MOLECULAR STRUCTURES IN BIOLOGY

Molecular Structures in Biology

Edited by

R. DIAMOND

MRC, Laboratory for Molecular Biology
Cambridge, UK

T.F. KOETZLE

Protein Data Bank, Brookhaven National Laboratory
New York, USA

KEITH PROUT

Chemical Crystallography Laboratory
University of Oxford, Oxford, UK

and

JANE S. RICHARDSON

Department of Biochemistry, Duke University
School of Medicine, North Carolina, USA

Oxford • New York • Tokyo
OXFORD UNIVERSITY PRESS
1993

Oxford University Press, Walton Street, Oxford OX2 6DP

Oxford   New York   Toronto
Delhi   Bombay   Calcutta   Madras   Karachi
Kuala Lumpur   Singapore   Hong Kong   Tokyo
Nairobi   Dar es Salaam   Cape Town
Melbourne   Auckland   Madrid

and associated companies in
Berlin   Ibadan

Oxford is a trade mark of Oxford University Press

Published in the United States
by Oxford University Press Inc., New York

© Contributors listed on p. ix, 1993

A catalogue record for this book is available from the British Library

Library of Congress Cataloging in Publication Data

Molecular structures in biology/edited by R. Diamond . . . [et al.].
p.   cm.
Includes index.
1. Proteins—Structure.   2. Nucleic acids—Structure.
I. Diamond, R. (Robert), 1929-   .
QP551.M63   1993   547.7—dc20   92-28936

ISBN 0 19 854771 4

Typeset by Colset Pte Ltd, Singapore
Printed in Great Britain on acid-free paper by
Biddles Ltd, Guildford and King's Lynn

Preface

In this book we present on paper ten articles that were prepared for the CD-ROM publication *Molecular Structures in Biology*, and as such I would like to preface it with some comments on the CD-ROM itself and the thinking that lead to its production.

It was the increasing complexity of information in the 1980s that caused some re-thinking of the process of publication and dissemination of information. Today's information is of better quality and is in much greater quantity than ever before. It was with this in mind that in the autumn of 1986 Adam Hodgkin, then of Oxford Electronic Publishing, suggested that I might give some thought to the use of CD-ROM for the publication of scientific information. The challenge was the publication and interpretation of highly complex, often numerically intense, information that needs to be viewed at a variety of levels of detail, and the possible subject areas that had been considered were as varied as language, economics, archeology, astronomy, and molecular biology.

The result of these deliberations was *Molecular Structures in Biology*, an experiment in the publication of numerically intense information held in the Protein Data Bank, machine readable and transmittable, together with a textual description in hypertext form, a 'picture bank' of predetermined graphical presentations, and software for the user to create interactive graphics, either on the instructions of an author or of the user's own choosing.

The CD-ROM, the data variant of CD audio, is compact, durable, physically robust, machine readable, high capacity, and has internationally agreed standards for data structure and format. It can contain a record of data as text, numerical data, bitmapped images, and single frame or motion video, together with programs to manipulate data either as an interactive graphic rendering or in the form of a database. Each CD-ROM has a storage capacity of 550 Mbytes, that is a quarter of a million pages of text or ten thousand colour pictures. As a distribution medium it comes into its own, and exceeds the value of paper, microfiche and microfilm only when the data volume is very large and machine readability is a premium requirement. Initially the production costs of a CD-ROM look very attractive, but there is the 'hidden' cost of software development which is almost open ended and is where compromises must be made.

An Editorial Board, with Jane Richardson and Bob Diamond representing molecular biology, computer graphics, and X-ray crystallography, Tom Koetzle representing the Protein Data Bank, and myself was formed to guide the publication in textual content and software design. Potential authors were selected by the editors with the generous assistance of a panel of international experts in the field. The authors faced a daunting task, to write for a publication which itself was going through a series of design changes, so that every letter from the editors seemed to change the instructions to authors, and at the same time to produce an article that was suitable for paper publication.

From the beginning of the project very generous assistance was received from IBM (UK) Ltd, first from the United Kingdom Scientific Centre at Winchester, England and then from IBM Software Technology.

In this book some of these texts are now presented to a much wider potential readership in conventional form on paper. They represent both the methodology of structural molecular biology and structure description by those active in the respective fields.

Oxford   Keith Prout
February 1993   for the Editors

Contents

Contributors

Charles L. Brooks III   Department of Chemistry, Carnegie Mellon University, Pittsburgh, Pennsylvania 15213, USA

U. Derewenda   Department of Chemistry, University of York, Heslington, York YO1 5DD, UK

Guy G. Dodson   Department of Chemistry, University of York, Heslington, York YO1 5DD, UK

Giulio Fermi   Laboratory of Molecular Biology, Medical Research Council, Hills Road, Cambridge CB2 2QH, UK

William N. Hunter   Department of Structural Chemistry, University of Manchester, Oxford Rd, Manchester M13 9PL, UK

Olga Kennard   University Chemical Laboratory, Lensfield Road, Cambridge CB2 1EW, UK

William N. Lipscomb   Gibbs Chemical Laboratory, Harvard University, Cambridge, Massachusetts 02138, USA

F. Scott Mathews   Department of Biochemistry and Molecular Biophysics, Washington University School of Medicine, St. Louis, Missouri 63110, USA

Madeleine H. Moore   Chemistry Department, University of York, Heslington, York YO1 5DD, UK

Joseph H.B. Pease   Department of Chemistry, University of California, Berkeley, California 94720, USA

Scott F. Sneddon   Central Research Division, Pfizer Inc., Groton, Connecticut 06340, USA

Raymond C. Stevens   Department of Chemistry, University of California, Berkeley, California 94720, USA

Mark B. Swindells   Biomolecular Structure and Modelling Unit, Department of Biochemistry and Molecular Biology, University College, Gower Street, London WC1E 6BT, UK

Janet M. Thornton   Biomolecular Structure and Modelling Unit, Department of Biochemistry and Molecular Biology, University College, Gower Street, London WC1E 6BT, UK

David A. Waller   Department of Biochemistry and Molecular Biology, University of Leeds, Leeds LS2 9JT, UK

David E. Wemmer   Department of Chemistry, University of California, Berkeley, California 94720, USA

Kurt Wüthrich   Institut für Molekularbiologie und Biophysik, Eidgenossische Technische Hochschule-Honggerberg, CH-8093 Zürich, Switzerland

1

Biological structures obtained by X-ray diffraction methods

David A. Waller and Guy G. Dodson

1.1 Introduction

This article is aimed at biochemists and molecular biologists without crystallographic training who wish to learn about the structural determination of macromolecules by X-ray analysis. Examples are taken mainly from the crystallography of proteins. X-ray analysis has had an immense impact on biochemistry and biological science. Our knowledge of the three-dimensional structure of proteins, nucleic acids, viruses, and other macromolecular assemblies has come almost entirely through this technique.

The first high-resolution protein structure determined by X-ray crystallography, myoglobin (Kendrew *et al.* 1960), gave an initial impression of great complexity. On further study it was revealed that the structure was composed of eight helices arranged around the haem group. The conformation of the α-helices agreed with the predictions of Pauling and Corey (1951). This first protein structure was found to contain a core of predominantly non-polar residues with polar residues distributed over its surface. The separation of non-polar side-chains into the interior and polar side-chains onto the exterior of the protein explained many of its physical and chemical properties and gave a reliable structural basis to direct future chemical and biochemical research.

The structure of myoglobin was determined by the technique of heavy atom isomorphous replacement (see Section 1.3.4.1). This technique still remains the most successful approach to determining structures. With the great expansion of protein crystallography, other methods have been developed based on homologies between structures (see Section 1.3.4.3).

Since the 1960s hundreds of macromolecular structures have been solved—proteins, nucleic acids, protein-nucleic acid complexes, viruses, and other larger molecular assemblies. It was from these analyses that the principles of protein architecture emerged. They showed that although proteins are superficially complicated structures, they are constructed from elements of defined structure—helices and sheets (referred to as secondary structure) whose arrangement gives the overall (or tertiary; see Glossary) structure. In addition these structures have defined catalytic sites of enzymes and explained the specificity of many protein reactions. One very important outcome of this now large body of research is the insight into the evolutionary relationships that have been shown to exist in molecules where this was unexpected, e.g. flavodoxin and the nucleotide-binding domain in lactate dehydrogenase (Rossmann *et al.* 1974). This is a molecular equivalent of palaeontology in which a three-dimensional structure proves to be a more persistent feature in a molecule than its sequence.

It is vital that the biochemist should be able to assess crystallographic structures and interpret the results sensibly; equally, biochemists should recognize that knowledge of three-dimensional structures gives unparalleled insight into the relationship between sequence, structure, and function, but that there are also limitations in crystallographic analysis. For example, many protein mechanisms involve structural changes that cannot yet be identified and studied in the crystalline state. The object of this article is to equip biochemists to read a crystallographic paper with good judgement. Therefore, in the following sections we will discuss:

(1) the nature of protein crystals and crystallization (Section 1.2);
(2) how a crystal structure is obtained by diffraction methods (Section 1.3);
(3) electron density and its interpretation in atomic terms (Section 1.4); and
(4) the factors relevant for assessing the quality of crystallographic structures (Section 1.5).

1.2 Crystals

By definition, a crystal is a solid, usually bounded by well-formed planes meeting at edges and corners, which has a regularly repeating internal structure. Examination will show that the crystal faces and edges are arranged with some degree of symmetry. The unit cell is the basic building-block from which the whole volume of the crystal is built up. The basic shape of the unit cell is described by the cell dimensions a, b, c, which are the lengths of the axes, and α, β, γ, which are the interaxial angles (α is the angle between b and c, etc.). The crystal lattice is a regular three-dimensional array of points upon which the contents of the unit cell may be considered to be arranged by infinite repetition. The space group of a crystal describes the symmetry of the unit cell. There are 230 possible space groups, which are tabulated in the *International tables for crystallography* (Hahn 1983). Because biological molecules contain chiral centres, they can only crystallize in those space groups that do not contain centres of symmetry or mirror planes. There are 65 such space groups.

1.2.1 Macromolecular crystals

Many proteins are functional in aqueous conditions and this means that it is usually quite straightforward to purify and solubilize them. However, protein crystallization implies that each protein molecule is positioned in a three-dimensional lattice with great precision. It is remarkable that in spite of the mobility of protein surfaces, they frequently form beautiful, well-organized crystals. Such crystals are different from small molecule crystals in one important respect, their unit cells typically contain 30-80 per cent by volume of solvent. This means that there are few contacts between molecules, which 'float' in baths of solvent; the crystals are stabilized by the few hydrogen bonds and van der Waals contacts between molecules. Protein crystals are therefore very fragile, and in the absence of surrounding liquid will rapidly lose solvent by evaporation, with a concomitant collapse of the crystal lattice. Although the high proportion of solvent poses experimental problems, it does mean that the crystal environment is reasonably similar to the natural environment and, consequently, proteins in crystals are often found to be functional.

Typically, protein crystals used in X-ray diffraction studies vary in size from 0.05 to 1 mm in their longest dimension. Their unit cell dimensions range from 25 Å to hundreds of angstroms. The virus that causes the common cold in man, human rhinovirus 14, has been crystallized in a cubic cell with a cell dimension of 445 Å (Arnold *et al.* 1984). A protein crystal will typically be made up of 10^{10}–10^{15} identical unit cells; this three-dimensional lattice will diffract X-rays. The large solvent volumes have made it possible to react the proteins in the crystal lattice. It was this property that led to the technique of heavy atom isomorphous replacement (see Section 1.3.4.1) which provided the key to solving macromolecular structures. In this technique the heavy atom is positioned in exactly the same place or places in the protein in all unit cells in the crystal, without disturbing the crystal lattice. Equally, substrate analogues, inhibitors, and ligands can often be introduced in the crystalline state; such studies have laid the basis for identifying catalytic sites and recognition surfaces.

1.2.2 Crystal growth

Unfortunately, the growing of crystals remains the major uncertainty in protein crystallography. In some ways this is not surprising given the delicate character of the forces that define and maintain the crystal lattice. The basic technique for protein crystallization involves bringing the protein to supersaturation and then leaving it undisturbed for days, weeks, or even months. Factors such as protein purity, buffer type, pH, temperature, ionic strength, and the presence of organic solvents or counter-ions may all affect the point of saturation. Thus it can be seen that in attempting to crystallize a new protein one is faced with a multidimensional problem and frequently insufficient protein to carry out an exhaustive survey of possible conditions. Examination of the existing literature on crystallizations has led to the development of strategies to approach this problem (Gilliland 1988).

The practicalities of protein crystallization have been extensively described and will not be further addressed here (Blundell and Johnson 1976; McPherson 1982; Wyckoff *et al.* 1985).

1.3 Determination of crystal structure (the X-ray image)

Normally the images of objects we see in everyday life are focused from scattered light by the eye onto the retina. The scattering from an individual molecule is, however, so small that it is not possible to detect. There are two problems with molecules; first, their dimensions are very much smaller than the wavelength of light, which means that their details cannot be resolved. Secondly, scattering will be very weak. X-rays, with their much shorter wavelength (about the order of a bond length) will be scattered by individual atoms in a molecule. However, again, scattering by an individual molecule will be so weak as to be undetectable. Moreover, X-rays cannot be focused in the normal sense, which means that it is not possible to use lenses to form an image. The periodic character of crystals means, however, that the scattering from each individual atom is coherently combined with that from all the other equivalent atoms in the crystal, giving sufficient intensity to be measured. A consequence of the periodic structure of the crystal is that the scattering from a crystal is not continuous, as it would be from an isolated object, in this case a molecule. Friedrich *et al.* (1912) first demonstrated that crystals diffract X-rays, giving individual diffraction spots. Bragg (1913) then showed that the diffraction of X-rays by crystals had to satisfy a law of reflection and that it was possible to index each reflection.

1.3.1 Bragg's law

Bragg's law states that:

$$n\lambda = 2d \sin \theta \tag{1}$$

where n = an integer;
λ = wavelength;
d = spacing in the crystal;
θ = angle made by X-ray and reflecting plane.

Note that angles of entry and exit are equal.

The reflected ray from one crystal unit cell will be parallel to the set of rays reflected from other unit cells in the crystal. The wavefronts of the reflected rays are different since, as Fig. 1.1 illustrates, the path length of the reflected rays has an increment, the path difference. This increment equals $2d \sin \theta$.

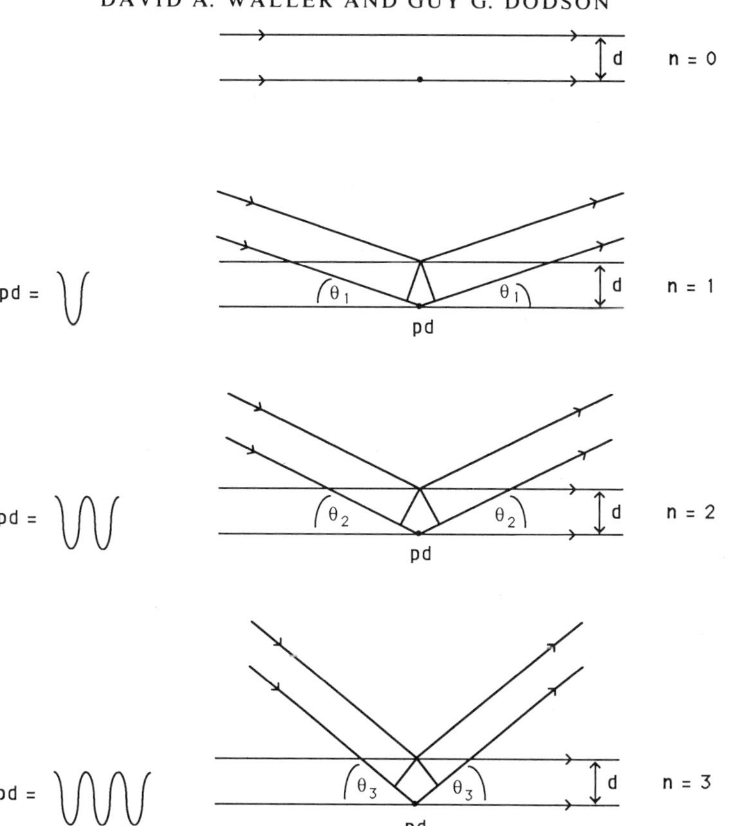

Fig. 1.1 X-ray reflection for zero, first, second, and third orders corresponding to path differences of 0, 1, 2, 3 λ. Note that the Miller indices *h, k, l* correspond to the orders of reflection as illustrated above.

When the increment is equal to an integral number of wavelengths ($n\lambda$), the wavefronts will line up exactly and interfere constructively. When the increment is not equal to an integral number of wavelengths, the wavefronts do not line up and the reflected rays will interfere destructively and cancel out.

The values of n: 0, 1, 2, 3 correspond to the zero, first, second, and third orders of reflection. In the zero order the waves obviously superimpose exactly. In the first order the waves are superimposed again, but the wavefronts from adjacent planes are displaced by one wavelength (λ). This corresponds to a relative difference in phase of 360° or 2π. Equivalently, for the second order the reflected rays overlap exactly but the wavefronts are displaced by two wavelengths (2λ) which corresponds to a phase difference

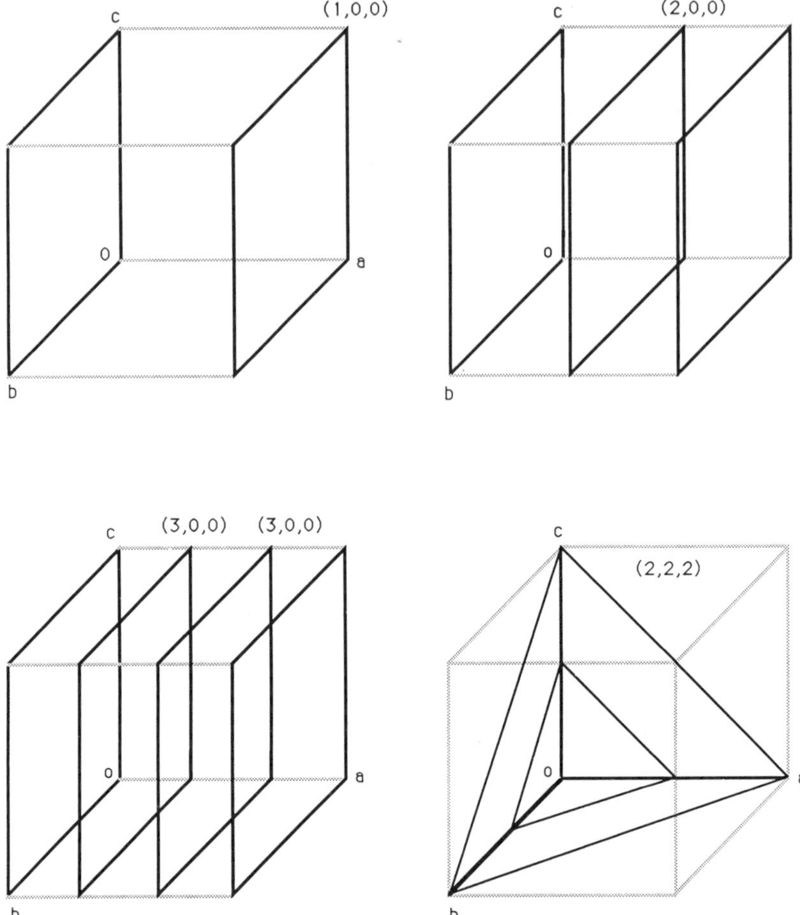

Fig. 1.2 A geometrical construction for the reflecting planes 1,0,0; 2,0,0; 3,0,0; and 2,2,2. The crystal unit cell directions are indicated *a, b, c*. The reflected planes are emphasized with bold lines.

of 720° or 4π, and so on. The use of the concept of crystal planes to explain reflection is common and is a useful construction (Stout and Jensen 1989). The planes are defined by the Miller indices, their geometry is illustrated in Fig. 1.2.

1.3.2 Diffraction from crystals

The periodicity of the crystal leads to the X-ray scattering from each molecule in the unit cell adding up with the scattering from all the molecules in the crystal in the directions where Bragg's law is satisfied. Thus the precise periodic structure of crystals creates conditions where the scattering from all the molecules in the crystal is equivalent to the scattering from one molecule amplified many millions of times, making it detectable as a reflection on X-ray film, Geiger counters, etc.

Bragg's law means that the X-ray reflections themselves will form a lattice, the spacings of which are reciprocally related to the crystal unit cell. It is a three-dimensional lattice in which the reflections are defined by indices h, k, and l, corresponding to the directions associated with the a, b, and c axes of the unit cell, respectively. The h, k, l indices, each in their own direction, correspond to the n of Bragg's law Thus, for example, the reflection with h, k, l: 10, 0, 30 means that for the h index the path difference in the adjacent crystal planes corresponds to 10 wavelengths, there is no path difference for the k index, and a path difference of 30 wavelengths for the l index. Typically, a macromolecular crystal will give rise to many thousands of X-ray reflections. Each of them can and must be indexed if we are to construct the image.

1.3.3 The phase problem

If it were possible to focus the X-ray reflections then the image of diffracting molecules could be constructed. In order to construct the image analytically the amplitude of the reflections together with the relative relationships of their wavefronts (the phases) are needed. The amplitude of the reflection can be measured directly from the observed intensity but the phase information is lost. The determination of the phase forms the heart of X-ray analysis — this is the so-called phase problem.

The expression for the electron density, ρ, from which atomic coordinates are derived is:

$$\rho(xyz) = \frac{1}{V} \sum_{h=-\infty}^{\infty} \sum_{k=-\infty}^{\infty} \sum_{l=-\infty}^{\infty} F_{hkl} \exp - 2\pi i(hx + ky + lz) \qquad (2)$$

where $\rho(xyz)$ = the electron density at the point xyz;
V = the unit cell volume;
F_{hkl} = the structure factor;
$\exp - 2\pi i(hx + ky + lz)$ = a three-dimensional wave.

The summations are over all available indices h, k, and l.

The expression $\exp - 2\pi i(hx + ky + lz)$ is a complex number which combines with its partner of opposite exponent to give a three-dimensional cosine wave with a frequency of h, k, and l in each of the directions associated with the a, b, and c axes, respectively. This function has an amplitude $|F_{hkl}|$. To construct the image of the diffracting object, the three-dimensional cosine waves have to be combined correctly. That is, the position of the wavefront for each reflection hkl must be known relative to some fixed origin. The relationship between the origin and the wavefront is defined by the phase angle.

The phase angles have values of 0 to 2π. At zero the wavefront is at the origin, at π the wavefront is displaced half a period from the origin.

In practice, the summations in equation 2 have finite limits on h, k, and l, determined by the availability of diffraction data. The larger the limits, the shorter the wavelength of the shortest included waves, and hence the finer the detail in the resulting map. This is the basis of the concept of resolution, discussed further below (Section 1.4).

1.3.3.1 The structure factor   The structure factor F_{hkl} is the sum of the atomic scattering in the unit cell for a reflection hkl.

$$F_{hkl} = \sum_{j=1}^{N} f_j \exp 2\pi i (hx_j + ky_j + lz_j) \tag{3}$$

Note that $f_j = $ the atomic scattering factor.

$$F_{hkl} = |F_{hkl}| \exp i\alpha_{hkl} \tag{4}$$

where $|F_{hkl}| = $ the structure factor amplitude, and $\alpha_{hkl} = $ the phase angle.

$$\alpha_{hkl} = \left(\tan^{-1} \frac{\Sigma f_j \sin 2\pi (hx_j + ky_j + lz_j)}{\Sigma f_j \cos 2\pi (hx_j + ky_j + lz_j)} \right) \tag{5}$$

The actual measured intensity, I, of a reflection is:

$$I = c|F_{hkl}|^2 \tag{6}$$

Thus we can see again the origins of the phase problem. The intensity measurement gives us the structure factor amplitude but the phase information, which is vital in the calculation of an electron-density map, is lost. The crystallographer is faced with the problem of determining the phase.

1.3.4 Solutions to the phase problem

In protein crystallography there are now two widely used techniqes for overcoming the phase problem — isomorphous replacement (which can be combined with anomalous scattering effects) and molecular replacement (see Section 1.3.4.3).

1.3.4.1 Isomorphous replacement   The isomorphous replacement method was developed by Perutz and colleagues in the 1950s (Green *et al.* 1954). It provided the breakthrough in the phase problem which allowed the structures of myoglobin (Kendrew *et al.* 1960) and haemoglobin (Perutz *et al.* 1960) to be solved. Today it is still the principal technique used to surmount the phase problem in the determination of a truly new protein structure. It involves producing a derivative crystal by soaking into a protein crystal a solution of a compound containing a heavy atom and collecting a set of diffraction data from that crystal. If the heavy atom has bound to the protein at a site or sites without affecting the protein structure itself, there will be differences in the intensities of the X-ray reflections which contain information about the phases. The positions of the heavy atoms can usually be calculated using a Patterson function (see Glossary) (Patterson 1934) whose coefficients are related to the intensity differences between the native and derivative data[1]. The native data from a crystal without heavy atoms attached, the derivative data from the crystal containing heavy atoms, and the knowledge of the position of the heavy atoms can be combined to calculate the phases for the native protein data.

A native data set used in conjunction with one derivative data set and heavy atom positions for that derivative will, in fact, give rise to an ambiguity in the form of two different possible phases for each X-ray reflection. This problem can be overcome by using a second isomorphous heavy atom derivative which will again give a pair of possible phases for each reflection. The true phase for each reflection can be obtained from the combination of the pairs of ambiguous phases, as the correct solution will be common to each of the ambiguous solutions. These phases can then be used together with the structure factor amplitudes for the calculation of an electron-density map. Unfortunately, experimental difficulties such as non-isomorphism, weak substitution by the heavy atom, crystal variations, etc. often mean that two heavy atom derivatives are not sufficient to determine the phases well enough to define the image. Frequently three, four, or five heavy atom derivatives are incorporated in the calculation of the phase.

1.3.4.2 Anomalous scattering   When X-rays are scattered by an atom, the phase of the scattered ray is delayed if the wavelength of the radiation is near a characteristic absorption edge. This effect is known as the Bijvoet effect (Bijvoet 1949). Normally in X-ray diffraction $F_{hkl} = F_{\overline{hkl}}$, this is known as Friedel's law (1913); if, however, the crystal contains a heavy atom with an absorption edge near that of the incident radiation, then generally F_{hkl} is not equal to $F_{\overline{hkl}}$. This is because the anomalous scattering by the heavy

---

[1] The nature of the coefficients in the Patterson function varies. The simple difference of amplitudes between native and derivative (F_P, F_{PH}) is only equal to the heavy atom scatter in special cases, i.e. when the phase can only have two values 0 or π, these are the so-called centric terms. Generally, the heavy atom contribution has to be constructed by using the isomorphous and anomalous scattering differences (see below). Alternatively the heavy atom contribution can be approximated by the difference in amplitudes. This procedure usually works very well. The refinement of the heavy atom positions deduced from the Patterson function is carried out either by limiting it to the centric terms ($F_H = F_{PH} - F_P$) or by using the experimentally determined F_H (derived from isomorphous and anomalous scattering) – the F_H lower estimate, or F_{HLE} method. If approximate protein phases are known from other derivatives, the F_H can be calculated and refinement carried out using experimental phases. The reader is directed to several excellent texts on the methods used in heavy atom refinement (Blundell and Johnson 1976; McPherson 1982).

atom combines differently in the directions corresponding to F_{hkl} and $F_{\overline{hkl}}$. If the heavy atom positions are known, then the anomalous scattering term can be calculated. This information can improve the accuracy of the isomorphous phasing by resolving the intrinsic ambiguity present in a single isomorphous phase set. In the best case, anomalous scattering associated with one isomorphous derivative can provide a set of phases good enough to produce an interpretable electron-density map.

In the absence of anomalous scattering, it is not possible to distinguish crystallographically between a set of coordinates xyz and $\overline{xyz}$ or a set of phases α and $-\alpha$. The coordinates xyz and $\overline{xyz}$ form a mirror image each of the other and the electron density generated by phases α is the mirror image to that generated by $-\alpha$. Of course, features in electron density and coordinates such as a right-handed α-helix allow one to determine the hand relatively easily. However, because the anomalous scattering has an absolute phase, it can be used to determine the handedness of the X-ray image directly.

The tunable radiation available from synchrotron sources (see Glossary) makes it possible to measure anomalous scattering effects because they vary with wavelength. These multiple observations make it possible to determine the phases using only a protein crystal containing a heavy atom. If the protein contains an intrinsic heavy atom like myoglobin (Fe), azurin (Cu), and streptavidin with selenobiotin (Se), not even one heavy atom derivative need be prepared.

1.3.4.3 Molecular replacement   The X-ray analysis of some 300 structures has revealed that many of them are closely related in their molecular organization. When two proteins have similar folds, i.e. similar tertiary structure, it is sometimes possible to use a crystallographically determined structure to solve the phase problem for the related but unknown structure (Rossmann 1990). Molecular replacement is particularly powerful for the analysis of the same enzyme from different species, e.g. the structure of phosphofructokinase from *Bacillus stearothermophilus* was solved using isomorphous replacement (Evans and Hudson 1979) and subsequently phosphofructokinase from *Escherichia coli* was solved by molecular replacement (Rypniewski and Evans 1989).

In molecular replacement it is assumed that the unknown molecule has a structure similar to some known structure. The structure factor amplitudes for a molecule of known structure are calculated for the cell of the related molecule of undetermined structure. Structure determination requires two steps. The first is to determine the orientation of the model structure in the cell of the unknown molecule by comparing the Patterson function calculated from the model in the unknown cell with that observed for the undetermined molecule. The possible orientations are explored systematically and when the two functions correlate, i.e. the calculated and observed Patterson functions match best, the angles of rotation are determined. This is referred to as the rotation function. Note that there is no phase information in this function. The second step is to determine the position of the correctly oriented model in the cell of the unknown molecule (this calculation is often referred to as the translation function). The model molecule is translated systematically in the unknown cell in steps of between 0.1 and 0.5 Å, depending on the resolution of data and size of unit cell. At each position the structure amplitudes are calculated and compared with those observed for the unknown molecule. A good match between the model and unknown molecules will lead to a detectable correlation between the calculated and observed structure amplitudes usually measured with the conventional crystallographic R factor (see Glossary). A typical R value for incorrect translation parameters is 0.55 (this depends on resolution, data quality, etc.); at the correct translation values the R value will fall to between 0.4 and 0.5 (again depending on resolution, data quality, etc.).

Because of the discrepancies that exist between the model and unknown molecules, and errors in the rotation and translation parameters, the calculated phases are not accurate. But in many cases these phases provide sufficient information to permit the refinement of the unknown molecule atomic parameters to commence.

1.3.4.4 Phase improvement   It is sometimes possible to improve the experimental phasing significantly by exploiting the existence of non-crystallographic symmetry (Bricogne 1976) and, more generally, by exploiting the contrast between the protein volumes and non-protein volumes in a crystal (Wang 1985). The electron density in the protein volume is characteristically higher than that in the solvent volume. When this information is imposed on the phasing it can lead to a useful and sometimes dramatic improvement in the clarity of the electron density for the protein. This latter technique is now being applied increasingly. (In some ways the density modification methods used in the symmetry averaging and solvent flattening procedures are halfway between direct methods of phase determination used in small molecule crystallography, which depend on the non-negativity of the electron density, and experimental methods used in macromolecular crystallography.)

1.4 Electron density and the determination of atomic positions

A crystal structure is the location of atoms in a crystal unit cell. The atoms are identified from a contoured map of electron density. Remember that it is electrons which scatter the X-rays and therefore when the image is constructed from the X-ray diffraction pattern it will reveal the positions of

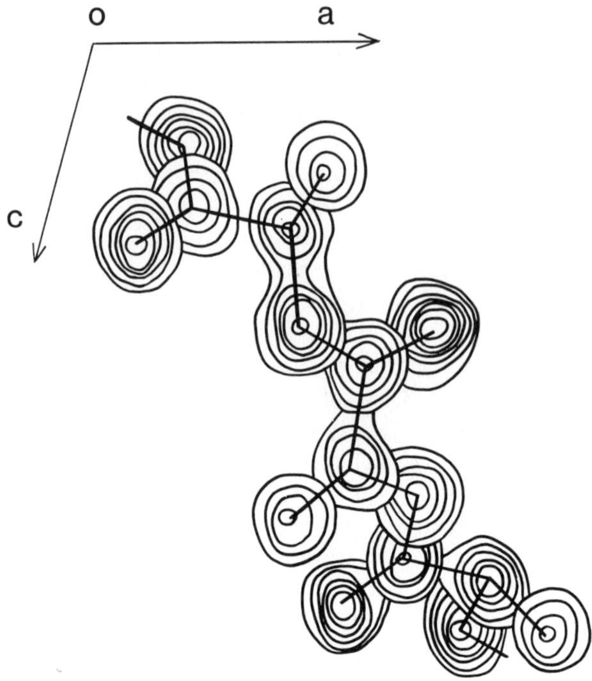

Fig. 1.3 The electron density for the atoms of a fragment of the antibiotic thiostrepton (Anderson *et al.* 1970). Note that each atom is individually resolved. The peaks for oxygen atoms contain more contours than those for nitrogen and carbon atoms, and the structure of the peptide bonds can be recognized.

electrons. Figure 1.3 shows the electron density for a small molecule in which the atoms are individually resolved. The bond lengths and angles can then be determined from the positions of the atoms (in this case to an accuracy of about 0.03 Å or better). Figure 1.4 illustrates the electron density for a protein, insulin. Here, at a resolution of 1.5 Å spacing, the atoms are almost resolved. Proteins do not generally diffract to 1.5 Å, a more typical limit is 2 Å spacing. The effects of this lower resolution on the definition of the X-ray image are seen in Fig. 1.5.

1.4.1 Interpretation of electron density

When the phases (obtained either by isomorphous replacement or molecular replacement) are available an electron-density map is calculated. At resolutions between 2 and 3 Å this map will reveal continuous threads of density which correspond to polypeptide backbone. Usually the side-chains are identified as extensions from the main chain, occurring every 4 Å or so. Features

(a)

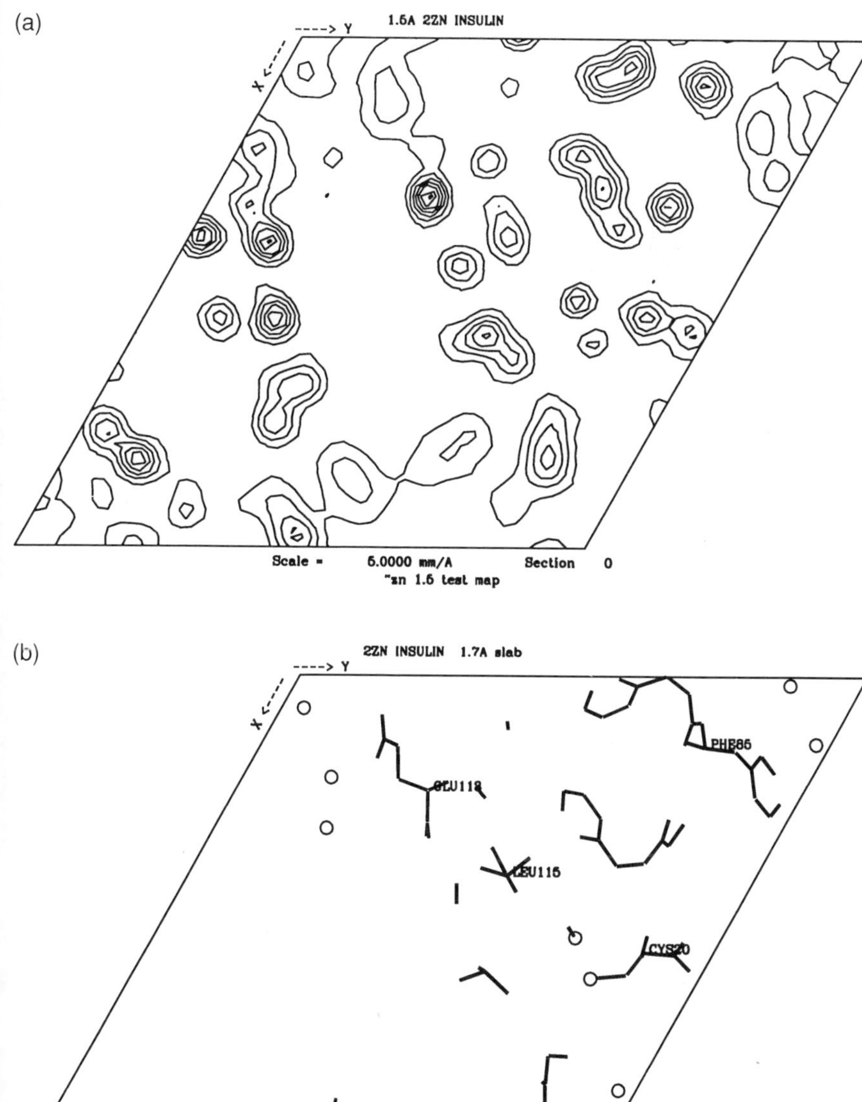

(b)

Fig. 1.4 (a) The section $z = 0$ for the 1.5 Å electron density map of 2 Zn insulin (Baker *et al*. 1988) computed with observed amplitudes and calculated phases. The contour levels are at 1 ς intervals (circa 0.15e/Å^3) from zero.
(b) The positions of atoms derived from the 1.5 Å electron density map above.

(a)

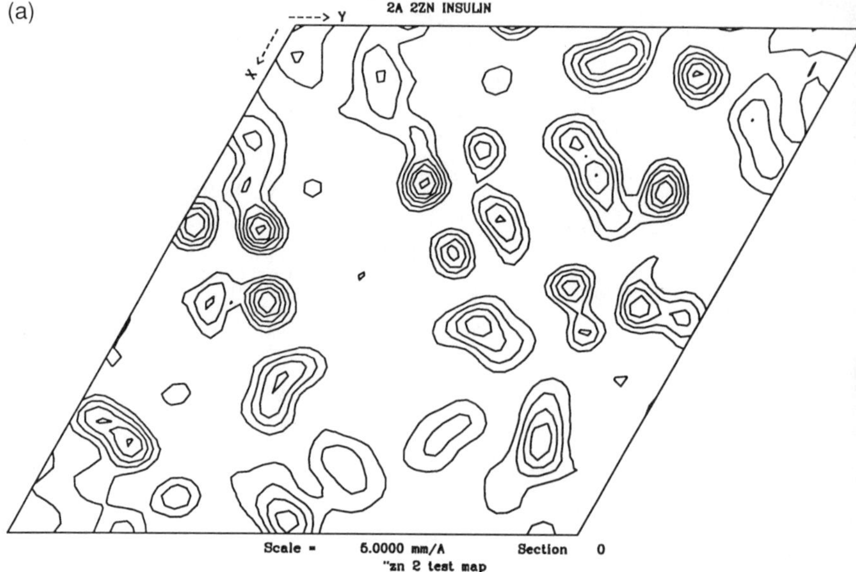

(b)

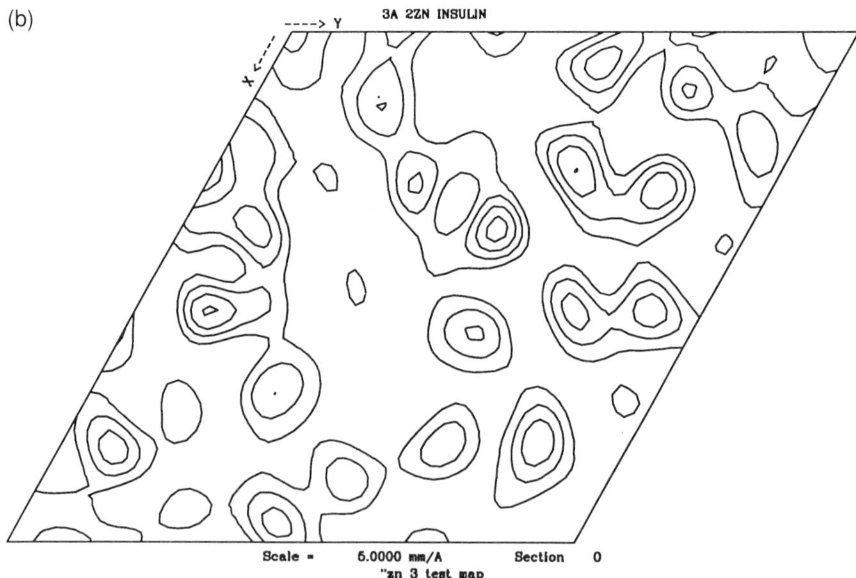

Fig. 1.5 The section $z = 0$ from a series of 2 Zn insulin maps computed with observed amplitudes and calculated phases at resolutions of 2 Å (a), 3 Å (b), 4 Å (c), and

(c)

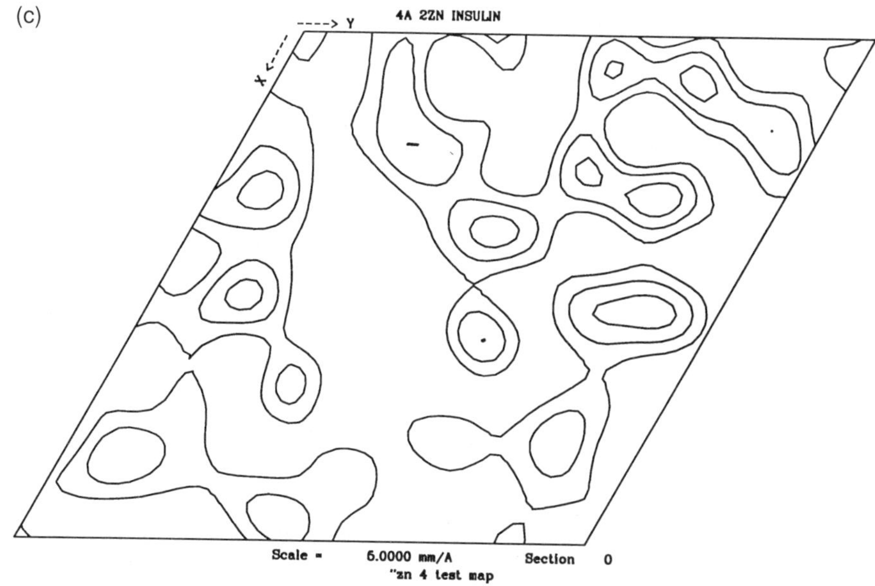

(d)

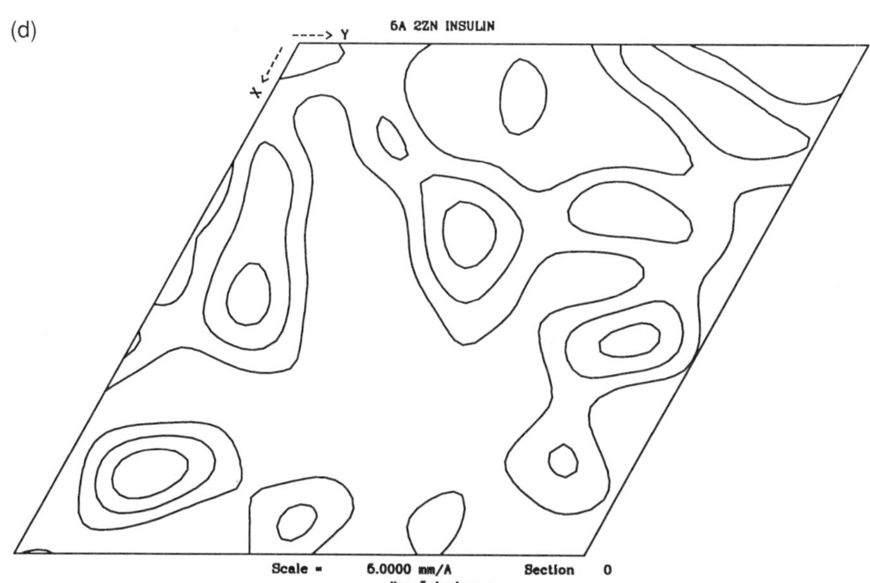

5 Å (d). Contours are drawn at 1 σ intervals. The maps do not show the same number of e/Å^3. The positions of atoms in this section are shown in Fig. 1.4b.

such as α-helix are usually readily identified and, if seen, confirm the general correctness of the phases. At resolutions of about 2 Å much more detail will be seen if the phases are good — carbonyl oxygen atoms appear as bumps and it will be possible to distinguish between small and large side-chains, e.g. alanine and tyrosine.

Because the X-ray image depends on experimental phasing with its inherent errors, the definition of the electron density is often degraded. It is from such imperfect maps that most protein structures are initially interpreted. Therefore the coordinates from these maps are inaccurate and often contain errors, for this reason it is necessary to refine atomic positions and their thermal parameters (see Glossary).

When a macromolecular crystal structure has been refined at a resolution of, say, better than 2.5 Å spacing, the details of helical and sheet structures and their arrangement together in the structure will be clear. Even the degree of order or mobility of the atoms can be estimated. Well-defined protein atoms will have high electron density with tight contours within which the stereochemistry of peptide bonds and amino-acid side-chains can be deduced readily. When the atoms are less well ordered either through mobility or disorder, then the electron density at the atomic positions is lower and more spread out, consequently the atomic positions are less well defined. The shape of the electron density can give an indication as to the motion of the atoms or the positions of alternative structures generated by disorder. In these features of the electron density we are seeing, trapped in the framework of the crystal, something of the intrinsic flexibility that characterizes proteins.

It is also possible to locate some of the water molecules that form a network over the protein surface. When the water molecules are well defined by their hydrogen bonding and van der Waals contacts to the protein, their electron-density peaks are clear and spherical. More mobile water molecules appear with more diffuse electron density, often elongated in the direction of motion or nearby alternative sites. In the extreme case, the electron density for the water smears out to broad, weakly featured density, merging into the background level.

1.4.2 Improving the electron density and atomic coordinates

The availability of powerful real-time interactive computer graphics systems has made the inspection and interpretation of electron-density maps very convenient. The electron-density map used in the initial interpretation will either be the isomorphously phased map obtained from the heavy atom derivatives or a map calculated with the phases obtained by molecular replacement. Once refinement has begun, however, it is desirable to use calculated phases, not the initial isomorphous phases which are relatively inaccurate. The electron-density map with coefficients $2F_o - F_c$ is very widely used once refinement has begun. It has the property of revealing both the atoms included in the phasing and those that are not. Atoms included in the phasing with the wrong coordinates reappear, generally speaking, very weakly or at zero density. The difference electron-density map with coefficients $F_o - F_c$ shows positive density for atoms not included in the phasing and negative density for wrongly placed atoms. In molecular replacement studies it is usual to examine the $2F_o - F_c$ map immediately, or after some cycles of refinement of the model structure, since of course there are no experimental phases. The refinement of the model coordinates is usually accompanied by rebuilding new or altered structure and deleting portions or residues that turn out to be incorrect in the model structure. Analysis of the electron-density maps can often appear straightforward, but when the coordinates contain significant errors the $2F_o - F_c$ map can be misleading.

1.5 Assessing macromolecular structures

Limited resolution and inaccurate diffraction data make it imperative to refine the atomic parameters with great care. It is important to extract as much genuine information from the analysis as possible; it is equally important not to overinterpret and to ensure that there are no detectable errors in the atomic parameters. With the power of modern computing and graphics systems it is generally now the case that only refined structures appear in the literature.

For completeness we list here the criteria that are useful in judging the quality of an X-ray structure determination.

1. The agreement, R (see Glossary), between observed and calculated structure factor amplitudes. It is expected that this should be about 0.20, values greater than 0.25 need to be justified. The resolution range that is used for R factor calculation should be noted, since omission of either high or low resolution terms will artificially reduce the R value.

2. The resolution of the X-ray diffraction data and the proportion actually measured need to be given. Generally, resolutions poorer than 3.0 Å will not reveal reliable atomic positions even though the R value is reported to be low. Atomic details can be reasonably defined at resolutions of 2.5 Å or better, provided the higher resolution data are well recorded.

3. The agreement between multiply observed symmetrically equivalent reflections (R_{sym}) needs to be reported and should be less than 10 per cent. Obviously, there will be higher values if the data are weak or there are any special problems with crystals (which are not infrequent).

4. The stereochemistry of the protein should be reported. The values of the peptide dihedral angles (shown in a Ramachandran plot (see Glossary)) provide a convenient way of describing main-chain geometry. Any deviations from expected values should be justified.

5. Error estimates. For a well-defined structure at 2.5 Å spacing or less

the estimate of error will be in the region of 0.30 Å for well-defined atoms [i.e. atoms with temperature factors <20 Å^2 (see Glossary)].

6. The agreement between expected and refined geometries should be reported. Typically, the root mean square deviations from expected values for bond lengths should be about 0.02 Å and for bond angles about 5°.

7. The accuracy of an isomorphous phase set can be judged from the quality of the heavy atom derivatives. As a rule of thumb, the agreement between the observed and calculated heavy atom contributions should be in the range 0.3–0.6. These high values are a consequence of the inaccuracies in the data, particularly the anomalous scattering term, problems of non-isomorphism, and complex heavy atom subsitution. A heavy atom may have four or more sites per asymmetric unit and there may be minor sites and local disorder, all of which are difficult to describe with normal parameters.

8. It is important for the quality of the phasing that the refinement of the heavy atoms is well behaved to a resolution of 2.5 Å or better. It is also important that the phasing power (mean heavy atom contribution divided by mean error) should be greater than 1.5.

References

Anderson, B., Crowfoot Hodgkin, D.M., and Viswamitra, M.A. (1970). The structure of thiostrepon. *Nature* **225**, 233–5.

Arnold, E., *et al.* (1984). Virion orientation in cubic crystals of the human common cold virus HRV14. *Journal of Molecular Biology* **177**, 417–30.

Baker, E.N., *et al.* (1988). The structure of 2 Zn pig insulin crystals at 1.5 Å resolution. *Philosophical Transactions of the Royal Society of London B* **319**, 369–456.

Bijvoet, J.M. (1949). Phase determination in direct Fourier synthesis of crystal structures. *Proceedings Koninklijke Nederlanse Akademie van Wetenschappen Series B* **52**, 313–14.

Blundell, T.L. and Johnson, L.N. (1976). *Protein crystallography*. Academic Press, London.

Bragg, W.L. (1913). *Proceedings of the Cambridge Philological Society* **17**, 43.

Bricogne, G. (1976). Methods and programs for direct-space exploitation of geometric redundancies. *Acta Crystallographica* **A32**, 832–47.

Evans, P.R. and Hudson, P.J. (1979). Structure and control of phospho-fructokinase from *Bacillus stearothermophilus*. *Nature* **279**, 500–4.

Friedel, G. (1913). Sur les symàtries cristallines que peut révéler la diffraction des rayons Rontgen. *Comptes Rendus Hebdomadaires des Séances de l'Académie des Sciences* **157**, 1533–6.

Friedrich, W., Knipping, P., and von Laue, M. (1912). Interferenz-Erscheinungen bei Rontgenstrahlen. *Proceedings of the Bavarian Academy of Sciences* p. 303. Reprinted (1952) in *Naturwissenschaften* **39**, 360–72.

Gilliland, G.L. (1988). A biological macromolecule crystallisation database: a basis for a crystallisation strategy. *Journal of Crystal Growth* **90**, 51–9.

Green, D.W., Ingram, V.M., and Perutz, M.F. (1954). The structure of haemoglobin IV: Sign determination by the isomorphous replacement method. *Proceedings of the Royal Society of London* **A225**, 287–307.

Hahn, T. (ed.) (1983). *International tables for crystallography*, Vol. A. D. Riedel, Dordrecht.

Kendrew, J.C., *et al.* (1960). Structure of myoglobin. A three-dimensional Fourier synthesis at 2 Å resolution. *Nature* **185**, 422–7.

McPherson, A. (1982). *Preparation and analysis of protein crystals*. Wiley, New York.

Patterson, A.L. (1934). A Fourier series method for the determination of the components of interatomic distances in crystals. *Physical Review* **46**, 372–6.

Pauling, L. and Corey, R.B. (1951). Atomic coordinates and structure factors for two helical configurations of polypeptide chains. *Proceedings of the National Academy of Sciences, USA* **37**, 235–40.

Perutz, M.F., Rossmann, M.G., Cullis, A.F., Muirhead, H., Will, G., and North, A.C.T. (1960). Structure of haemoglobin: a three-dimensional Fourier synthesis at 5.5 Å resolution, obtained by X-ray diffraction. *Nature* **185**, 416–22.

Ramachandran, G.N., Ramakrishnan, C., and Sasisekharan, V. (1963). *Journal of Molecular Biology* **7**, 95–9.

Rossmann, M.G. (1990). The molecular replacement method. *Acta Crystallographica* **A46**, 73–82.

Rossmann, M.G., Moras, D., and Olsen, K.W. (1974). Chemical and biological evolution of a nucleotide binding domain. *Nature* **250**, 194–9.

Rypniewski, W.R. and Evans, P.R. (1989). Crystal structure of unliganded phosphofructokinase from *Escherichia coli*. *Journal of Molecular Biology* **207**, 805–21.

Stout, G.H. and Jensen, L.H. (1989). *X-ray structure determination*. Wiley, New York.

Wang, B.C. (1985). Resolution of phase ambiguity in macromolecular crystallography. *Methods in Enzymology* **115**, 90–112.

Wyckoff, H.W., Hirs, C.H.W., and Timasheff, S.N. (1985). *Methods in Enzymology* **114**, Section II.

2

Biopolymers: an NMR survey

Kurt Wüthrich

During the last decade a method for the determination of the complete three-dimensional structure of proteins was developed, which uses nuclear magnetic resonance (NMR) spectroscopy for the data collection and distance geometry, or possibly other mathematical techniques, for the structural interpretation of the NMR data (Wüthrich *et al.* 1982; Wüthrich 1986). NMR is the only approach, in addition to diffraction techniques with single crystals, that is available for protein structure determination at atomic resolution. The structural information obtained by NMR is in many ways complementary to that from X-ray crystallography, thus widening our view of protein molecules with regard to a better grasp of the relations between structure and function. The NMR method can be applied with closely similar strategies to other biopolymers, such as DNA, RNA, and polysaccharides. However, since the potentialities of the method are intimately linked with the intrinsic structural properties of the molecules studied, the information on the molecular conformation obtainable with these other classes of compounds may be more limited than with proteins. Much interest of the NMR approach is also focused on studies of intermolecular aggregates between proteins and other small or large molecules.

In this publication the present short survey and two more extensive chapters are devoted to NMR methodology for biopolymer structure determination. For ease of presentation the method can be described as consisting of several independent, successive steps: (i) sample preparation, (ii) recording of the NMR data and spectral analysis to yeild NMR parameters, (iii) sequence-specific resonance assignments, (iv) collection of the conformational constraints needed as input for the structure determination, (v) structure calculation from the NMR data and structure refinement. In practice, these different steps are intimately linked, and one usually goes through repeated cycles of NMR data analysis and structure determination. The diagram in Fig. 2.1 outlines how these different steps are related in a protein structure determination by NMR, and Fig. 2.2 displays a structure determined by NMR. The arrows in the lower right of Fig. 2.1 indicate the circular nature of the procedures of data collection,

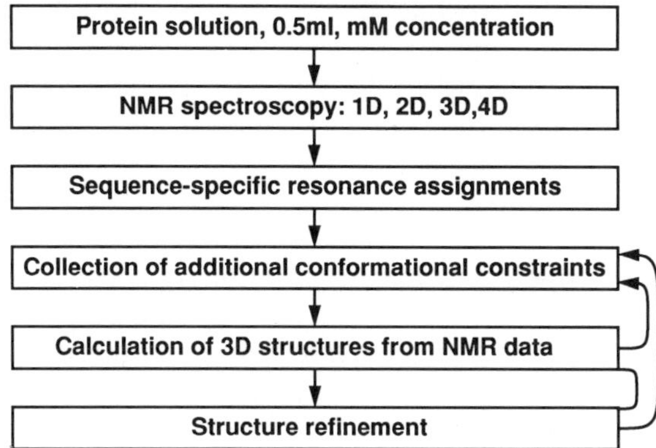

Fig. 2.1 Diagram outlining the course of a protein structure determination by NMR (see text for further explanations).

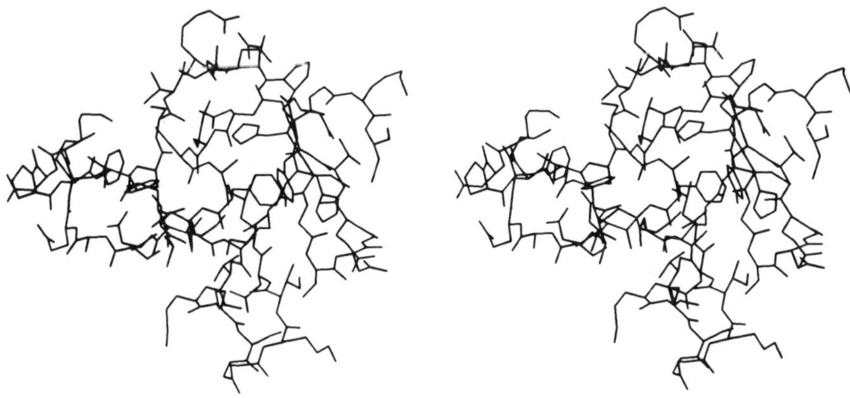

Fig. 2.2 Stereo view of a structure of the protein bull seminal proteinase inhibitor IIA (BUSI IIA), which was determined with the distance geometry program DISGEO (Havel and Wüthrich 1984) from NMR data recorded in aqueous solutions, and subsequently refined by energy minimization (Williamson *et al*. 1985). The disulphide bonds are not shown.

structure calculation, and structure refinement.

Chapter 3 on NMR methodology by Pease and Wemmer presents a detailed account of the techniques used for obtaining sequence-specific resonance assignments (Billeter *et al*. 1982; Wagner and Wüthrich 1982; Wider *et al*. 1982). Other reviews describe the mathematical basis and the practical implementation of the techniques used to compute three-dimensional structures from NMR data (Braun 1987; Clore and Gronenborn 1989; Kaptein *et al.* 1988; Wüthrich 1986, 1989). In addition to optimal sample preparation and the recording of high quality NMR spectra, these are clearly the two pivotal elements of the method.

2.1 Information from structure determination by NMR

Complementarity of the structural information obtained from X-ray diffraction techniques or from NMR results from the fact that the time scales of the two types of measurements are widely different in that, in contrast to the need for single crystals in diffraction studies, the NMR measurements are recorded with solutions of the compounds of interest. The most obvious consequence is that NMR can be applied to biopolymers for which no single crystals are available, thus yielding new structural data that cannot be obtained by any other method.

With regard to *de novo* structure determination, the NMR method is of prime interest for those working with polypeptides and proteins having molecular weights up to approximately 20 000. For molecules in this class the method provides stringent tests of the purity of the protein preparations (Otting *et al.* 1987; Widmer *et al.* 1988), the amino acid composition, and the amino acid sequence (Strop *et al.* 1983; Wagner *et al.* 1986), and can yield a three-dimensional structure in solution at high resolution (Kline *et al.* 1988). Comparisons of corresponding structures in single crystals and in non-crystalline states are then highly relevant, because the solution conditions for NMR studies can often be chosen so that they coincide closely with the natural, physiological environment of the protein (Wüthrich 1986). Extensive similarities between crystal and solution structures have been observed for globular proteins, while major conformational rearrangements between the two environments have been documented for non-globular polypeptides. Quite generally, significant differences between crystal and solution are observed for the protein surface, which is usually most directly related to the functional properties of the molecule and often includes the active site.

The use of NMR for structure determination with small proteins does not preclude studies of interest for larger systems. For example, individual domains excised by enzymatic or chemical cleavage from large proteins, or expressed separately by recombinant methods may be accessible for structure determination by the NMR method (Qian *et al.* 1989). Or, on a more general vein, conclusions relating to larger molecules, for which only a crystal structure is available, may be drawn from systematic comparisons of the molecular surface of small proteins in crystals and in solution.

An important facet of the method has not yet been mentioned, and is not

Fig. 2.3 Secondary structure of the DNA-binding domain comprising the N-terminal residues 1-51 of the *E. coli lac* repressor determined by NMR in aqueous solution (Zuiderweg *et al.* 1983). The locations of three α-helical segments are indicated below the amino acid sequence.

included in the survey of Fig. 2.1. Once sequence-specific resonance assignments have been obtained by the sequential assignment strategy, the determination of the polypeptide secondary structure is usually achieved without much additional work (Fig. 2.3). This secondary structure identification is based on the recognition of characteristic patterns of conformational NMR constraints between polypeptide backbone protons, and does not require support by computational techniques (Billeter *et al.* 1982; Wüthrich *et al.* 1984; Pardi *et al.* 1984). As a consequence, one frequently encounters preliminary reports on NMR structures of proteins which include the sequence-specific resonance assignments and a description of regular secondary structure elements (Wagner *et al.* 1981; Zuiderweg *et al.* 1983; Williamson *et al.* 1984; Kline and Wüthrich 1985). The pattern recognition approach for secondary structure identification has also found interesting applications in investigations on the early stages of protein folding (Dyson *et al.* 1988a; 1988b).

2.2 Acknowledgments

The author's research on NMR structures of biopolymers is supported by special grants of the ETH Zürich and by the Schweizerischer Nationalfonds (project 31.25174.88).

References

Billeter, M., Braun, W., and Wüthrich, K. (1982). Sequential resonance assigments in protein ¹H nuclear magnetic resonance spectra: Computation of sterically allowed proton-proton distances and statistical analysis of proton-proton distances in single crystal protein conformations. *Journal of Molecular Biology* **155**, 321–46.

Braun, W. (1987). Distance geometry and related methods for protein structure determination from NMR. *Quarterly Reviews in Biophysics* **19**, 115–57.

Clore, G.M. and Gronenborn, A.M. (1989). Determination of three-dimensional structures of proteins and nucleic acids in solution by nuclear magnetic resonance spectroscopy. *Critical Reviews in Biochemistry and Molecular Biology* **24**, 479–564.

Dyson, H.J., Rance, M., Houghten, R.A., Lerner, R.A., and Wright, P.E.(1988a). Folding of immunogenic peptide fragments of proteins in water solution. I. Sequence requirements for the formation of a reverse turn. *Journal of Molecular Biology* **201**, 161–200.

Dyson, H.J., Rance, M., Houghten, R.A., Wright, P.E., and Lerner, R.A.(1988b). Folding of immunogenic peptide fragments of proteins in water solution. II. The nascent helix.*Journal of Molecular Biology* **201**, 201–17.

Havel, T.F. and Wüthrich, K. (1984). A distance geometry program for determining the structures of small proteins and other macromolecules from NMR measurements of intramolecular proton-proton proximities in solution.*Bulletin of Mathematical Biology* **46**, 673–98.

Kaptein, R., Boelens, R., Scheek, R.M., and van Gunsteren, W.F. (1988). Protein structures from NMR. *Biochemistry* **27**, 5389–95.

Kline, A.D. and Wüthrich, K. (1985). Secondary structure of the α-amylase polypeptide inhibitor tendamistat from *Streptomyces tendae* determined in solution by ^{1}H nuclear magnetic resonance. *Journal of Molecular Biology* **183**, 503–7.

Kline, A.D., Braun, W., and Wüthrich, K. (1988). Determination of the complete three-dimensional structure of the α-amylase inhibitor tendamistat in aqueous solution by nuclear magnetic resonance and distance geometry. *Journal of Molecular Biology* **204**, 675–724.

Otting, G., Marchot, P., Bougis, P.E., Rochat, H., and Wüthrich, K. (1987). Monitoring the purification by high-performance liquid chromatography of cardiotoxins from *Naja mossambica mossambica* using phase-sensitive two-dimensional nuclear magnetic resonance. *European Journal of Biochemistry* **168**, 603–7.

Pardi, A., Billeter, M., and Wüthrich, K. (1984). Calibration of the angular dependence of the amide proton-Cα proton coupling constants, $^3J_{HN\alpha}$, in a globular protein: Use of $^3J_{HN\alpha}$ for identification of helical secondary structure. *Journal of Molecular Biology* **180**, 741–51.

Qian, Y.Q., *et al.* (1989). The structure of the *Antennapedia* homeodomain determined by NMR spectroscopy in solution: Comparison with prokaryotic repressors. *Cell* **59**, 573–80.

Strop, P., Wider, G., and Wüthrich, K. (1983). Assignment of the ^{1}H nuclear magnetic resonance spectrum of the proteinase inhibitor IIA from bull seminal plasma by two-dimensional nuclear magnetic resonance at 500 MHz. *Journal of Molecular Biology* **166**, 641–67.

Wagner, G. and Wüthrich, K. (1982). Sequential resonance assignments in protein ^{1}H nuclear magnetic resonance spectra: Basic pancreatic trypsin inhibitor. *Journal of Molecular Biology* **155**, 347–66.

Wagner, G., Kumar, A., and Wüthrich, K. (1981). Systematic application of two-dimensional ^{1}H nuclear-magnetic-resonance techniques for studies of proteins. 2. Combined use of correlated spectroscopy and nuclear Overhauser spectroscopy for sequential assignments of backbone resonances and elucidation of polypeptide secondary sequences.*European Journal of Biochemistry* **114**, 375–84.

Wagner, G., *et al.* (1986). Sequence-specific ^{1}H-NMR assignments in rabbit-liver metallothionein-2. *European Journal of Biochemistry* **157**, 275–89.

Wider, G., Lee, K.H., and Wüthrich, K. (1982). Sequential resonance assignments in protein ^{1}H nuclear magnetic resonance spectra: Glucagon bound to perdeuterated dodecylphosphocholine micelles. *Journal of Molecular Biology* **155**, 367–88.

Widmer, H., Wagner, G., Schweitz, H., Lazdunski, M., and Wüthrich, K. (1988). The secondary structure of the toxin ATX Ia from *Anemonia sulcata* in aqueous solution determined on the basis of complete sequence-specific ^{1}H-NMR assignments. *European Journal of Biochemistry* **171**, 177–92.

Williamson, M.P., Marion, D., and Wüthrich, K. (1984). Secondary structure in the solution conformation of proteinase inhibitor IIA from bull seminal plasma by nuclear magnetic resonance. *Journal of Molecular Biology* **173**, 341–59.

Williamson, M.P., Havel, T.F., and Wüthrich, K. (1985). Solution conformation of proteinase inhibitor IIA from bull seminal plasma by ^{1}H nuclear magnetic resonance and distance geometry. *Journal of Molecular Biology* **182**, 295–315.

Wüthrich, K. (1986). *NMR of Proteins and Nucleic Acids*. Wiley, New York.

Wüthrich, K. (1989). Protein structure determination in solution by nuclear magnetic resonance spectroscopy. *Science* **243**, 45–50.

Wüthrich, K., Wider, G., Wagner, G., and Braun, W. (1982). Sequential resonance assignments as a basis for determination of spatial protein structures by high resolution proton nuclear magnetic resonance. *Journal of Molecular Biology* **155**, 311–19.

Wüthrich, K., Billeter, M., and Braun, W. (1984). Polypeptide secondary structure determination by nuclear magnetic resonance observation of proton-proton distances. *Journal of Molecular Biology* **180**, 715–40.

Zuiderweg, E.R.P., Kaptein, R., and Wüthrich, K. (1983). Secondary structure of the *lac* repressor DNA-binding domain by two-dimensional ^{1}H nuclear magnetic resonance in solution. *Proceedings of the National Academy of Sciences of the USA* **80**, 5837–41.

3

NMR Resonance Assignments

Joseph H. B. Pease and David E. Wemmer

3.1 Introduction

NMR (see Glossary) spectra contain information about the structure of a molecule through the chemical shift (which is sensitive to the local electronic environment), spin-spin coupling constants (which are sensitive to dihedral angles), and through relaxation [nuclear Overhauser effect (NOE)], which is sensitive to the positions of nearby spins (see Glossary for definitions of chemical shift and NOE). However, before any of this information can be put to use in determining the structure of a molecule (see Chapter 2), it must first be determined which resonances come from which spins. The process of associating specific spins in the molecule with specific resonances is called the sequence-specific assignment of resonances. For many years this was the major hurdle in interpretation of NMR data for biological molecules. However, over the past decade tremendous progress has been made through the introduction of multidimensional NMR experiments and through advances in spectrometer hardware. These improvements were combined with development of a systematic approach for analysis of the spectra, to make an efficient algorithm for obtaining assignments.

In this chapter we describe in some detail the process of assigning nuclear magnetic resonance spectra of biological macromolecules: proteins, nucleic acids, and carbohydrates. Historically, most NMR work has been done with protons (the common form of the hydrogen nucleus) since they are present at many sites in the molecules, at high abundance for each site, and are one of the most sensitive nuclei to detect. Protons continue to be the most important nuclei, since they are most directly related to structural information of interest; however, methods have also developed for assignment of essentially any other NMR-active nucleus (with spin $\frac{1}{2}$). In addition, these other spins are becoming more useful in 'filtering' proton spectra to simplify them, and for defining some of the connectivities within the molecules. This will be discussed in more detail below.

3.2 NMR methods

We will assume a basic familiarity with NMR spectroscopy, particularly Fourier transform (FT) NMR. There are numerous texts that treat this topic in detail, providing a useful starting point for those who are not comfortable with the concepts (Farrar and Becker 1971; Harris 1983; Derome 1987). The resonance from each spin (labelled i) in a molecule can be characterized by a chemical shift (δ_i) and a set of coupling constants to all other spins around it (J_{ij}) giving rise to a multiplet. Although the 'normal' one-dimensional (1D) spectrum is a useful reference for thinking about spectral regions in which particular types of multiplets occur, there is very little that can be done with 1D spectra for resonance assignments. Instead, essentially all contemporary work on biopolymers is done with two-dimensional (2D), or three-dimensional (3D) NMR. First, we will describe the basic experimental methods for the most commonly used of these multidimensional experiments, and discuss their information content. The application to particular classes of biomolecules will then be discussed in separate sections.

3.2.1 Basics of multidimensional NMR experiments

In a 1D FT NMR experiment the radiofrequency pulse tips the spins, and hence the macroscopic magnetic moment that arises from them, from their equilibrium position along the applied magnetic field (which defines the z axis). They then rotate in the x-y plane and are detected during the following time-period. The signal to noise (S/N) ratio is improved by

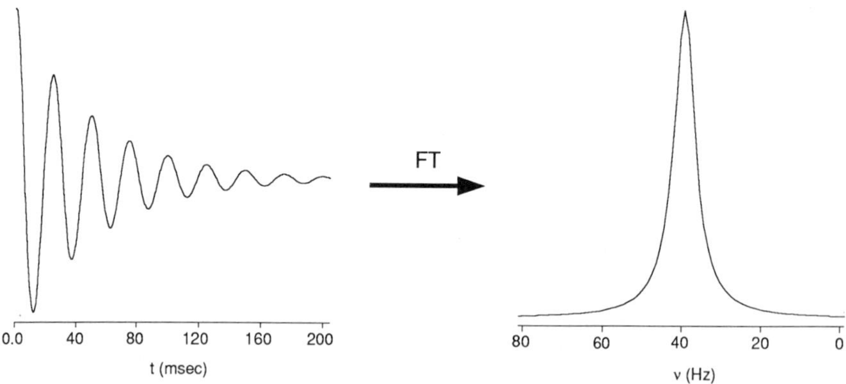

Fig. 3.1 One-dimensional Fourier transform (FT). On the left, a free induction decay (FID) which is the signal amplitude as a function of time. On the right, the FT of the FID, which is the signal amplitude as a function of frequency.

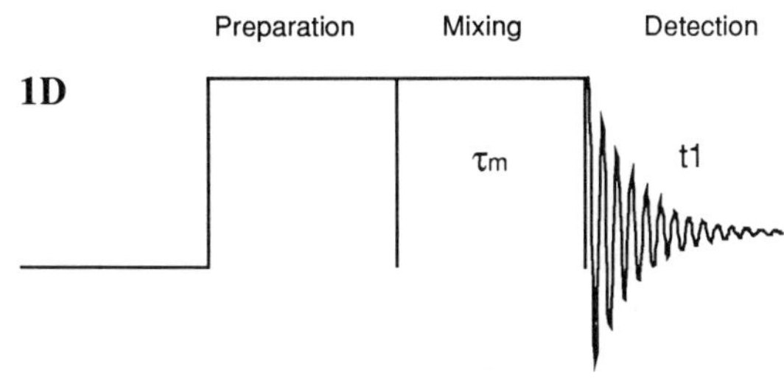

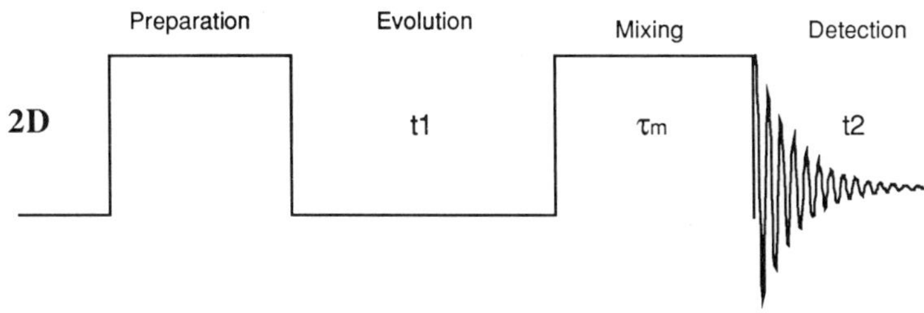

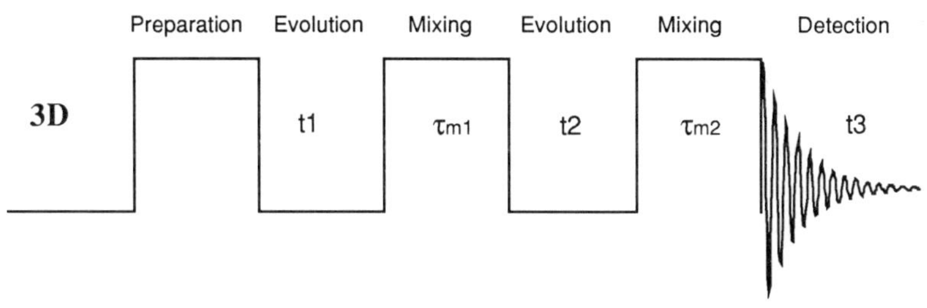

Fig. 3.2 1D, 2D, and 3D general experiments. Preparation, the spins are put into a known non-equilibrium state. Evolution, the time when spins are frequency labelled so that the origin of cross-peaks is known. Mixing, the time when the desired transfer of information takes place. Detection, the time when the signal is recorded. In the 1D experiment, the signal is acquired during t_1 and is a function of τ_m. In the 2D experiment, the value of t_1 is incremented to obtain a second time-domain and the acquired signal is a function of both t_1 and τ_m. Similarly for the 3D experiment, both t_1 and t_2 are incremented, to obtain a second and third time domains, and the signal acquired is a function of t_1, τ_{m1}, t_2, and τ_{m2}.

co-adding the signal after each pulse. The spectrum is obtained by Fourier transforming the time-domain signal to give an amplitude as a function of frequency ν, as shown schematically in Fig. 3.1. There are two important time constants which affect the behaviour of spins in all NMR experiments: T_1 and T_2, also known as spin-lattice and spin-spin relaxation times, respectively. T_1 is the time constant for the magnetization to return to its equilibrium value M_0 along z. A period of several times T_1 should be left between signal acquisitions to obtain proper amplitudes for all peaks. T_2, on the other hand, is the time constant for magnetization to return to its equilibrium value of zero in the x-y plane. A direct manifestation of T_2 is that individual lines in the spectrum will have a width $1/\pi T_2$.

To extend this to a 2D FT experiment, a second time-period is added. The experiment then is composed of sections: preparation, evolution, mixing, and detection, as shown in Fig. 3.2. The evolution period t_1 is increased in equal increments, and for each t_1 a complete signal is collected as a function of the detection time, t_2. In order to analyse the data, two Fourier transformations must be done, the first over t_2, giving a set of spectra with lines whose amplitudes are a function of the time-periods t_1 and τ_m (see Glossary). To determine the oscillation frequencies in t_1, the amplitude at each frequency point ν_2 is taken as a function of t_1 and is Fourier transformed over t_1. This then gives a two-dimensional spectrum which is an amplitude as a function of two frequency variables ν_1 and ν_2. These can be displayed as stacked plots, but are more convenient in a contour representation, in which lines of equal amplitude are drawn, as in a topographic map (see Fig. 3.3). The useful information from these spectra arises from the fact that the spins do not evolve independently throughout the experiment. If two spins, with frequencies ν_i and ν_j, exchange information during the mixing period of the experiment, then cross-peaks will arise away from the diagonal (at the intersection of horizontal and vertical lines through ν_i and ν_j), as is illustrated in Fig. 3.3. It is this correlation information which is crucial to the systematic assignment of resonances.

The further extrapolation to three dimensions by addition of another evolution and mixing period before detection should be relatively clear: preparation, evolution, mixing, evolution, mixing, and detection (Fig. 3.2). This is essentially just the concatenation of two 2D experiments, but of course the spins now retain information from the first section of the experiment into the second and third parts. To obtain the spectrum in this case three successive Fourier transforms must be done to give amplitude as a function of three frequencies. In this case cross-peaks lie off of the body diagonal of the cube of frequency space representing the spectrum, and arise from exactly the same correlation mechanisms as in 2D spectra. Representation of these data is difficult since intrinsically four dimensions are required. At present this is done either by storing the spectrum in a computer and

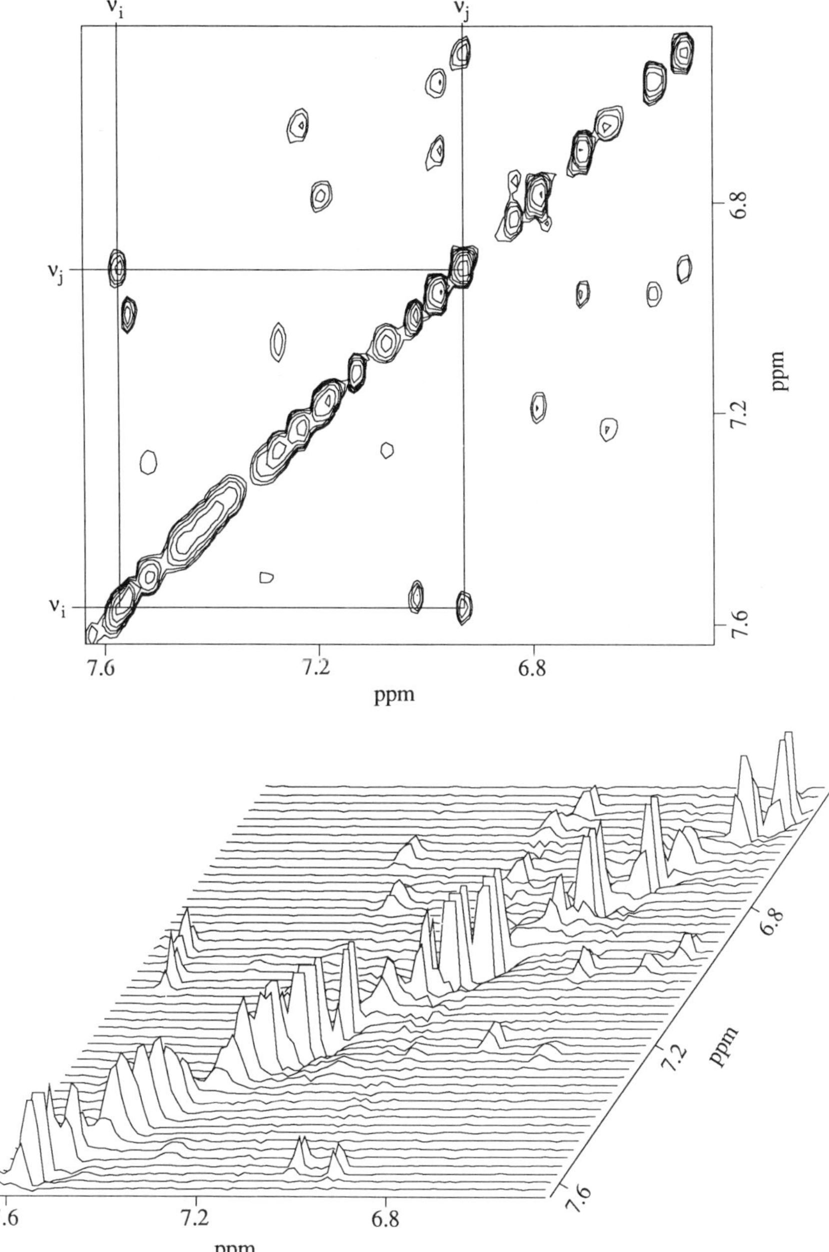

Fig. 3.3 Contour (top) and stacked (bottom) plots of the aromatic region of a protein. The contour plot is useful for finding cross peaks whereas the stacked plot is useful for determining whether there are any ridges or baseline rolls. The diagonal is similar to the 1D NMR spectrum and the cross-peaks represent correlations between spins. Note how cross-peaks appear on either side of the diagonal: at the intersection of horizontal and vertical lines through ν_i and ν_j.

drawing stereo-graphic representations of it, or by making stacks of two-dimensional contour plots where each plot represents a single plane of the 3D volume. This will become more concrete as examples are given below.

The wide variety of 2D and 3D NMR experiments in the literature arises from the different mixing mechanisms that can be used. There are two fundamentally different mixing mechanisms (causing transfer of information) found in NMR: the first uses coherence (through-bond scalar couplings) and the second uses relaxation (dipolar coupling through space). In both categories there are several variations of the basic experiment, which often have considerable practical importance. In the limited space of this chapter it will not be possible to describe all of these, but representative examples will be given. Most of these experiments can be done in a way that either retains all of the phase information (phase-sensitive mode) or ignores it (magnitude mode). In almost all cases, superior quality data are obtained by retaining the phase information, and henceforth we will assume that this has been done. The details of how this is accomplished are described in the literature, and will not be repeated here (Drobny *et al.* 1979; States *et al.* 1982).

3.2.1.1 COSY   The basic experiment based on coherence transfer through bonds is called COSY (for COrrelated SpectroscopY) (Jeener 1971; Aue *et al.* 1976). The pulse sequence, which is the simplest of all 2D experiments, is shown in Fig. 3.4. In this experiment the spins undergo rotations about one another, in addition to rotation about the applied magnetic field. During the mixing period of this experiment, the second pulse, J-coupled spins (through bond) exchange coherence and, hence, information about their rotation frequencies. The entire effect is well understood and can be calculated precisely for any combination of spins and couplings. For the COSY experiment a cross-peak between two spins, i and j, will occur at the positions (δ_i, δ_j) and

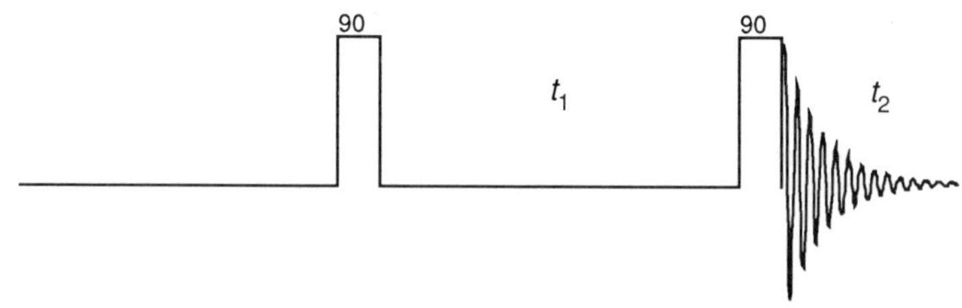

Fig. 3.4 The 2D correlated spectroscopy (COSY) pulse sequence. The 90s represent 90-degree pulses. The mixing period for this experiment is the last pulse. Cross-peaks are observed between spins which are J coupled. For protons, cross-peaks generally appear between protons two and three bonds apart.

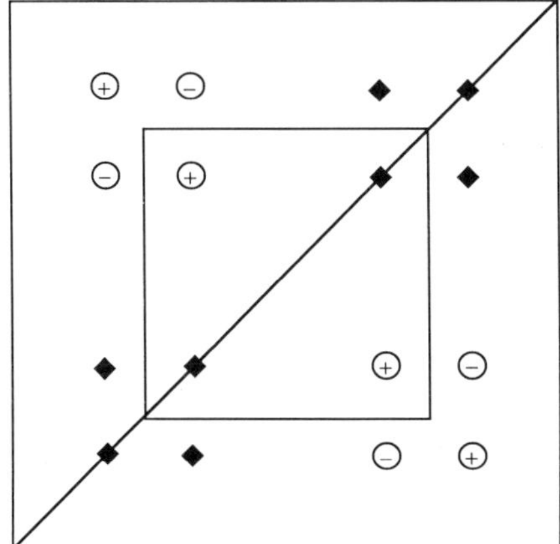

Fig. 3.5 A simulated 2D COSY contour plot showing the antiphase character of the cross-peaks. + represents positive intensity and − negative intensity. The box traces the centres of all of the peaks. The diagonal is dispersive (derivative-like).

(δ_j, δ_i) in the spectrum if, and only if, the two spins are directly coupled to one another.

In all of the biopolymers that will be discussed here, the primary structure logically breaks up the molecule into groups of coupled spins — there are normally one or two groups per residue. An important part of the assignment process is to identify which resonances belong to the same group. The existence of a COSY cross-peak identifies a pair of spins within a group. By looking for cross-peaks that have one frequency in common, the coupling pattern can be extended through the entire coupled group. This experiment has been used extensively for defining protons within individual residues for many types of biopolymers.

The nature of the transfer process is such that the cross-peaks arising from the coherence transfer are antiphase in character, that is half of the multiplet is 'up' and the other half is 'down', as shown schematically in Fig. 3.5. From the nature of this fine structure it is clear that destructive cancellation of the multiplet components will occur when the width of an individual component is larger than the splitting (Fig. 3.6). Thus, to be considered 'coupled' the coupling constant J_{ij} must be the same order of magnitude as the linewidth. There is no sharp cut-off, cross-peaks from couplings as small as $\frac{1}{5}$ of the linewidth (or even less) can be detected, but their amplitude is much lower than that for spins with couplings equal to or larger than the linewidth, and it is really the signal to noise ratio that limits their detection.

In COSY experiments peaks also occur on the diagonal, arising from

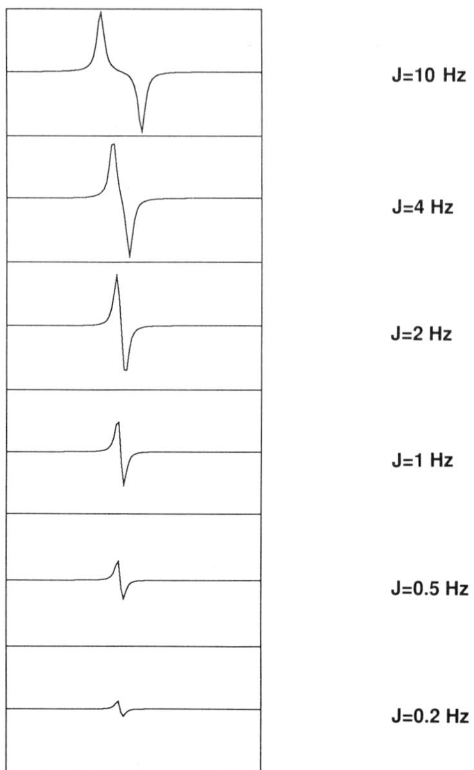

Fig. 3.6 Antiphase line intensity as a function of J coupling. The linewidth is 2 Hz. Note how the line intensity drops dramatically once the coupling is less than the linewidth. Alternatively, cancellation can be due to an increase in the linewidth, which is often the problem with larger biomolecules.

coherence that remained on the same spin during both parts of the experiment. In the normal COSY the diagonal signals are out of phase (dispersive, or derivative-like) relative to the cross-peaks and hence are much broader than the absorption signal. This sometimes interferes with detection of cross-peaks near the diagonal. A very useful modification of the COSY is to select only coherence that was transferred at some point in the experiment, which is achieved by applying a 'double quantum filter' (Rance *et al.* 1983). This is done by adding an extra pulse in the mixing period, and rotating its phase (direction in the *x-y* plane) through a cycle while alternately adding and subtracting the result from the averaged signal. The rotations are such that by the end of a cycle all coherence from spins that do not have coupling partners (singlets in the spectrum) and all coherence that remained on the same spin during the period t_1 adds to zero, but that which was transferred adds coherently. In the final spectrum, then called a double quantum filtered COSY (or DQF COSY), all peaks have antiphase character, so the region near the diagonal is not obscured. This version is now the 'usual' COSY experiment for biomolecular applications, even though there is a factor of 2 loss in sensitivity relative to the ordinary COSY. When run at high digital resolution, the cross-peak lineshapes provide detailed information about the couplings within the multiplets involved in the cross-peak. More sophisticated filters can also be applied selecting either for groups of N or more spins (an N-quantum filter) (Rance and Wright 1986), or for specific coupling patterns (spin topology filters) (Ernst *et al.* 1987). However, these more complex filters have not been as useful in practice as the DQF (see Glossary), and hence will not be discussed further.

3.2.1.2 RELAY, TOCSY, and MQ   It is usually chemical shift degeneracies (more than one proton resonating at the same frequency) that cause problems with the assignment of resonances in large molecules. Because the grouping of spins is an important aspect of the assignment process, techniques have been developed which circumvent the degeneracy problem. The simplest of these is the relayed coherence transfer experiment (RELAY) (Eich *et al.* 1982). In this method, the coherence transfer which occurred in the COSY experiment is repeated by addition of another pulse after a fixed second evolution period (Fig. 3.7). Since this period is of fixed length 2τ (to maintain the experiment as 2D), there is no information from chemical shifts during it, and in fact possible decreases in cross-peak amplitude due to chemical shift rotations can be avoided by inserting a refocusing (180° pulse) in the centre of this period. The RELAY spectrum has not only regular COSY-type cross-peaks (from directly coupled spins), but also cross-peaks

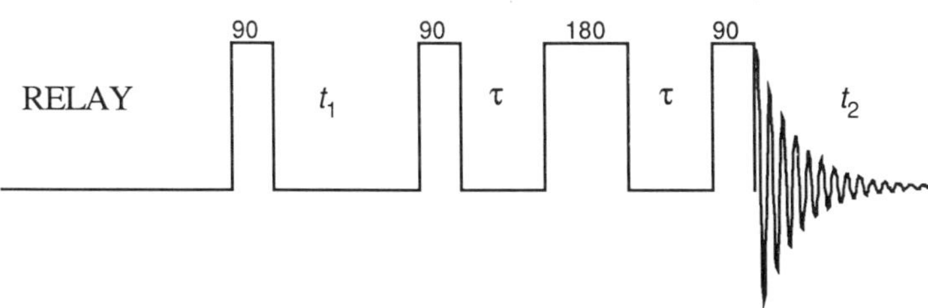

Fig. 3.7 The 2D relayed coherence transfer spectroscopy (RELAY) pulse sequence. The value of τ is adjusted to maximize the desired RELAY. Normal COSY cross-peaks as well as RELAY cross-peaks are observed. RELAY cross-peaks are between spins coupled to a common spin (see Fig. 3.8).

between spins that are not directly coupled, but which have a common coupling partner. This is shown schematically in Fig. 3.8 for a complex system of protons within an amino acid. The advantage of this approach is that when a degeneracy in chemical shift occurs for two separate spin systems (see Glossary), the RELAY has redundant information which often allows unambiguous identification of the coupled spins.

The transfer of coherence in COSY and RELAY experiments is driven by the spin-spin couplings, and the rate of transfer is dependent on the value of coupling constants in the system. This means that the amplitude of RELAY cross-peaks depends on the value of the delay (2τ) in the experiment. This dependence is complex, but has been calculated for a number of amino-acid and nucleic-acid spin systems (Bax and Drobny 1985; Chazin and Wüthrich 1987). Fortunately, many of the coupling constants involved are not strongly conformation dependent, so these calculations can be used for selection of the optimum delay. The T_2 value is also an important consideration. It can be estimated from a 1D spectrum and, luckily, the dependence on the precise value is not strong (Bax and Drobny 1985).

In cases where chemical shift crowding is severe (especially common for polysaccharides), it can be advantageous to repeat this process once (or even several times) again to give multiple RELAY information. The basis for coherence transfer is the same for each step, but with each additional step cross-peaks will occur for pairs of spins with one more intervening spin allowed. Thus in a double relay experiment a system such as $H_a-H-H-H_b$ will give a cross-peak between protons a and b. In general,

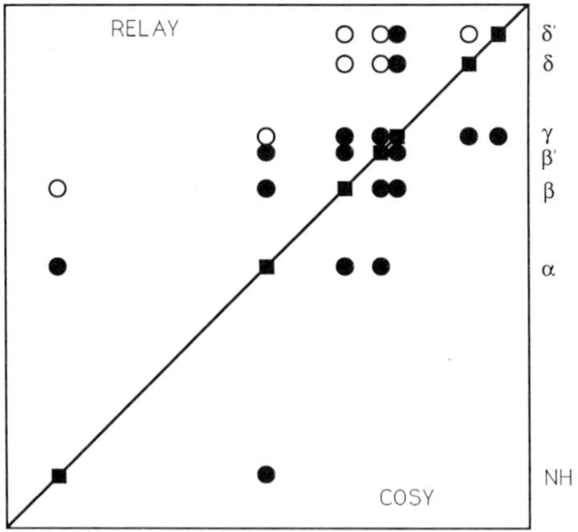

Fig. 3.8 Half of a simulated RELAY (above diagonal) and half of a simulated COSY (below diagonal) spectrum of a Leu in a protein. The filled circles represent COSY-type cross-peaks and the open circles RELAY-type cross-peaks. The protons are labelled on the right.

it is not possible to transfer all the coherence from one spin to the next, and so the amplitude of remote cross-peaks will decrease with each successive step required to generate them. During all delays in the experiment spin-spin relaxation also occurs, which reduces the total amount of magnetization in the system, again reducing cross-peak amplitudes. In practice, for macromolecules this latter effect usually dominates, and limits the number of RELAY steps to two or three. The multiple RELAY experiment has the advantage that successive transfers are done in discrete steps, unlike TOCSY (discussed below). The delays for the individual steps can be optimized independently, which is important if coupling constants for different steps are quite different from one step to the next.

A more recent addition to the repertoire of experiments for identifying extended couplings is TOCSY [for TOtal Correlation SpectroscopY (Braunschweiler and Ernst 1983); also known as HOHAHA for HOmonuclear HArtman HAhn spectroscopy (Davis and Bax 1985; Bax and Davis 1985a)]. In this case the stepwise mixing is replaced by a period of 'isotropic mixing', so called because the effects of all Zeeman couplings (chemical shifts) are removed during this time. The isotropic mixing is achieved by applying a sequence of pulses which effectively averages out chemical shifts, Fig. 3.9. Although the real behaviour is more complex, this can be simply thought of as a sequence of 180° pulses, each of which refocuses (and hence removes on average) the chemical shift. During this multiple pulse sequence the spins behave as though they are all strongly coupled, and coherence that begins on one spin can be spread to all others that are part of the coupled group. (See Glossary for definition of isotropic mixing.) In the 2D experiment, coherence is frequency-labelled during the first evolution period, then is detected on all members of the coupled group, providing information equivalent to all orders of RELAY combined in one experiment. This is indicated schematically in Fig. 3.10.

In TOCSY the mixing between spins does not occur instantaneously, and

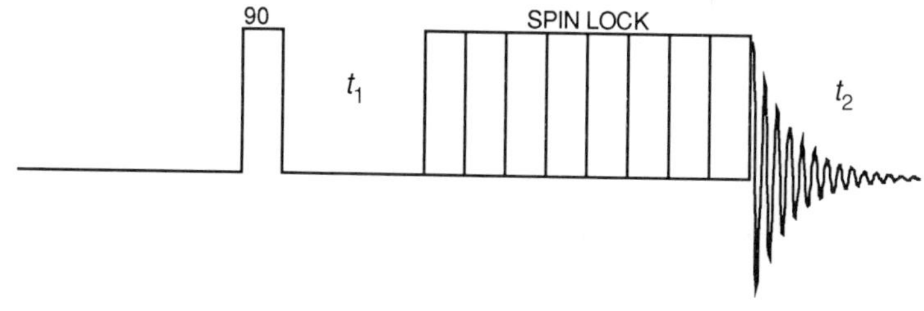

Fig. 3.9 A homonuclear Hartmann Hahn spectroscopy [HOHAHA (TOCSY)] pulse sequence. Isotropic mixing occurs during the spin lock sequence giving rise to cross-peaks between all spins in a coupled spin system.

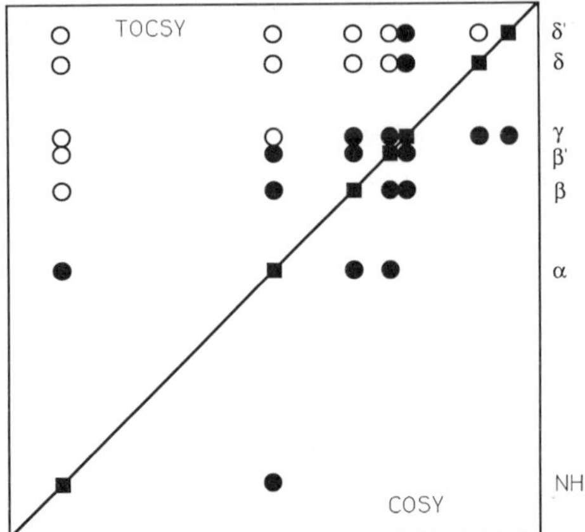

Fig. 3.10 Half of a simulated TOCSY (HOHAHA) (above diagonal) and half of a simulated COSY (below diagonal) spectrum of a Leu in a protein. The filled circles represent COSY-type cross-peaks and the open circles TOCSY-type cross-peaks. Note how cross-peaks in a TOCSY are observed between all of the protons, which is similar to the information obtained from all orders of a RELAY. The protons are labelled on the right.

the number of spins intermediate between the initial spin and the detected spin can be adjusted by modifying the length of the mixing period of the experiment. The evolution during the mixing period is a function of the coupling constants for all of the spins involved, and is quite complex. The behaviour for some amino-acid spin systems has been calculated and can be used to choose appropriate mixing times (Remerowski *et al.* 1989). TOCSY has an additional advantage that part of the magnetization transfer is in phase and thus cross-peak cancellation does not occur at higher molecular weights. Because of the higher information content relative to RELAY experiments, the TOCSY has become the preferred method for obtaining extended connectivity information.

A final method for obtaining extended connectivity information is multiple quantum spectroscopy (MQ). In any pulse sequence with two or more pulses there are coherent states created which involve multiple spins. In the cases discussed above these are never detected, and hence can be ignored in the description of how the experiments work. However, if the pulse sequence is rearranged slightly, then these can be detected and used as an alternate source of information. The MQ pulse sequence is shown in Fig. 3.11. As with the other experiments, the initial step is the creation of coherence. However, in this case it is multiple quantum coherence, and so the usual single-pulse 'preparation' must be replaced by at least two pulses. Evolution then occurs

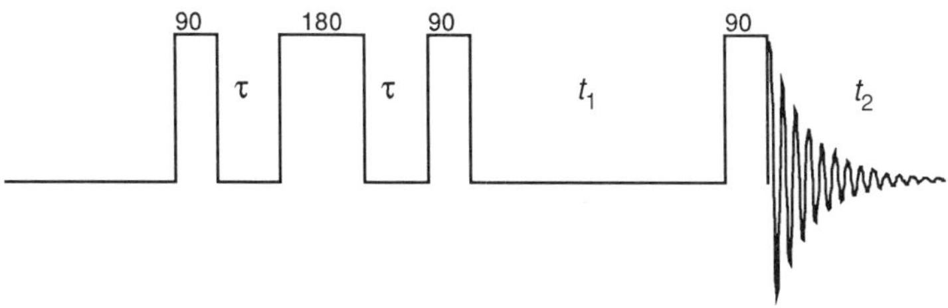

Fig. 3.11 The 2D multiple quantum spectroscopy (MQ) pulse sequence. The phases of the pulses are selected to pick the order of coherence. τ is chosen to maximize the transfer into multiple quantum coherence.

at the multiple quantum frequency during t_1, and then another pulse converts the multiple quantum coherence back into ordinary coherence which is detected. The multiple quantum frequencies are sums and differences of the involved spins' chemical shifts, and there is some change in the multiplet structure depending on the number of spins involved. By cycling the phases of pulses in appropriate ways, coherences from a specific number of spins can be selected (Ernst *et al.* 1987). The MQ frequencies can be detected either on the spins that were involved directly in generating it (direct connectivities), giving COSY-like information, or on spins coupled to those involved in generating it (remote connectivities), giving RELAY-like information. Both of these occur in a single spectrum, and can be distinguished by the phase characteristics of the lines (Braunschweiler *et al.* 1983; Rance and Wright 1986).

3.2.1.3 NOESY and ROESY   In all of the experiments discussed above, information transfer between spins was mediated by the through-bond spin-spin couplings. The basic experiment in the second approach for information transfer (through space-dipolar relaxation mediated) is Nuclear Overhauser Effect SpectroscopY (NOESY) (Jeener *et al.* 1979; Kumar *et al.* 1980; Macura and Ernst 1980). In this case the z component of magnetization is used, rather than the x-y, and the experiment must be modified to achieve this. To create a frequency-labelled z-population distribution, two pulses must be used for preparation, as shown in Fig. 3.12. The spins are tipped into the x-y plane by the first pulse, then are rotated by chemical shifts, and finally the component perpendicular to the pulse direction is tipped back to z by the second pulse. Relaxation is then allowed to occur during the mixing time, τ_m, and the amount of z-magnetization at each frequency is then detected by application of the final pulse.

During the mixing period cross-relaxation occurs if the spins are close to one another in space, for protons they must be less than about 5 Å apart,

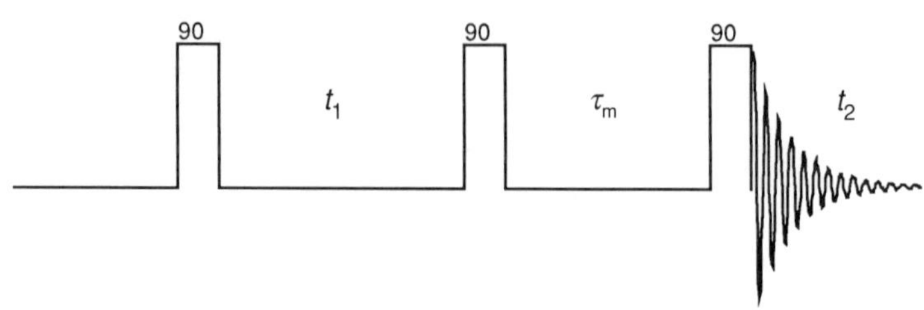

Fig. 3.12 The nuclear Overhauser effect spectroscopy (NOESY) pulse sequence. The phases of the pulses are cycled to minimize coherence transfer. Cross-peaks are observed between protons close in space (<5 Å). The maximum distance to give an observable cross-peak depends on the value of τ_m.

and for other spins they must be even closer. The magnitude of the effect (and hence the intensity of the cross-peaks) depends on the rate of cross-relaxation relative to the rate of longitudinal relaxation (through all mechanisms, not just dipolar). In addition, there is also a strong dependence on correlation time. For a pair of like spins (e.g. two ^{1}Hs) the effect is positive at short correlation times (the maximum enhancement of a line's intensity is 50 per cent) and negative at long correlation times (the maximum loss of amplitude is 100 per cent), going through zero amplitude when the correlation time is approximately the inverse of the NMR frequency. For pairs of unlike spins (e.g. a ^{1}H$-^{13}$C pair) the effect is largest for very short rotational correlation times (in small molecules), and drops to almost zero when the correlation time reaches the inverse of the NMR frequency. This occurs at about 1 kD molecular weight in water solutions. The long correlation time limit is applicable for most biomolecules, and in this limit the initial rate of cross-relaxation between a pair of like spins, i and j, scales as τ_c/r_{ij}^6. By adjusting the mixing time in the experiment, one can then effectively adjust the maximum distance between spins for which cross-peaks will be seen. In this limit there is a useful relationship between the volume of a cross-peak and the distance between spins. For long mixing times magnetization may be transferred several steps along a chain of spins (a process known as spin diffusion), in which case observable peaks no longer indicate spins that are very close in space. With judicious choice of mixing time this problem can usually be avoided.

The second form of this experiment is the rotating frame Overhauser effect spectroscopy [ROESY (Bax and Davis 1985b); also known as CAMELSPIN (see Glossary) (Bothner-By *et al.* 1984)]. Again, cross-peaks come from relaxation through dipolar couplings, but this time in the rotating frame. Spins are tipped into the *x-y* plane and are frequency-labelled (as for NOESY), but then are 'spin-locked' in one direction by application of a long

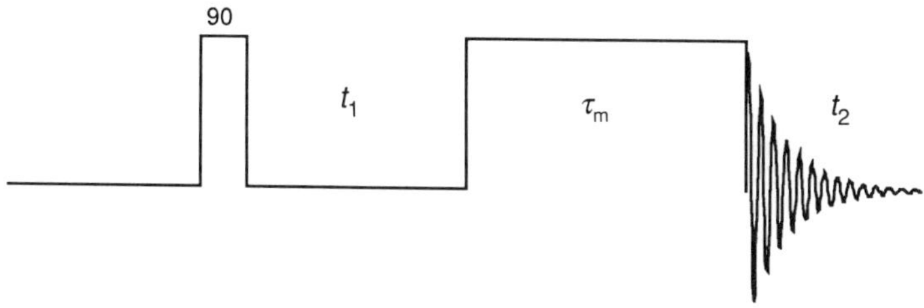

Fig. 3.13 The rotating nuclear Overhauser effect spectroscopy (ROESY) pulse sequence. The mixing period is one long pulse.

pulse, as shown in Fig. 3.13. During this locking pulse spins undergo cross-relaxation, but since the 'flip-flops' of spins occur relative to the rotating field, the correlation time dependence is different. In ROESY the behaviour is always similar to the short correlation time limit for NOESY: the diagonal and cross-peaks are opposite in phase, and the maximum amplitude effect is between 38.5 per cent ($\omega\tau_c \ll 1$) and 67.5 per cent ($\omega\tau_c \gg 1$) (Bothner-By *et al.* 1984). This experiment is most important for small molecules that have τ_c values near the zero in the NOE curve, but may also be helpful for macromolecules since spin diffusion is reduced (due to the lower maximum effect) and is more easily identified (since spin diffusion peaks have opposite signs from direct ROE peaks). In addition, it is helpful when both exchange and cross-relaxation are present (Otting and Wüthrich 1989).

3.2.1.4 HETCOR and HMQC In ordinary COSY experiments on protons, the combination of pulses and delays gives rise to coherence transfer.

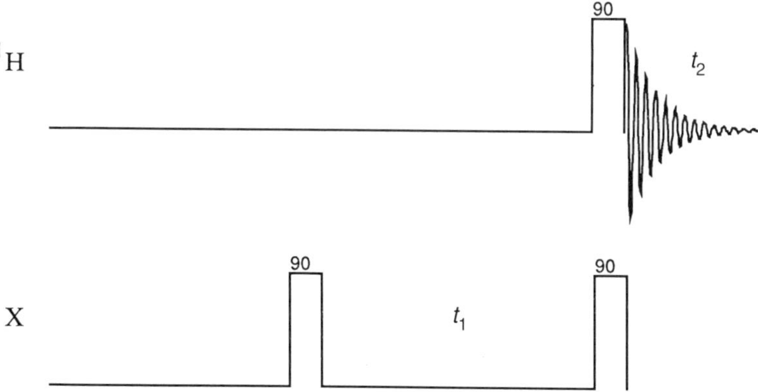

Fig. 3.14 The heteronuclear correlated spectroscopy (HETCOR) pulse sequence. Cross-peaks are observed between a heteroatom and the proton(s) it is coupled to. The most common heteroatoms are ^{13}C, ^{15}N, and ^{31}P.

When two different kinds of spins are considered, the most common being
^{1}H$-$X, where X $= ^{13}$C, ^{15}N, or ^{31}P in biomolecules, individual pulses can
be applied to either the ^{1}Hs or the Xs separately, since they have different
resonance frequencies. However, the same kind of coherence transfer can be
achieved as in a COSY, but now going from ^{1}H to X or X to ^{1}H. In the
simplest experiment, HETCOR (see Glossary) (Maudsley and Ernst 1977),
a single pulse on one of the spins begins the frequency-labelling for time t_1,
then pulses are applied to both types of spins in the mixing period, and finally
the signal is detected (Fig. 3.14). The coherence transfer leads to antiphase
character of the multiplet lines, as for COSY. For directly bonded ^{1}H$-$X
pairs, the coupling constant is almost independent of the other bonds to X,
and extra delays can be introduced to refocus the multiplet to be in phase.
An alternative approach, which has found more practical use, is to use the
^{1}H$-$X double quantum frequency in the first dimension and the proton
frequency in the second, called HMQC (Fig. 3.15) (Müller 1979). From such
experiments the X frequency is derived directly. The most sensitive way to
carry out the experiment is to detect the signal from the protons, since it is
at higher frequency (this can be up to 1000 times more sensitive than direct
detection of X). For X $= ^{13}$C or ^{15}N at their natural abundance, sensitivity
presents a problem since only 1 per cent or 0.3 per cent of the protons will
be attached to an X, the rest will be attached to inactive isotopes ^{12}C or ^{14}N.
Isotope enrichment can sometimes be used to solve the sensitivity problem.
The unwanted proton signals can be suppressed very efficiently since the

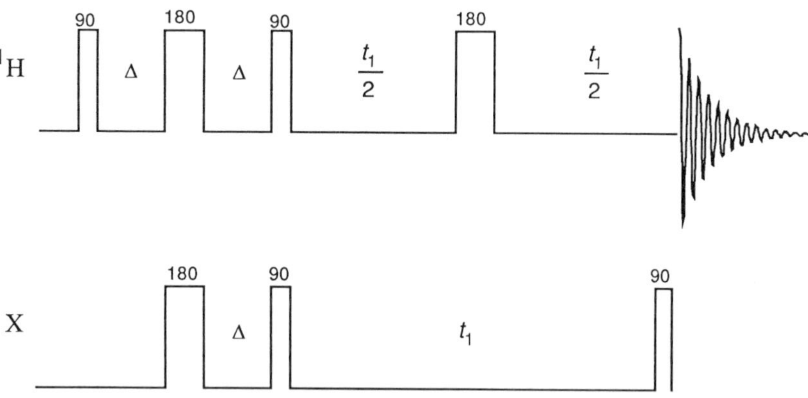

Fig. 3.15 The heteronuclear multiple quantum coherence spectroscopy (HMQC)
pulse sequence. The value of Δ is adjusted to maximize the transfer into multiple
quantum coherence. Cross-peaks are identical to those observed from the HETCOR.

phase of proton signals attached to an active X will change with the phase of the X pulse, whereas the phase of proton signals on inactive isotopes remains constant. Co-addition of results from experiments with appropriate cycles of phases can be used to keep only the active ^{1}H$-$X signals. Since the ^{1}H$-$X couplings for ^{13}C and ^{15}N are large (≥ 100 Hz), antiphase cancellation does not occur and the efficiency of transfer is very high even when lines are rather broad (high molecular weight). This is an especially advantageous aspect of this experiment for these nuclei. For ^{31}P, polarization transfer usually involves protons that are three or four bonds away, and hence have much smaller and more variable coupling constant values. The polarization transfer can still be done, but with lower efficiency (and hence S/N ratio). ^{31}P has the advantage that it has 100 per cent natural abundance.

3.2.2 3D NMR

From the discussion of 2D NMR it should be clear that in 3D NMR spectra there are correlations of three different frequencies, generated through the two different mixing times of the experiment. The mixing mechanisms are the same as in 2D NMR, that is 3D NMR experiments are essentially combinations of 2D NMR experiments run end-to-end (Fig. 3.2). There are many different 2D experiments which give rise to even more 3D ones. Of these, a relatively small number has been used so far, the important types being represented by homonuclear (all proton) NOESY-TOCSY and NOESY-NOESY, and heteronuclear TOCSY-HMQC and NOESY-HMQC. These spectra can be thought of as a two-dimensional spectrum corresponding to the first experiment listed, e.g. the NOESY of a NOESY-TOCSY, with the

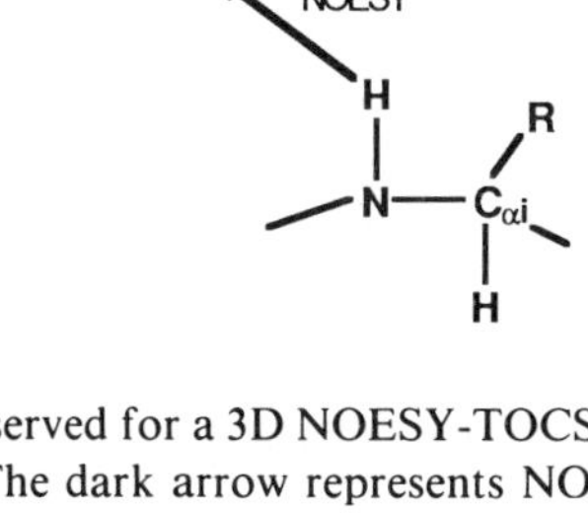

Fig. 3.16 The transfer observed for a 3D NOESY-TOCSY between two amino acids as described in the text. The dark arrow represents NOESY type transfer, and the light arrows TOCSY transfer.

detection done on all atoms connected by the second type of experiment, e.g. TOCSY in this case, breaking the 2D peak up into a set of peaks in the third dimension. For example, an NOE from an NH proton of residue i to an α of another residue j (a cross-peak at $\nu_{\text{NHi}}\, \nu_{\alpha\text{Hj}}$ in the NOESY) would also be detected at the other frequencies from protons connected to αH_j by the TOCSY, such as NH_j and βH_j in the third dimension, Fig. 3.16. The single frequency of detection αH_j could be degenerate with other spins (for large molecules this becomes very likely), but when the detection is done on the whole group, NH_j, αH_j, and βH_j (or even further out along the side-chain), the identification of the residue becomes clear. Since neither the NOESY or TOCSY gives rise to terribly strong cross-peaks, the intensities of the 3D peaks will be even weaker. Still, this experiment has been used successfully for proteins and carbohydrates. If the third dimension is generated by transfer to a heteroatom, as in NOESY-HMQC, then the large $^1\text{H}-\text{X}$ couplings make this transfer very efficient. In this case there is only a single detection of the NOESY cross-peak, instead of the multiple detection of the NOESY-TOCSY, but the resolution is increased by spreading the NOESY peaks into the third dimension by the chemical shift of the X atoms. For proteins this has been done with ^{15}N labels at the amide position.

3.3 Protein assignments

The idea for systematic assignment of proton resonances in proteins seems to have evolved from 1D studies on basic pancreatic trypsin inhibitor (BPTI) by Wüthrich and his co-workers (Dubs *et al.* 1979). They realized that in β-strand regions the αH from one residue is quite close to the NH of the following residue, and hence these should show a relatively strong NOE. The connection from any amide to the αH of the same residue could then be found through decoupling experiments. The NOE $\rightarrow$ coupling identification could be followed stepwise along the chain. At about the same time that NOESY and COSY were first applied to proteins to follow the two types of connectivity much more efficiently than 1D experiments could, it was realized that the covalent linkage of a protein along the backbone makes the idea of sequential connections much stronger (Wüthrich *et al.* 1982).

Through an analysis of known protein structures it was shown that, regardless of the local conformation of the backbone, there is a very high probability that at least one proton among the NH, αH, or βHs from one residue will be near (less than 3.5 Å, or more importantly within the allowed NOE range) to the amide of the following residue (Billeter *et al.* 1982; Wüthrich *et al.* 1984). Thus it should always be possible to step from residue to residue through the protein using a combination of NOESY and COSY spectra. If the COSY and NOESY data are of high quality (cross-peaks can be distinguished from noise easily) and there are no degeneracies in chemical shifts, then this kind of tracing can be done from one end of the molecule to the other. However, most often one or both of these conditions are not met, especially the lack of chemical-shift degeneracy for large molecules. A degeneracy can be thought of as a branch point — that is there are two possible spin systems (residues) that could be identified as the neighbour at that particular step. One obvious way to resolve the problem is to continue along both possible assignment paths until several amino acids are connected. If the side-chains of these amino acids can be identified, then the possible assignments can be checked against a known primary sequence. By the time a few amino acids are connected sequentially, it is usually possible to unambiguously identify one of the possible branches as the correct one. This process is then continued until all resonances are assigned. While this process is fairly simple in principle, there are many technical details that make it difficult in practice. We will now discuss the problem of identification of amino acids, and then the types of sequential NOEs that are typically observed. This method assumes that the primary sequence of the protein is known. In several cases there have been errors in the chemical sequencing, which were found through disagreement with the NMR, and could be partially corrected from the NMR data (for example, Wemmer *et al.* 1986).

3.3.1 Identification of amino acids

The primary information for recognizing the type of amino acid associated with each resonance comes from couplings established in COSY (normally DQF) spectra. As much of the spin system as possible for each amino acid must be identified. This includes an amide proton (NH), an α-proton (αH), one or more β-protons (βHs) and, for long side-chains, γHs, δHs, and even ϵHs (see Fig. 3.17). In general, the proton resonances from different types of protons on side-chains can be roughly grouped based on chemical shift. (Fig. 3.18). For example, the amide protons usually resonate in the region of 6–11 p.p.m. (see Glossary), αHs in the region 3.5–6 p.p.m., and CH_3s in the region of 0.5–2.5 p.p.m. However, the chemical shifts are dependent on environment, both local conformation of the backbone and also contacts with side-chains from residues distant in the primary structure but close in space. For this reason the ranges of chemical shifts must always be interpreted loosely — in special cases resonances from the specified types of protons could occur well outside the given range. The most extreme examples are proteins containing haeme groups, giving rise to large ring current effects, and those containing paramagnetic metal ions, which can give rise to huge shifts through hyperfine interactions with the unpaired electron. Recently there have been some tabulations of chemical shifts found for each type of residue in different proteins (Gross and Kalbitzer 1988). It is clear from these studies that chemical shift is a useful guide, but should

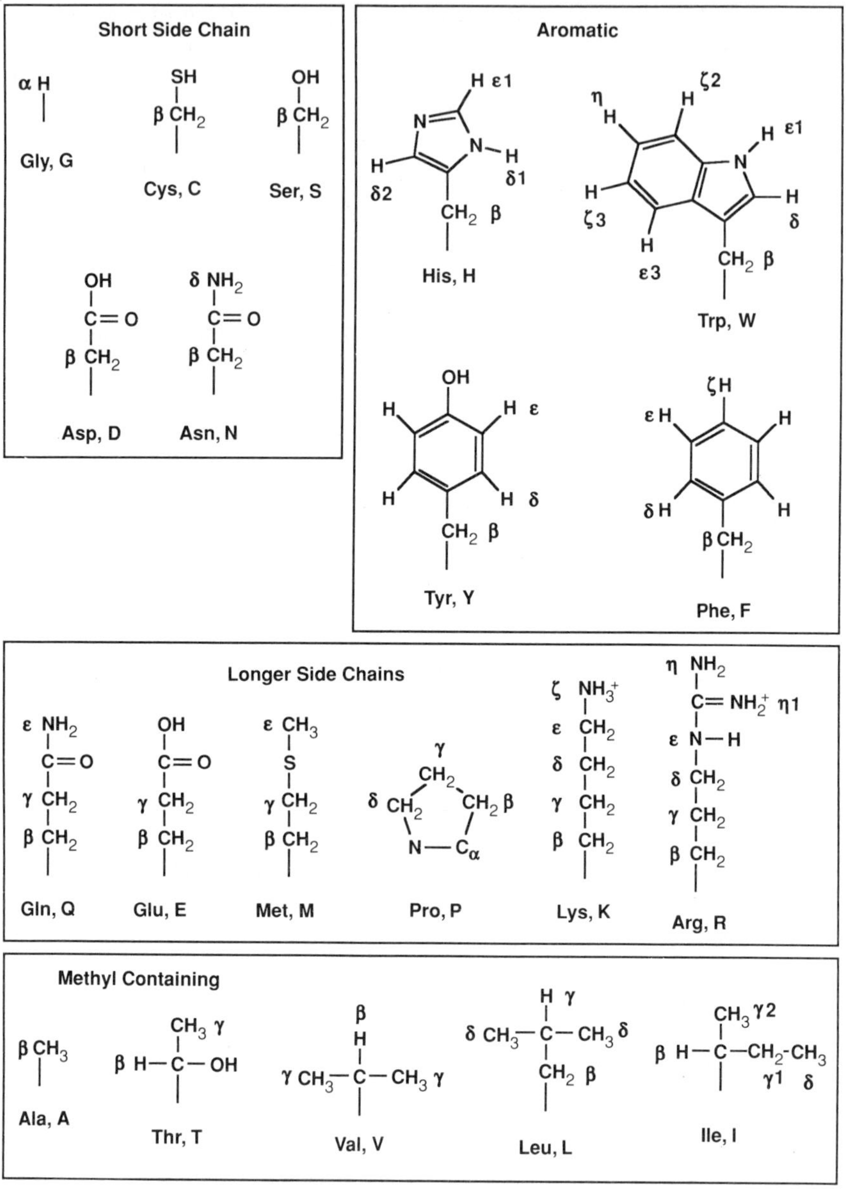

Short Side Chain

α H

Gly, G

SH
β CH$_2$

Cys, C

OH
β CH$_2$

Ser, S

OH
C=O
β CH$_2$

Asp, D

δ NH$_2$
C=O
β CH$_2$

Asn, N

Aromatic

H ε1
N
N—H
δ1
δ2
CH$_2$ β

His, H

η
H
ζ2
H
H ε1
N
ζ3
H
ε3
δ
CH$_2$ β

Trp, W

OH
H
H ε
H
H δ
CH$_2$ β

Tyr, Y

ζH
ε H
H
δ H
H
β CH$_2$

Phe, F

Longer Side Chains

ε NH$_2$
C=O
γ CH$_2$
β CH$_2$

Gln, Q

OH
C=O
γ CH$_2$
β CH$_2$

Glu, E

ε CH$_3$
S
γ CH$_2$
β CH$_2$

Met, M

γ
CH$_2$
δ CH$_2$ CH$_2$ β
N — C$_\alpha$

Pro, P

ζ NH$_3^+$
ε CH$_2$
δ CH$_2$
γ CH$_2$
β CH$_2$

Lys, K

η NH$_2$
C= NH$_2^+$ η1
ε N—H
δ CH$_2$
γ CH$_2$
β CH$_2$

Arg, R

Methyl Containing

β CH$_3$

Ala, A

CH$_3$ γ
β H—C—OH

Thr, T

β
H
γ CH$_3$—C—CH$_3$ γ

Val, V

H γ
δ CH$_3$—C—CH$_3$ δ
CH$_2$ β

Leu, L

CH$_3$ γ2
β H—C—CH$_2$-CH$_3$
γ1   δ

Ile, I

Fig. 3.17 The 20 common amino acids. The amino acids are grouped as in the text. The proton positions are labelled and both the three-letter and one-letter amino acid codes are shown.

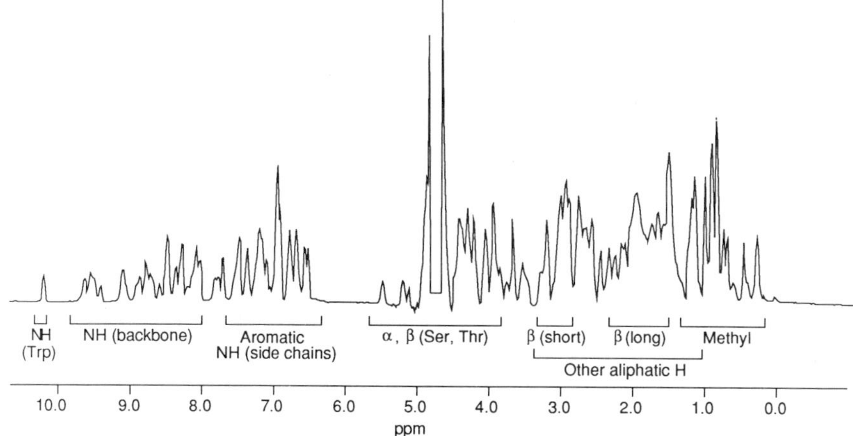

Fig. 3.18 A 1D proton NMR spectrum of a 48 amino-acid peptide. The chemical shift regions of various protons are indicated. The H_2O peak at 4.8 p.p.m. has been erased.

not be used as an absolute identifier. A similar study has shown that chemical shifts of the same protons on different residues do correlate with secondary structure to some degree, but again the scatter is rather large (Pardi *et al.* 1983a; Szilagyi and Jardetzkey 1989).

The amide protons of solvent-exposed residues exchange fairly rapidly with protons from solvent water. Thus, when a sample is dissolved in ^2H_2O (D_2O) solution these amide protons are replaced by deuterons, eliminating these amide proton resonances from the spectrum. Hence, to observe all amide resonances from a protein, experiments must be done in ^1H_2O solutions. There are a number of methods for suppressing the *c.* 100 M proton signal from water in order to detect the *c.* mM signals from a protein. The most common way of achieving this is to saturate the solvent signal during the delays in the experiment. This can cause saturation of αH resonances that have the same chemical shift as water, although with special additions to the pulse sequence [SCUBA (Brown *et al.* 1988) and preTOCSY (Otting and Wüthrich 1987)] these can usually be recovered. (See Glossary for definitions of preTOCSY and SCUBA.) When the solvent signal suppression is done carefully, the NH to αH cross-peaks for each residue (except Pro which has no NH) can be seen in a characteristic spectral window, commonly called the fingerprint region, roughly 6–11 p.p.m. for amides and 3.5–6 p.p.m. for αHs. The strength of a cross-peak depends to a large extent on the ratio of coupling constant to linewidth. If linewidths are uniform throughout the protein (usually the case), then differences in amplitude arise just from differences in J values. In extended chain regions (as in β-sheets) the NH-αH coupling is near its maximum of about 9–10 Hz, and strong COSY peaks will be seen. However, for helical regions the J value is considerably smaller, about 3–4 Hz, and cross-peaks will be correspondingly weaker. The fingerprint region also contains COSY peaks between amino and guanidino protons on Lys and Arg residues to the side-chain ϵHs and δHs, respectively. The nearby 6–8 p.p.m. region also contains weak COSY cross-peaks between the side-chain amides of Asn and Gln residues, as well as aromatic-residue cross-peaks.

Each NH-αH correlation should line up in αH chemical shift with cross-peaks from the αH to the βHs (except Gly which has two αHs). For long side-chains the βHs should then correlate further to γHs, etc. However, there are several problems that occur in trying to follow these patterns. First, even when solvent suppression is done well, some resonances near the water chemical shift (c. 4.7 p.p.m. at ambient temperature) will be obscured by the residual water signal. This can be avoided by working in D_2O solution, but at the expense of losing information about many of the amide protons. If conditions of temperature, pH, spectrometer settings, etc. are carefully matched, then data from separate H_2O and D_2O experiments can be combined to provide complete information. The second problem is that typically there will be degeneracies in αH chemical shifts, making association of NH and βHs belonging to the same residue ambiguous. Thirdly, some of the needed COSY connectivities for following the full side-chain system are very near the diagonal in crowded regions. These latter two problems are best solved by using TOCSY data together with the COSY to obtain extended correlation information. Particularly useful in this respect are the correlations from NHs to βHs and γHs directly. Finally, for some types of residue there are no observable couplings between some side-chain protons and the βHs (such as aromatic amino acids, Fig. 3.17). In this case NOESY experiments often provide useful information; however, care must be used when interpreting data since there is no guarantee that the cross-peaks seen are intra-residue. When good correlation information is available, the type of amino acid may be classified. The different types of amino acids are discussed below.

3.3.1.1 Short side-chains: G, S, C, D, and N   Glycine can usually be identified in several ways. First, it is the only amino acid with two α-protons, and hence is the only amino acid that can give rise to two NH to αH COSY cross-peaks in the fingerprint region. Typically, the two α-protons will not have the same chemical shift. The large geminal coupling (see Glossary) between the two αHs, and different couplings from them to the NH give rise to characteristic splitting patterns in the two COSY cross-peaks. The two NH to αH cross-peaks should also align with strong αH to α'H COSY cross-peaks, which do not correlate with any other resonances. For proteins with large linewidths the splitting pattern becomes less clear, but the presence of two COSY correlations to the same amide remains diagnostic for Gly. In some cases the two α-protons can be at the same chemical shift, making it difficult to identify in COSY spectra. However, in DQ spectra such degenerate α-protons give rise to characteristic peaks.

Normally the thiol, hydroxyl, and carboxylic protons on Ser, Cys, and Asp, respectively, exchange so rapidly with solvent (even in H_2O solution) that no couplings or NOEs are seen from them to the βHs. Thus, each of these amino acids should have the same coupling pattern, having one NH, one αH, and two βHs (typically at somewhat different chemical shifts). The chemical shifts of the βHs are in the downfield region, 2.5–4.0 p.p.m., but none of these residues has sufficiently different intrinsic chemical shift to distinguish it from the others. These must be assigned through sequential connectivities following the primary structure. The coupling pattern of Asn is also identical to these, since the side-chain amide to βH couplings are too weak to be seen. However, it is frequently possible to distinguish Asn from the others by the observation of an NOE from one of the side-chain amides to the βHs. As noted above, observation of such an NOE must be considered with caution since it could be inter-residue.

3.3.1.2 Methyl-containing: A, T, V, I, and L Methyl resonances are distinctive through their three-proton intensity, low multiplicity, and characteristic chemical shift. Together these make it possible to pick out cross-peaks involving methyl groups. The Met-S-CH$_3$ really does not fit in this category well, since it has no couplings to the γ-protons, and is therefore discussed below (see Section 3.3.1.3 for a discussion of longer side-chains). For Ala and Thr a coupling is seen from the 3.5–5.0 p.p.m. region to the methyl at 0.5–2.5 p.p.m. For Ala the coupling is αH to βCH$_3$, while for Thr it is βH to γCH$_3$. These are distinguished by the presence of an αH to βH coupling in the 3.5–5.0 p.p.m. region for Thr, and/or by the existence of a TOCSY and/or NOESY connectivity from the αH to γCH$_3$. For methyls in longer side-chains, couplings connect the methyls in the 0.5–2.5 p.p.m. region to protons in the 1.0–2.5 p.p.m. region. Val residues are distinguished by the characteristic αH to βH to two γCH$_3$ pattern, confirmed by direct TOCSY and/or NOESY αH to γCH$_3$ peaks. In Leu there is an extra methylene group, the βCH$_2$, which is normally very similar in chemical shift to the γCH. The region of the COSY that contains the β-γ cross-peaks is crowded by peaks from other long side-chains and is also near the diagonal, making it difficult to clearly identify these peaks. TOCSY can be used to follow from the αH out to the δCH$_3$s, but if the β and γHs are degenerate the coupling pattern looks identical to that observed for Val. The methyls are also usually further from the αH, so direct αH to methyl

NOEs are not seen as often as for valine. Only Val and Leu have two methyls that couple to the same proton. In Ile the γCH_3 couples to the single βH, while the δCH_3 couples to the two protons of the γCH_2. A coupling should be observed between the βH and the two (usually distinct) γ-protons. All of these protons are in the upfield region (0.5–2.5 p.p.m.), giving a rather characteristic, but complex, pattern. Again, TOCSY and/or NOESY data are often important for establishing the connectivity between particular βCH_3 and δCH_3 groups, since the intervening couplings are in a crowded region.

3.3.1.3　Longer side-chains: E, Q, M, P, R, and K

There are two significant differences that distinguish these amino acids from the short side-chain group discussed in Section 3.3.1.1. First, the side-chains contain more protons, out to the γ position for Glu, Gln, and Met (as far as couplings go), δ for Pro and Arg, and to the ϵ for Lys. Secondly, since the γ position is a CH_2 in all cases, the βHs are upfield from the typical position observed for the shorter side-chains, usually appearing in the 1.0–2.5 p.p.m. region. The cross-peaks from βHs to the γHs are very often near the diagonal, in a crowded region, and thus are difficult to identify clearly in a COSY. However, RELAY and/or TOCSY experiments can indicate the position of such resonances through αH to γH cross-peaks. In favourable cases, further resonances can be resolved and can then be used to distinguish Glu, Gln, and Met from Pro, Arg, and Lys. For Glu the γH position is often somewhat downfield of the βHs, due to the carboxyl group. For Gln and Met it is very similar to the βH position, and for Pro, Arg, and Lys it is often slightly upfield of the βH resonances. With further steps out along the side-chain, the geminal protons are more likely to become equivalent in chemical shift, making the δHs of Arg and the ϵHs of Lys into simple, sharp triplets. The exception is Pro, where looping of the side-chain back to join the backbone amide nitrogen usually makes the δHs inequivalent. In those cases where couplings can be distinguished all the way out along the sidechain, Pro, Arg, and Lys form distinct patterns. Gln, as for Asn, has side-chain amide protons which often show NOEs to the γHs. Met is a special case, there is no measurable coupling from the γHs to the methyl group, even in the free amino acid. Hence the methyl resonance is a distinctive singlet, usually appearing in the 2.0 p.p.m. region. It can be connected to the rest of the side-chain only through NOEs, which is sometimes difficult if the γHs are at almost the same chemical shift, as is often the case.

3.3.1.4　Aromatic side-chains: H, F, Y, and W

All of the aromatic groups are attached to the β-carbon of the side-chain (Fig. 3.17). The pattern of couplings and chemical shifts of the αH and βHs are very similar to the short side-chain amino acids, discussed above (Section 3.3.1.1). The nearest protons on the side-chains are four bonds away, making couplings to the βHs weak. In His and Trp the *c.* 1 Hz coupling to the ring δH is observable in COSY spectra for favourable cases, but antiphase cancellation becomes quite severe at high molecular weights. For Phe and Tyr such cross-peaks are not seen, even for free amino acids.

The aromatic ring protons also show distinctive coupling patterns, which can be identified in COSY spectra. For His the ϵ-proton is quite far downfield of the δ2-proton, and there is only a weak coupling between them (*c.* 1 Hz coupling constant). Since there are no other couplings to these protons, this coupling gives rise to weak, but visible cross-peaks between two peaks which basically appear as singlets. The positions are strongly pH-dependent in the range near the pK of the imidazole group. For Trp the δH is a singlet, while the remaining ring protons form two doublets (slightly downfield in the free amino acid) and two triplets (slightly upfield, with the singlet). The three-bond couplings among these, *c.* 8 Hz, give rise to three clear COSY cross-peaks when there is no accidental degeneracy in chemical shift. In water solutions the indole ring NH is also observable (typically in the 10–11 p.p.m. region), and the coupling to the δH is sufficiently large to observe a COSY cross-peak. In NOESY spectra there is an additional cross-peak between the indole ring NH and the ζ2H, which allows unambiguous identification of all the ring protons. For Phe and Tyr there is a special consideration, whether the ring flips rapidly about the twofold axis or not. Frequently the protons on either side of the ring are made equivalent, on the average, by rapid flipping about the βC to γC bond. In this case Tyr has just two inequivalent sets of protons, each a doublet (with intensity two protons), giving rise to one COSY cross-peak. Similarly for Phe there will be one doublet (intensity two), one triplet (intensity two), and one triplet (intensity one), giving rise to two COSY cross-peaks. However, in the interior of a protein there may be strong steric hindrance of such a flipping motion, and the two sides of the ring may be completely inequivalent. In this case Tyr will give rise to two pairs of coupled protons (two COSY cross-peaks), and Phe will give a set of five coupled protons (four COSY cross-peaks). For Phe particularly, it is common that the protons are close in chemical shift, and accidental (not caused by symmetry) degeneracy in chemical shifts may occur, confusing the analysis.

Since the connection of the aromatic group to the backbone cannot usually be done with couplings, it must instead be done with NOEs. Examination of the structure of the amino acids shows that normally one of the ring protons is near to either the αH or the βH of the side-chain (intra-residue). These NOEs are best observed in D_2O solution, to avoid overlap with the many amides that occur in the same chemical-shift region as the aromatics. There may be present both intra-residue and inter-residue NOEs, but the intra-residue ones can usually be distinguished by the fact that they should involve the δ-protons of the ring, whereas inter-residue NOEs usually involve those further out along the chain. The ultimate verification of such identifications is through proper fitting of the aromatics into the correct backbone sequential assignment pattern.

It should be mentioned that for exposed residues the exchange of side-chain hydroxyl (Thr, Ser, and Tyr) and carboxyl (Asp and Glu) protons is usually fast, making these resonances unobservable. However, when the hydroxyl is protected against exchange by hydrogen bonding, these protons have been identified through couplings to other side-chain protons (Feng *et al.* 1989). Similarly, the exchange of the backbone amides, side-chain amides of Asn and Gln, the indole NH of Trp, and imidazole NH of His is usually so rapid that they can be seen only in H_2O solution; but, again, internal hydrogen bonding can greatly slow the exchange of these. Thus if unusual coupling patterns are seen for some resonance, the possibility that they come from these exchangeable protons must be considered. In addition, the chemical-shift ranges given above are quite approximate, and individual residues can occur well outside the ranges given if they are in an unusual environment (e.g. near an aromatic ring).

3.3.2 Sequential connectivities by NOE

In most cases to date, the first stage of analysis has been the identification of spin systems, as described above. However, except for amino acids with unique spin systems that occur only once in the protein sequence, these identifications do not lead to sequence-specific assignments. To accomplish this it is necessary to identify contacts between neighbouring amino acids. As noted earlier, NOEs occur whenever spins are sufficiently close in space. For assignment purposes the mixing time in the NOESY experiment is usually adjusted so that the distance limit for direct NOEs is about 3.5 Å. The observed cross-peaks in a NOESY spectrum can be classified as: (1) intra-residue; (2) neighbouring residue; and (3) distant (in sequence) residue contacts. The major task in achieving sequential assignments is to sort these correctly, and to identify the correct spin system involved for each cross-peak. As pointed out above, a statistical analysis of protein structures has shown that there is a high probability that some NOE between neighbouring residues will occur, and will involve the amide proton (Billeter *et al.* 1982). The NOESY spectrum breaks up fairly easily into regions, containing the relevant groups of protons: NH, αH, and βH. The complexity arises from the fact that in addition to the sequential NOEs (between neighbours) these regions also contain distant NOEs. To understand the expected patterns of NOEs it is worthwhile to examine the regular secondary structures that occur in proteins, and to understand the patterns of sequential NOEs expected from each of them (Wüthrich 1986).

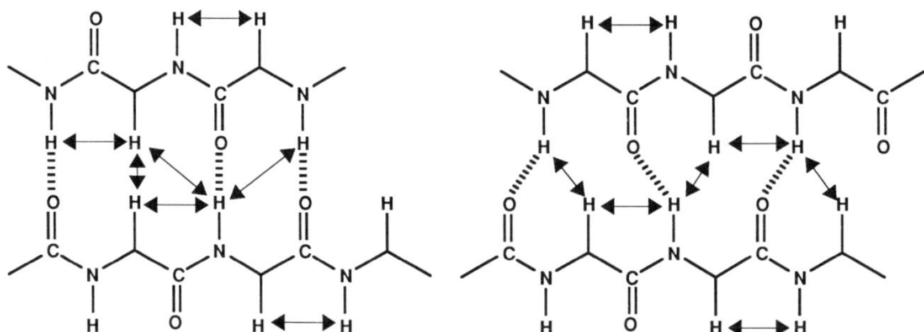

Fig. 3.19 Antiparallel (left) and parallel (right) b-sheet. Observed NOEs are shown by the arrows, and the hydrogen bonds by the dashed lines.

3.3.2.1 $d_{\alpha N}$, d_{NN}, and $d_{\beta N}$ The major elements of secondary structure are: (1) β-sheets (antiparallel and parallel); (2) helices (α and 3_{10}); and (3) turns (types I, II, and III, plus others), each of which has a somewhat different pattern. First, in β-sheets the chain is fully extended, as shown in Fig. 3.19, corresponding to the $\phi = -119°$ (parallel) and $-139°$ (antiparallel)], $\psi = 113°$ (parallel) and $135°$ (antiparallel)] region of a Ramachandaran plot. In this case the NH of each residue in the sheet is quite close to the αH of the preceding residue (c. 2.2 Å), and hence a strong NOE is expected between them. On the other hand, the distance between successive amides is quite large (c. 4.2 Å), and cross-peaks between them will not be seen. The orientation of the side-chain determines the distances from an NH to the βH, so cross-peaks between them cannot be predicted on the basis of secondary structure alone. Thus the major region to be searched for sequential connectivities for β-sheets is the NH to αH. However, it is worth noting that there are two other types of NOEs that are expected from sheet regions, which are similar to the sequential ones. First, there may be cross-strand NH to NH and NH to αH NOEs. These are somewhat longer distances than the sequential NOEs, but often appear with sufficient intensity to be seen quite clearly. In addition, for antiparallel sheets there will also be αH to αH cross-strand NOEs, which do not influence the typical sequential assignments, but are diagnostic for antiparallel sheet. These cross-strand NOEs are used for main-chain-directed (MCD) assignments (see Section 3.3.3) and for structural analysis.

In helical regions (both α-and 3_{10}-helix) the local conformation brings the NHs from successive residues close together (to c. 2.8 Å), but the distance from an NH to the preceding αH is then long (c. 3.5 Å), Fig. 3.20. The pattern of an NH having NOEs to both the preceding and following NH is repeated throughout the length of a helix. In addition, the side-chain orientation is constrained by the body of the helix, leading to NOEs from an NH to a preceding βH for many residues. This pattern is distinct from that in sheet regions. Again, in helical secondary structure there are protons that are

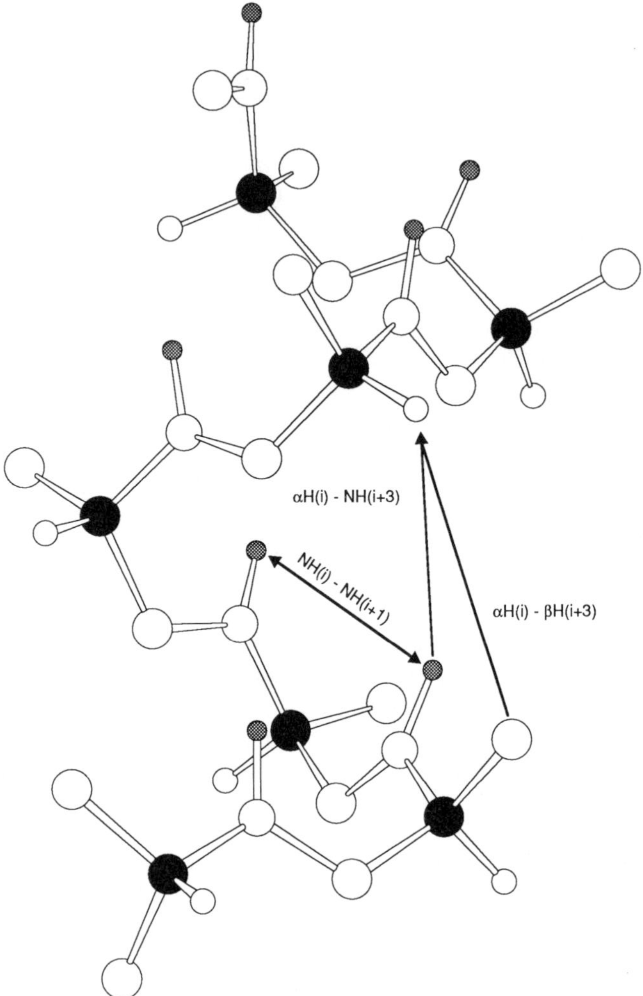

Fig. 3.20 α-Helix. Both sequential and longer-range NOEs are indicated. The α-carbons are black and the amide protons are shaded.

sufficiently close in space to give rise to non-sequential NOEs. In an α-helix the αH points out from the helix and toward the C-terminal end of the helix, bringing it close to the NHs from the following turn of helix, three or four residues further along the chain. In a 3_{10}-helix the tighter winding also brings an NH of a residue within NOE range of the NH two residues up or down the helix, one helical turn. These distances are longer than those to the neighbouring NH, but still care must be taken during assignments not to confuse these.

For turn regions, patterns of short inter-proton distances are only very local. For type I turns there are short NH to NH distances from residue 2 to 3, and from 3 to 4, with no particularly short NH to αH distances. In a type II turn, however, there is a short NH to αH from 2 to 3, and a short NH to NH from 3 to 4. For irregular sections of proteins the pattern of sequential NOEs will also be irregular. However, the allowed region of the Ramachandran plot near the sheet-like conformations is larger than that near the helices, suggesting that NH to αH NOEs should be seen more frequently than NH to NH.

In general, it is important to use all available information to identify sequential cross-peaks. For example, for a particular amide chemical shift there may be both NH-αH and NH-NH types of connectivities, but to different residues. Independent connectivities can often be found in the NH to βH region. In cases where all the residue types can be identified, the sequence can be searched to see if only one of the possibilities is consistent. This is not always possible. However, the intensities of the cross-peaks can also be examined. If the NH-αH is strong, then it is probably sequential, and since we know that some long-range NH-NH NOEs are seen in sheets, such a peak is not contradictory. In addition, since this implies an extended conformation, it would be expected that the residue in question should have a strong intra-residue NH-αH COSY peak as well. If several lines of evidence are consistent with one interpretation, then it will usually be correct. In the sequential method as many residue identifications are made as possible, and as tentative sequential connectivities are identified they are checked against the primary sequence. Clearly, since there are many ambiguous residue identifications, it may be necessary to connect several residues before either a clear confirmation from or contradiction of the primary sequence is evident. However, the final verification must come from completing the assignments so that *all* cross-peaks are consistent with both the primary structure and the rules for sequential connectivities.

A general notation for designation of distances used in protein assignments has been developed. The letter d is used to designate distance; the protons observed are indicated by subscripts α, β, or N; and the residues involved are indicated by numbers in parentheses after. For example, $d_{\alpha\beta}(2, 3)$ would be the distance from the αH of residue 2 to the βH of residue 3. (For further information, refer to the Glossary definitions for $d_{\alpha N}$; $d_{\beta N}$; and d_{NN}.) For discussions of secondary structure it is frequently useful to consider distance *separations* for a fixed number of residues, e.g. $d_{\alpha N}(i, i + 3)$. The presence of a short (NOE range) distance is often shown graphically. In such an assignment chart the sequence of the protein is shown with a bar along one or more of these distance designators. The bar indicates that the corresponding distance is sufficiently short to give an observable NOE between the designated pair of residues. Sequential connectivities must always connect

Table 3.1 Short inter-proton distances for various secondary structures

Sequential	Long range	Intra-residue
α-helix		
$d_{NN} = 2.8$	$d_{\alpha\beta}(i, i+3) = 2.5\text{--}4.4, 4.0, 3.1^*$	$d_{N\alpha} = 2.7$
$d_{\alpha N} = 3.5$	$d_{\alpha N}(i, i+3) = 3.4$	$d_{N\beta} = 2.0\text{--}3.4$
$d_{\beta N} = 2.5\text{--}4.1, 3.8, 3.0^*$		
Type I β-turn: 4 residues, NH(4) hydrogen bonds to O(1)		
$d_{NN}(2,3) = 2.6$	$d_{\alpha N}(1,4) = 3.1\text{--}4.2$	$d_{N\alpha}(2,2) = 2.7$
$d_{NN}(3,4) = 2.4$		$d_{N\alpha}(3,3) = 2.8$
$d_{\alpha N}(2,3) = 3.4$		$d_{N\beta}(2,2) = 2.0\text{--}3.4$
$d_{\alpha N}(3,4) = 3.2$		$d_{N\beta}(3,3) = 2.1\text{--}3.5$
$d_{\beta N}(2,3) = 2.9\text{--}4.4, 4.1, 3.4^*$		
Type II β-turn: 4 residues, NH(4) hydrogen bonds to O(1) and third residue is Gly		
$d_{NN}(3,4) = 2.4$	$d_{\alpha N}(2,4) = 3.3$	$d_{N\alpha}(2,2) = 2.7$
$d_{\alpha N}(2,3) = 2.2$		$d_{N\alpha}(3,3) = 2.2$
$d_{\alpha N}(3,4) = 3.2$		$d_{N\beta}(2,2) = 2.0\text{--}3.4$
		$d_{N\beta}(3,3) = 3.2\text{--}4.0$
β-sheet		
$d_{\alpha N} = 2.2$	$d_{\alpha\alpha}(i, j) = 2.3 \text{ (anti)}$	$d_{N\alpha} = 2.8$
$d_{\beta N} = 3.2\text{--}4.5, 4.2, 3.6 \text{ (anti)}^*$	$d_{\alpha N}(i, j) = 3.2 \text{ (anti)}$	$d_{N\beta} = 2.6\text{--}3.8 \text{ (anti)}$
	$d_{NN}(i, j) = 3.3 \text{ (anti)}$	$d_{N\beta} = 2.4\text{--}3.7 \text{ (para)}$
	$d_{\alpha N}(i, j) = 3.0 \text{ (para)}$	
Pro–X		
$d_{\delta N} = 2.8\text{--}5.7$	$d_{\delta N} = 2.9 \text{ (α-helix)}$	
X–Pro		
$d_{N\delta} = 1.9\text{--}4.8$	$d_{\alpha\delta} = 2.1\text{--}3.8$	$d_{\alpha\delta} = 2.1 \text{ (β-sheet)}$
		$d_{N\delta} = 2.1 \text{ (α-helix)}$

* Distance to 1, 2, and 3 β-protons, respectively.
Anti, antiparallel β-sheet; para, parallel β-sheet.
All distances in angstroms, and only distances <3.5 Å are shown.

residues i and i + 1, but other NOE patterns are seen which indicate secondary structure, and these are used in the MCD approach (discussed below). Table 3.1 shows the typical short distances seen for common secondary structures. In addition, Fig. 3.21 shows a 'map' of sequential connectivities seen for a protein, which uses this graphical representation.

3.3.3 Main-chain-directed assignments (MCD)

This approach was developed after the original sequential method, and uses the same fundamental information, but in different orders and ways (Englander and Wand 1987). The MCD method was built upon the realization

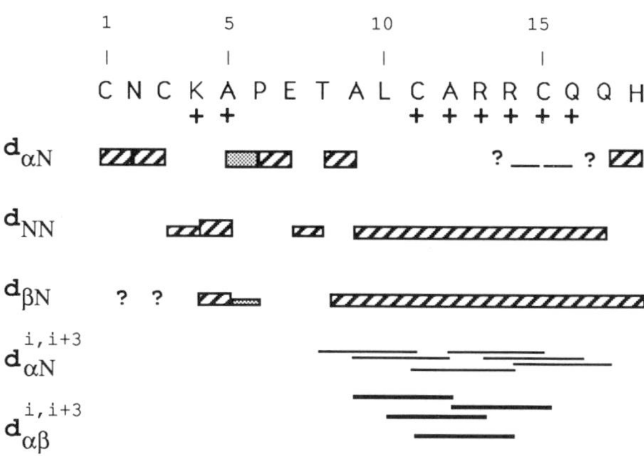

Fig. 3.21 A chart of the sequential and secondary structure NOEs observed for the 18 amino-acid peptide, apamin. The striped bars indicate that the NOE was observed, and the shaded bars indicate that the NOE was observed to the Pro δ-proton(s) (Pro has no amide proton). The size of these bars is related to the observed NOE intensity. The bottom two sets of lines are the observed longer-range NOEs indicative of α-helix. The +s are possible NOEs but the chemical shift degeneracy makes it impossible to tell if they are present. The ?s are the observed slowly exchanging amide protons, presumably due to hydrogen bonding. Clearly, this peptide has an α-helix from residues 8 to 17.

that in elements of regular secondary structure there are extended patterns of connectivities through different sets of protons. The usual problem with assignments is degeneracy in chemical shift, but the correctness of a particular set of assignments can be checked using the independent connectivity patterns. The appropriate connectivities involve only the backbone atoms (NH, αH, and βHs), and hence the method does not require complete identification of the spin systems as a first step in the process. When side-chains are identified, this information can be used to verify the assignments, or it can be used during the process to enhance the ability to find correct assignments in more complex systems.

3.3.3.1 Connectivity patterns   The three types of extended secondary structure are shown in Figs 3.22 and 3.23. In each of these there is a repeated pattern of sequential NOEs, together with some longer-range NOEs, which are structure specific. As described above, these structure-specific NOEs are used to confirm assignments and to remove ambiguities arising from degeneracies. For example, in an α-helix the NH-NH connectivities are still a primary identifier of neighbours in the sequence (supplemented by NH-βH). However, rather than trying to identify the spin system of each

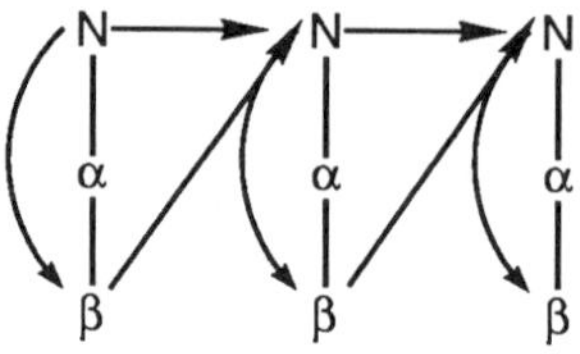

Fig. 3.22 This pattern is cylindrical and is confirmed by the longer-range helical NOEs, as described in the text.

amino acid in order to resolve ambiguities, instead connectivities of the αH(i)-NH(i + n) and αH(i)-βH(i + n) type, with n equal to 3 or 4, are used. The presence of such 'loops' indicates that the primary connectivities are correct, and since these loops occur frequently, one per residue beyond the first turn, a complete set provides a stringent verification.

For both types of β-sheet there are other types of extended patterns. In antiparallel sheet, these involve cross-strand αH-αH, αH-NH, and NH to NH connectivities (Fig. 3.23). Such NOEs are used to cross to the opposite strand, step along it, and finally return to the original strand. Again, these two independent pathways connecting resonances will be present only if the assignments of the intervening protons in each path are correct. For parallel sheets the principle is the same, but there are no cross-strand αH-αH NOEs, instead just cross-strand αH-NH and NH-NH NOEs are used. In both types of sheets there are several different patterns which can be used to confirm the direct sequential path.

In the MCD approach a major concern is how often the various patterns do correspond to the proper secondary structure, and how well degeneracies in the primary sequential connectivities can be resolved. A number of studies have been made, using both known protein structures and synthetic data (Wand and Nelson 1988). The results depend significantly on the accuracy of the input data, and on the particular MCD pattern which is involved. The accuracy of the input data simply means how well the chemical shift of a particular cross-peak can be determined in each dimension. In the past these positions have usually been determined by hand, 'picking' the tops of cross-peaks. It has been found that there is significant error in doing this, both from human error in selecting the centre of a peak, and from shift in the apparent centre of a cross-peak due to roll in the baseline (which is not seen in contour plots but can be observed in stacked plots, Fig. 3.3). Lower accuracy in the positions of peaks leads to more possible combinations of cross-peaks that could be assigned to the same spin system. This means, in turn, that there are more false connectivity pathways, whether sequential or MCD, and hence it is more difficult to sort them out. In addition, it has been

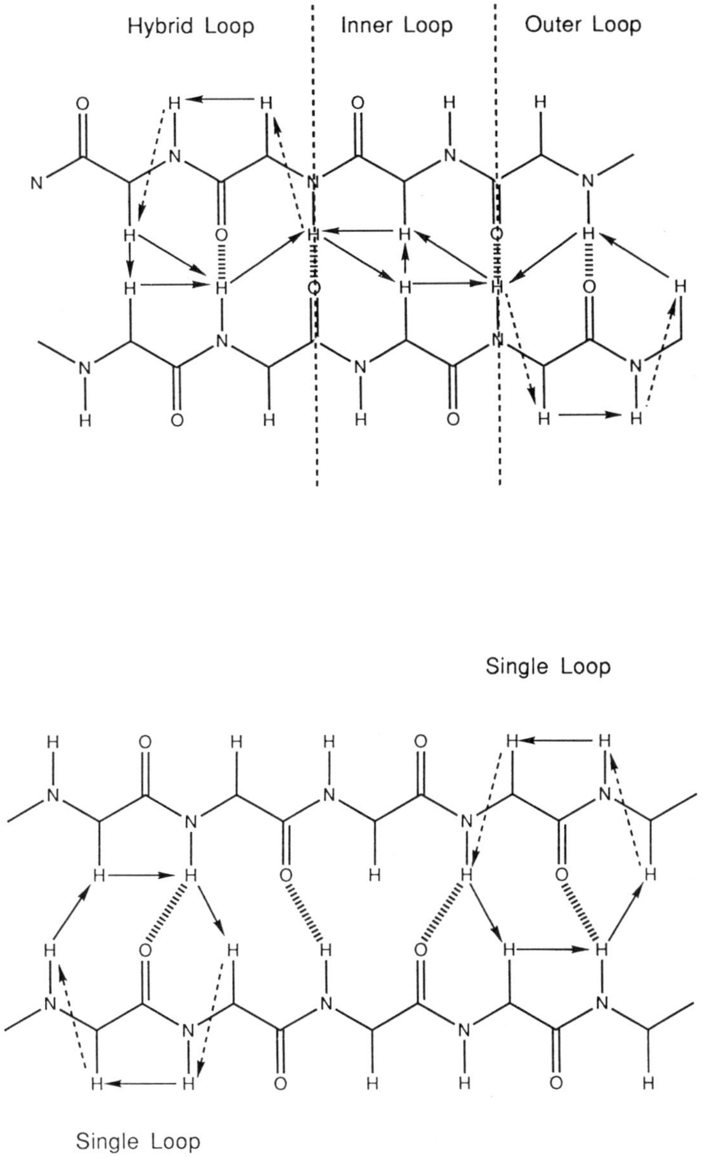

Fig. 3.23 The sheet MCD patterns. There are three patterns for antiparallel sheet (top) and one for parallel sheet (bottom). Note how the patterns intertwine (not drawn) to substantiate the assignments.

found that for the smaller MCD patterns, e.g. the inner loop of β-sheet, that there are non-sequential residues which also give rise to this pattern. This reduced fidelity also increases the problem of identifying the correct resonance assignments. However, when the extended MCD patterns are used with accurate data, the correct sequential assignments can indeed be identified. The MCD approach has the great advantage that it follows a sufficiently well-defined algorithm which can be carried out by a computer in an automated fashion. This, combined with automated 2D peak-picking, is already becoming a powerful tool for resonance assignment determination.

3.3.3.2 Including spin systems   The MCD approach requires identification of groups of αH, βH, and NH protons that belong to the same residue, but does not require complete identification of the side-chain. However, for 3D-structure determination, side-chain contacts are as important as backbone contacts. It should be noted that when the MCD backbone assignments are obtained, specific αH, βH, and NH sets are identified with specific amino acids in the sequence. To extend the main-chain assignments obtained through MCD, the same methods are used as in the sequential method, i.e. COSY, RELAY, and TOCSY, supplemented by NOESY. This process is somewhat easier since the features sought in each spin system are known, eliminating some of the degeneracy problems. It is also easier if 3D NMR data are available.

3.3.4 Stereo-specific assignments

Stereo-specific assignment of stereo-related proton and methyl pairs is extremely helpful for defining the local conformation in peptides. These assignments were originally obtained by using a combination of short intra-residue NOEs and vicinal J couplings (Zuiderweg *et al.* 1985; Hyberts *et al.* 1987). (See Glossary for definition of vicinal coupling.) This method has had limited success. More recently a second approach has been developed which uses all of the available structure information (Weber *et al.* 1988). This latter approach involves calculating several structures assuming no chirality for carbons with non-stereo-specifically assigned substituents. These structures are then analysed to determine if each carbon has a preferred chirality, hence defining the stereo-specific assignment.

3.3.5 Heteronuclear methods

There are now a wide variety of ways in which isotope substitution has been used to obtain information for resonance assignments. Some of these are direct (putting in a label so that only specific types of resonances can be seen, or not seen), while some are indirect (putting in extra spins, which can be used to affect changes in the proton spectrum in a predictable way). These have already been used in a wide variety of different systems, and cannot be covered comprehensively in the space available. However, several examples of the most important of these methods will be given.

3.3.5.1 Proton filtering There are two mechanisms for affecting proton spectra through incorporation of other spins. In the first method proton resonances are eliminated, or their multiplicity is reduced through substitution of some protons with deuterons. This can be done for specific residues by chemical exchange methods, or through biosynthetic incorporation of specifically deuterated precursors, or through growth of organisms on a uniformly partially deuterated medium. When residues are used which are perdeuterated in either all or specific positions, then resonances (and cross-peaks) involving the exchanged protons will disappear from the ^{1}H spectra (1D, 2D, and 3D). The residue identifications are easily obtained by comparing these spectra with those from the fully protonated material. Multiple substitutions can also be useful. For example, putting deuterated His, Trp, and Phe into a protein leaves only cross-peaks from Tyr in the aromatic region. This is also useful for residue-type identification, but neither method leads in any straightforward way to sequential assignments. Uniform deuteration does not help with even residue identification directly, but when *c.* 70 per cent deuterium is incorporated, the multiplicity of peaks is reduced, and the T_2 values of remaining protons are increased, making cross-peaks sharper in COSY and NOESY spectra, aiding interpretation (LeMaster and Richards 1988). In addition, the lower proton density reduces the spin diffusion problem in NOESYs.

Another mechanism for affecting proton resonances is to use other stable isotopes, the most common being ^{13}C and ^{15}N. In fact both are present at natural abundance, and simple heteronuclear correlation experiments to attached protons (one-bond heterocorrelation, HETCOR; or heteronuclear multiple quantum coherence, HMQC) can be done without enrichment. Biosynthetic incorporation of these isotopes to a level of over 50 per cent makes these experiments quite easy and sensitive when done with proton detection (1 mM solutions give good spectra in a few hours). By providing a different spreading parameter in one dimension (compared to COSY, etc.) these experiments can provide additional information not available in the 'proton only' experiments. Biological incorporation can be done in a residue-selective manner, by feeding specific, selectively labelled amino acids. In order to prevent scrambling of the label it is sometimes necessary to use auxotrophic organisms. In the HMQC experiments on such selectively labelled molecules only the labelled residues will give cross-peaks, providing residue identifications. For example, the amide proton to amide nitrogen 'fingerprint' of a protein provides one cross-peak per residue (except two for Gly and none for Pro). If the α-amino group is labelled only in Leu, then the same experiment will yield a subset of cross-peaks, one for each Leu residue. This is an important method for verifying residue identifications in large proteins where the proton COSY and TOCSY cannot always be followed far into the side-chain and sequential connectivities often have degeneracies.

Alternatively, these labels can be used to 'filter' the proton spectrum. The presence of the ^{13}C or ^{15}N can be used to select only the protons attached to them. With appropriate phase cycling and pulse sequences this means that the spectra will contain only cross-peaks for which magnetization either started on a proton attached to the heteroatom or, in the alternative version, was detected on such a proton. Thus, for the example of a Leu-labelled protein above, selection of ^{15}N-^{1}H pairs during the preparation phase of a NOESY experiment will give a spectrum in which all cross-peaks involve a Leu amide proton in the t_1 dimension. Again, this provides a great enhancement in resolution — clearly sequential connectivities which are observed in such a NOESY must be from Leu residues. This enhancement comes at the cost of having to generate a large number of samples (one per amino acid) in order to get complete information. It is possible to combine several different labelled amino acids into a single sample and still have sufficient resolution for assignments. Although ^{15}N-^{1}H pairs have been used most frequently in such experiments, α^{13}C-^{1}H pairs, etc. are also valuable.

3.3.5.2 Direct experiments When labels are present at a fairly high level, direct experiments involving only the heteroatoms can also be performed. For proteins generated with 50–70 per cent ^{13}C, carbon-carbon couplings can be identified with a COSY experiment, or a ^{13}C-^{13}C double quantum spectrum, usually called INADEQUATE (a misnomer in this case). It is known that the distribution of carbon chemical shifts is dominated to a larger extent by covalent bonding than for protons, and hence each type of amino-acid residue gives rise to a characteristic, complex pattern of coupled spins with a particular chemical-shift distribution (Oh *et al.* 1988). Tracing the full pattern identifies the residue type. The residue-type assignments generated in this way can then be carried over to the protons through HETCOR experiments. Strong couplings only occur for directly bonded carbons (hence only within residues), so these couplings cannot be used directly for sequential assignments. However, if double labelling is done with ^{13}C and ^{15}N, then the combination of homonuclear ^{13}C COSY or DQ with a heteronuclear ^{13}C-^{15}N HETCOR can give sequence-specific assignments (Stockman *et al.* 1989). The first such experiments have only been performed recently, so there are few applications of the method. When the isotopic labelling is straightforward, it seems clear that this may develop into a powerful approach.

3.3.5.3 3D applications Finally, these isotopes can also be used in 3D NMR experiments. As mentioned above, the heteroatom shift can be used

as a spreading parameter into the third dimension. The combination of
TOCSY-HMQC and NOESY-HMQC seems to be a powerful method
for obtaining the resolution necessary for sequential assignments in large
proteins. Since the proton chemical shifts are detected at the nitrogen
chemical shift, it is straightforward to make the correct association of NH
and αH within each residue. The sequential assignment approach is to start
at a particular ^{15}N-NH-αH position of the TOCSY-HMQC spectrum, and
then in the NOESY-HMQC search at the same ^{15}N and NH chemical shift
for a proton of the neighbouring residue; in an extended sheet, for example,

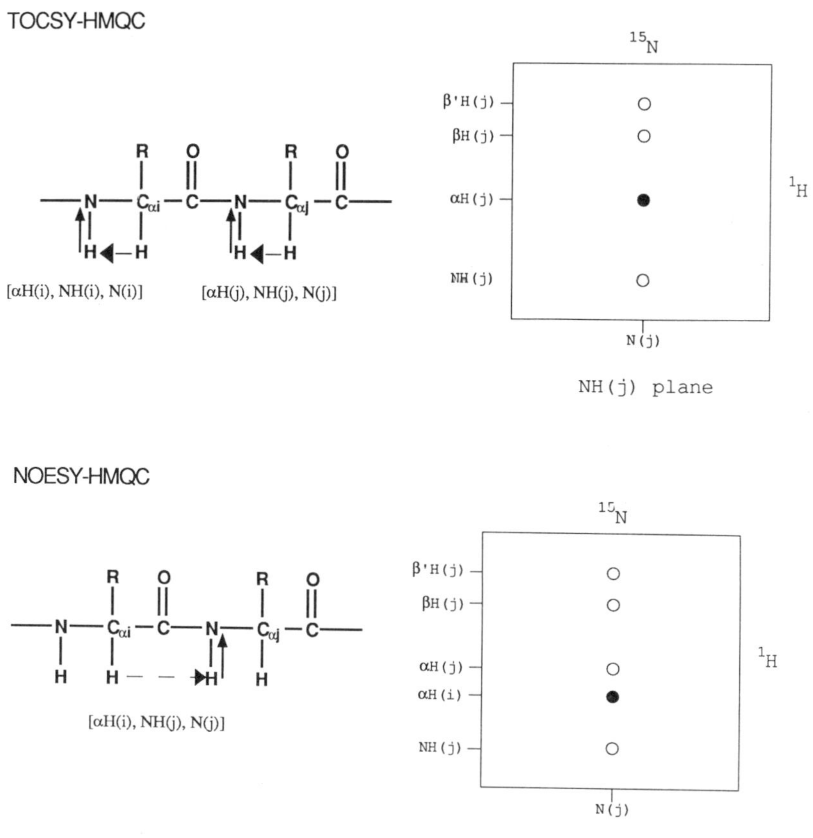

Fig. 3.24 The assignment strategy using both TOCSY-HMQC and NOESY-HMQC
3D NMR experiments, as described in the text. Top, larger arrow-head represents
TOCSY transfer. Bottom, dashed arrow represents NOESY transfer. Closed circles
represent desired peaks. Each 2D spectrum represents the same plane (the NH(j)
chemical-shift plane) from the 3D volume.

one would find an αH. The TOCSY-HMQC is then searched for a peak with the same αH chemical shift in order to identify the ^{15}N and αH chemical shifts of that neighbouring residue, Fig. 3.24. There may be degeneracy in the αH chemical shift, and the identification of the correct neighbouring residue still depends on either identification of the residue type (made easier through the ^{15}N sorting of the TOCSY-HMQC) or through identification of a second sequential connectivity, such as NH to βH. Since 3D experiments are time-consuming, these experiments require substantial enrichment of the isotope.

3.3.6 Mutants or natural variants

A final approach to assignments is to look at point mutants, whether they are natural or created through site-directed mutagenesis. In an ideal case, the spectra of the wild-type and mutant protein could be compared and only two differences would be seen: disappearance of the substituted amino acid's spin system in going from the wild-type to mutant spectrum, and appearance of the spin system of its replacement. In reality, however, the substitution of one amino acid for another will result in shifts for a substantial number of resonances in the spectrum (25 per cent or more of fingerprint-region cross-peaks commonly shift by more than one linewidth) (Wittekind *et al.* 1989). Shifts occur for amino acids nearby in the primary sequence and nearby in space, as expected, but also sometimes in sites displaced by a substantial distance from the site of change. The comparison of wild-type and mutant spectra is not trivial, but by following cross-peak patterns carefully throughout the spectra it is often possible identify at least most of both the substituted and original spin systems. When this can be done it provides an absolute assignment (assuming that the site and type of mutation is known). As with the substitution of specific labelled amino acids, to make a significant impact on assignment of the full protein requires a tremendous amount of work in generating mutants, but when they are available they provide a good check on the sequential assignments obtained by traditional methods.

3.4 Nucleic acid assignments

Although there are many technical differences in the way proteins and nucleic acids are assigned, there are many fundamental similarities. It is again necessary to identify NOEs between neighbouing residues, which enable one to step along the backbone of the polymer. When degeneracies occur in the chemical shifts of the spins involved, these can be resolved through correct identification of the type of residue involved and a knowledge of the primary sequence. In some ways this process is simpler for nucleic acids than for proteins since there are (usually) only four types of

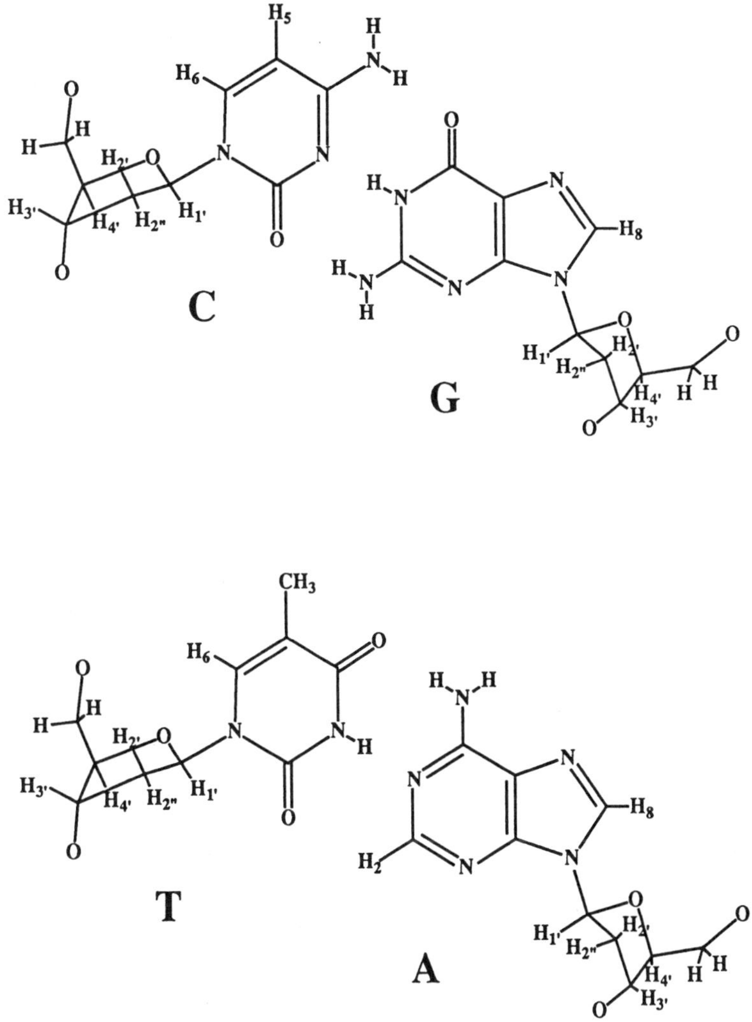

Fig. 3.25 The four nucleotides. U is equivalent to a T except that the CH_3 is replaced with H. Deoxyribose sugars are shown. A ribose sugar has an OH in place of the deoxyribose C2″ H.

bases (Fig. 3.25), and the regions of the spectrum where specific types of resonances occur are better defined, Fig. 3.26. Conversely, the limited spectral dispersion also leads to crowding for large nucleic acids.

3.4.1 Identification of bases and sugars The process of identifying the protons that belong to an individual base or sugar is done with COSY

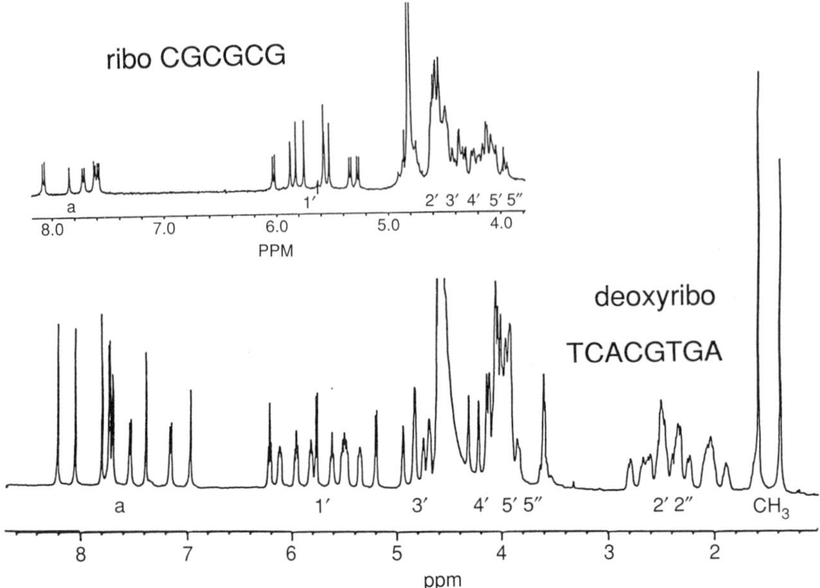

Fig. 3.26 DNA and RNA chemical-shift regions. Note how much more crowded the RNA spectrum is. The imino and amino protons, not shown, resonate from 11 to 16 p.p.m. and from 6 to 9 p.p.m., respectively.

or TOCSY spectra. The individual residues are somewhat like aromatic amino acids in that the sugar and base form separate spin systems, with no observable couplings between them. The association of specific bases with sugars must be done using a NOESY spectrum. Like the amide protons of amino acids, there are also acidic protons in nucleic acids which can exchange with solvent. Each normal, Watson-Crick base pair has one imino proton and either two or four amino protons. All of these exchange so rapidly that they can only be observed in H_2O solution. Unlike proteins, however, there are no couplings from these exchangeable protons to any of the non-exchangeable ones, making it necessary to use NOEs to assign them.

3.4.1.1 DNA   DNA is made up of the four types of bases A, T, G, and C, shown in Fig. 3.25. In A and G the aromatic protons are not coupled and occur as relatively sharp singlets in the spectrum. For T there is a weak four-bond coupling (*c.* 1 Hz) between the aromatic H-6 and the methyl group, which gives rise to easily identified COSY cross-peaks for small oligomers. For C there is a strong coupling between the H-5 and H-6 protons (*c.* 8 Hz), giving very clear cross-peaks even for fairly large oligomers. Each residue has a base attached to a deoxyribose sugar, also shown in Fig. 3.25. The sugar protons also resonate in characteristic frequency ranges (see Fig. 3.26) and coupling between these protons can be observed in both COSY and TOCSY spectra. The coupling constants within the sugar depend on its conformation: in DNA the most frequently observed form is C-2′ *endo*. The C-1′ protons are furthest downfield in chemical shift and normally couple to both the C-2′ and C-2″ protons. With the C-2′ *endo* conformation the C-2″ to C-3′ couping is very weak, but the C-2′ to C-3′ is observable. The C-3′ couples further to the C-4′ proton, but in a crowded region of the spectrum, and similarly, the C-4′ can, in principle, be connected to the C-5′ and C-5″ protons. The combination of relatively small coupling constants and crowded spectral regions makes it difficult to connect the full spin systems of each sugar from COSY data alone. TOCSY spectra are somewhat better since the most crowded regions can be avoided by looking at relayed cross-peaks.

The exchangeable protons are usually observed as sharp lines when they are hydrogen bonded, slowing their exchange with the solvent. No couplings involving any of the amino or imino protons are large enough to be observed, and hence assignment of these relies on NOEs alone, as discussed below.

3.4.1.2 RNA   RNA resonance assignments are more complicated than DNA assignments. First, the second pyrimidine base is U, instead of T, which from the coupling viewpoint is identical to C (both have a H-5 to H-6 coupling). The purine bases (A and G) can be distinguished from pyrimidines (U and C), but the type of base in either group cannot be determined from the coupling data. Secondly, the replacement of the 2″ proton of DNA by the 2′OH in RNA causes a change in chemical shift of the 2′ proton, moving it into the crowded region near the C-3′ and C-4′ protons (Fig. 3.26). Finally, introduction of the 2′OH causes a change in the typical sugar conformation to C-3′ *endo*, for which the coupling between the C-1′ and C-2′ protons is very small (usually too small to give a cross-peak), making the grouping of sugar resonances more difficult. However, in spite of these problems, it is still possible to systematically assign RNA (Chou *et al.* 1989). For transfer RNA (tRNA), there are many modified bases in addition to the normal ones described above. The modifications provide new, distinct groups of spins, some of which can be identified through their coupling patterns. Not much work has been done on the non-exchangeable proton systems of these, but the presence of such modifications will make the work easier when these systems are studied. The exchangeable protons have the same basic behaviour as in DNA.

3.4.2 Sequential connectivities

3.4.2.1 Helical regions   Many nucleic acids occur in a duplex form with standard Watson–Crick base pairs. In either A- or B-form helices the imino protons in the centre of the base pairs are stacked along the direction of the helix axis, with a separation of 3.5–4.0 Å. NOEs (either 1D or 2D) from one of these imino protons then identify the two neighbours in the sequence, except for the terminal base pairs which have only one (Roy and Redfield 1981). In a self-complementary oligomer, if there are no degeneracies in chemical shift, this information is sufficient to assign all of the resonances. For general duplexes in which the strands have different sequences some additional information is needed to associate resonances with the correct ends of the molecule. This can be done using NOEs from the imino protons to the nearby amino and C-2H protons within the same base pair. For a G-C base pair there are two protons from amino groups nearby, while for an A-T (or A-U in RNA) there is one amino group and one aromatic proton, the C-2H. The amino groups of A and G are normally very broad due to chemical exchange, arising from rotation of the amino group. The C aminos are significantly sharper than those of A and G, but are still broader than the aromatic resonance of the A C-2H. Thus examination of the imino to amino and aromatic cross-peak region can resolve the identity of the base pairs, and hence the absolute assignments. In addition, when there are chemical-shift degeneracies, the correct assignments can be determined by requiring that the assignments correctly follow the primary sequence of the molecule, assuming that it is available from chemical sequencing.

The CH and NH protons in nucleic acids form separate groups, with no through-bond couplings between them. Thus, unlike the case for proteins, it is possible to assign all of the backbone atoms working in D_2O solution, reserving work in H_2O for assigning imino and amino protons, as discussed above. In the regular helical regions of the nucleic acid (either A- or B-form), the bases stack from the 5′-end of the molecule in a right-handed sense. This places the base of any residue partially over the sugar of the preceding residue such that there are observable NOEs between the aromatic proton and the sugar C-1′H and C-2″H, Fig. 3.27. The intra-residue distances from the aromatic proton to these same sugar protons are also within NOE range. Thus it is possible to carry out a sequential walk using the base and sugar protons. Often the resolution is best in the region containing cross-peaks between the C-1′H and aromatic, so this is used first in analysis of the spectra. The inter-residue aromatic to C-1′H distance in B-DNA is about 3.6 Å, while the intra-residue distance is in the range of 3.7–3.9 Å. These are on the long end of the observable NOE range so that the cross-peaks are typically fairly weak. However, spin diffusion involving other sugar protons increases their intensity. While this is a problem for quantitative analysis,

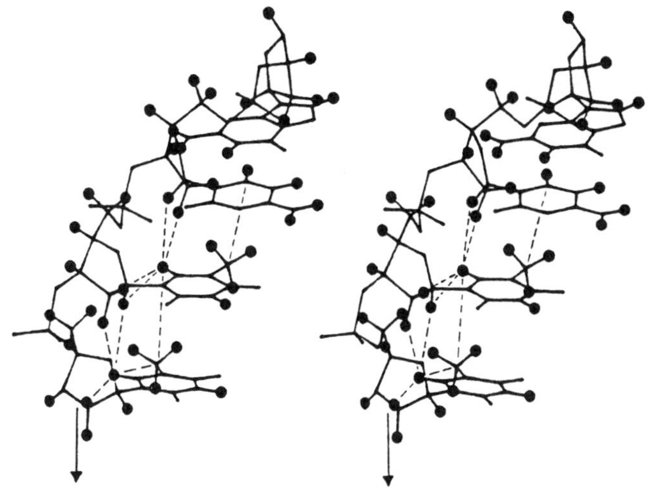

Fig. 3.27 Stereo view of the helical sequential connectivities observed for nucleic acids (DNA shown). The connectivities are from the aromatic protons to the C1′H and C2″H.

it does not cause any problems for assignments since the only combinations of aromatic and sugar protons that are at all close are the desired intra-residue and sequential ones. The aromatic to neighbouring C-2″H distance in B-DNA is about 2.2 Å, giving a strong cross-peak, while the intra-residue distance is about 3.5 Å; for C-2′H the inter-residue is about 3.9 Å while the intra-residue is about 2.3 Å. Again, spin diffusion will tend to equalize the cross-peak intensities, but does not corrupt the connectivities. The connectivities in the aromatic to C-2′H, C-2″H region can be used to check the C-1′H region assignments, or to provide an independent path when degeneracies cause ambiguities there. For pyrimidine bases there are additional cross-peaks from the protons at the base 5 position (C-5H for cytosine, CH_3 for thymidine) to the preceding aromatic and sugar protons, again providing an independent assignment pathway. The remaining sugar protons of each residue can be assigned using any combination of the COSY, TOCSY, and/or NOESY. There are generally a sufficient number of cross-peaks; however, the C-4′H, C-5′H, and C-5″H have very similar chemical shifts, and for oligomers with more than 12 or so residues it may be difficult to resolve many of these.

The assignment of duplex RNA oligomers follows a very similar process; however, there are some differences in detail because the sugars are typically C-3′ *endo*, altering the separation of protons, and the replacement of the C-2″H with an OH dramatically alters the chemical shift of the C-2′H. Since

A-form RNA is also right-handed, the same basic features of stacking apply to it as were the case in DNA. The distance from the aromatic proton to the preceding sugar C-1'H increases to 4.6 Å, with the intra-residue remaining at about 3.7 Å. The C-2'H sequential distance is 2.1 Å, while the intra-residue value is about 3.9 Å. In the C-3' *endo* sugar, however, the C-1'H to C-2'H distance is reduced to about 2.5 Å. While the sequential aromatic to C-1'H distance is very long, the distance to the C-2'H is very short and spin diffusion will again lead to observable intensity even though the 4.6 Å distance is outside the range normally considered observable. The path using only the aromatic to C-2'H cross-peaks is good in principle but, as mentioned before, the addition of the 2'OH in RNA shifts the C-2'H peak into a crowded region which also contains the C-3'H and C-4'H resonances. For small oligomers these can usually be sorted out, but for larger ones this is difficult.

3.4.2.2 Non-helical regions   Unlike proteins, in which sequential connectivities are present regardless of the local conformation of the backbone, in nucleic acids it is possible for essentially all protons on neighbouring residues to be far apart. Clearly, this means that in such segments it is not possible to obtain assignments through sequential connectivities using the NOE, as discussed above. Fortunately, in many nucleic acids long stretches of non-helical, unpaired bases do not occur. In short stretches there may be some specific structure formed, such as a hairpin loop, or internal bulge, in which some base stacking does occur. In these cases it is often possible to determine a sufficiently large number of sequential connectivities to make complete assignments, even if some bases cannot be connected. There should always be some NOEs connecting the base to its own sugar, and then consistency with the sequence and determination of base type from either COSY or NOESY data can eliminate most of the possible assignments. This approach has been successful for a number of hairpin loops and also for helices containing an extra residue. It should be noted that, unlike regular helical regions in which all NOEs involving the pyrimidine H-6 or the purine H-8 are either intra-residue or sequential, some tertiary NOEs may be seen, and these must not be confused with sequential ones. For short segments of unpaired bases, distinguishing these is usually not a problem; however, as cases arise with longer segments of this type caution must be used. One approach to avoiding these problems is to use a through-bond connectivity, as discussed below. Unfortunately this may not be feasible for large molecules.

3.4.3 Through-bond sequential assignment methods

Nucleic acids have phosphate groups linking the individual residues, each of which contains a ^{31}P spin. The 100 per cent natural abundance of this

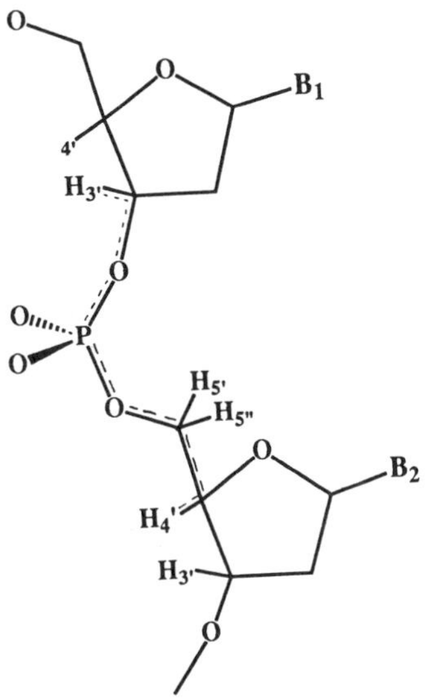

Fig. 3.28 Through-bond sequential connectivities for nucleic acids. The dashed line indicates the sequential connectivity.

isotope and the relatively high gyromagnetic ratio make it attractive for study. The three-bond couplings from the phosphate to the HC-3′ and HC-5′,5″ are sufficiently large to be seen as splittings in the spectrum, although those protons occur in crowded spectral regions. The four-bond coupling to the HC-4′ is also large enough be be detected, at least for fairly small oligomers. Since the couplings go to residues on both sides of each phosphate, they provide a link that can be used to establish sequential connectivities (Fig. 3.28). This has been done in two ways: first, through direct correlation with the phosphorous resonance (Pardi *et al.* 1983b; Marion and Lancelot 1984; Frey *et al.* 1985), requiring reasonable resolution in the ^{31}P spectrum, and secondly, by using a ^{31}P RELAY step (Frey *et al.* 1985).

The direct method is applied by obtaining correlations between coupled ^{31}P-^{1}H pairs. This can be done either with the HETCOR experiment, or with the double quantum version, HMQC, and can be done with either direct ^{31}P detection or ^{1}H detection for better sensitivity. In any case cross-peaks are obtained from each ^{31}P resonance to the H-5′, H-5″, and H-4′ on one side, and the H-3′ on the opposite side (Fig. 3.28). To complete the linkage using couplings along the backbone, it is necessary to find the H-3′-H-4′ couplings, which can be done with COSY, RELAY, or TOCSY spectra. With each of these couplings identified, it is possible to walk along the backbone: $^{31}P_i$-H-4$_i'$-H-3$_i'$-$^{31}P_{i+1}$. In practice this approach suffers from two problems. First, it relies on having reasonable resolution in the ^{31}P dimension of the heterocorrelated experiment. The phosphorous resonances are generally only spread over a range of 1 p.p.m., and the linewidth contributions from chemical-shift anisotropy limits the improvement with increasing field, so that problems can be expected for large oligomers. In addition, when linewidths reach the magnitude of the coupling constants the cross-peaks will become more difficult to observe. When two ^{31}P resonances overlap completely it is not possible to directly associate the correct proton frequencies. A similar problem can occur in the crowded H-3'-H-4' cross-peak region of the COSY. As is generally the case with assignments, if there is only one such degeneracy then each of the possible assignment choices at the degenerate step can be followed and tested against the primary sequence for consistency. Usually with one degeneracy there will only be one possible assignment scheme that will explain all of the data; however, for larger molecules in which multiple degeneracies occur it may not be possible to resolve the ambiguities.

The second approach, ^{31}P RELAYED spectroscopy, removes the observation of the ^{31}P altogether. In this experiment, a first step transfers magnetization from a ^1H to a coupled ^{31}P as in any heterocorrelated experiment. However, rather than detecting it on the phosphorous, the process is repeated with a ^{31}P to ^1H step. While some of the magnetization returns to its original source, some of it transfers to the following sugar. This gives rise to extra cross-peaks in what appears otherwise to be a COSY spectrum, but in this case between the H-3' and H-4's of neighbouring residues. These correlations being established through having a coupling to a common ^{31}P as in the direct method. Again, this approach suffers from having to use a crowded spectral region. In addition, doing successive ^1H-^{31}P and ^{31}P-^1H transfer steps limits sensitivity, especially as linewidths increase.

In spite of limited resolution and sensitivity these approaches have been used to assign DNA oligomers up to about 12 base pairs. Since these methods use only through-bond couplings, as long as the resolution is sufficient, the ambiguity in the interpretation of NOEs as intra-residue, inter-residue sequential, or inter-residue non-sequential is avoided. This makes the coupling approach especially attractive for non-standard conformations where the NOEs can be confusing.

3.4.4 Isotope-labelling schemes

Stable isotope labelling has been used in some nucleic acid systems, but to a lesser extent than for proteins. Two types of RNA that occur naturally, tRNA and 5S RNA, have been labelled with ^{15}N and ^{13}C by feeding the appropriately labelled bases to strains of bacteria. Experiments with ^{15}N-labelled uridine in tRNA were the first to use indirect detection for a large molecule (Griffey *et al.* 1982). The selective splitting introduced through labelled U allowed unambiguous identification of A-U base pairs. Experiments with isotope-selected and directed NOEs were also carried out. Similarly, 5S RNA has been labelled with ^{15}N, aiding resonance assignments both for the intact RNA and for fragments recombined with unlabelled material (Gewirth *et al.* 1987). Smaller RNA oligomers are more difficult to prepare chemically than DNA, and only the simple exchange methods used for DNA, discussed below, have been applied to them so far. Labelling in DNA has found less use than in RNA, primarily because the incorporation must be done chemically rather than through biosynthetic incorporation. While labelling is straightforward in principle, the relatively large number of synthetic steps needed to prepare the labelled nucleotide, and then to protect and activate it appropriately for chemical synthesis, has limited its use. On the other hand, there are some simple exchange reactions which can provide some useful information, although not to the same degree as by complete synthesis. First, it is well known that the H-8 aromatic position of the purines is mildly acidic and will exchange with solvent at high pH and elevated temperatures (*c.* 1 day at pH 8 and 70°C). When purine and pyrimidine aromatics overlap in a spectrum, deuteration with this method can remove some ambiguities (Brush *et al.* 1988). In addition, by heating in the presence of bisulphite, exchange will also occur at the cytosine H-5 position. Such exchange-labelled samples, used in various combinations, can aid interpretation of the normal proton spectra. It is likely that more generalized labelling methods will be needed to simplify the more complex systems.

3.5 Carbohydrates

Many of the general ideas for the analysis of carbohydrates follow closely those used for systems already discussed. The polysaccharide chains may be branched, unlike proteins and nucleic acids which are strictly linear, and may be attached to other biopolymers such as proteins or lipids. In some cases the primary structure of the polysaccharide is known (including each branch point and linkage stereochemistry), and the main point of the NMR study is to determine the three-dimensional conformation. However, unlike the usual case for either proteins or nucleic acids, there are many carbohydrates for which the primary structure is not known and their determination is the central goal of the work, although information about the three-dimensional conformation is always welcome. In either case, the process for determining resonance assignments is very similar in philosophy to that for proteins or nucleic acids: first, identify the spin systems corresponding to individual residues, using through-bond couplings, and then determine proximity of residues using Overhauser effects or heteronuclear coupling.

3.5.1 Identification of spin systems

Each oligosaccharide is made up of sugar, or modified sugar, residues linked together by ether bonds. There are a number of residues that occur frequently in certain classes of carbohydrate: simple sugars, such as glucose, galactose, mannose, fucose, and sialic acid; and derivatized ones, such as *N*-acetylglucosamine, *N*-acetylgalactosamine, and *N*-acetylneuraminate. However, in other types of oligosaccharide there are many other derivatized sugars, such as sulphated forms which occur in heparan. In each sugar there is an effectively linear chain of protons, numbered beginning at the anomeric carbon. Almost all of the carbons in each sugar have an oxygen attached, leading to a rather poor spread in chemical shifts, similar to the ribose ring resonances in RNA, ranging from 3–6 p.p.m. (Fig. 3.29). The anomeric protons generally have the best dispersion in chemical shift, ranging from about 4.5–6 p.p.m. (Fig. 3.29). The sugar at the reducing end will generally have both the α and β anomeric forms present, leading to two sets of resonances for most of its protons.

Identification of groups of spins corresponding to individual sugars is done with COSY, RELAY, or TOCSY spectra. Since the chemical-shift range of 3–4.5 p.p.m. is very crowded, with many couplings within it, the

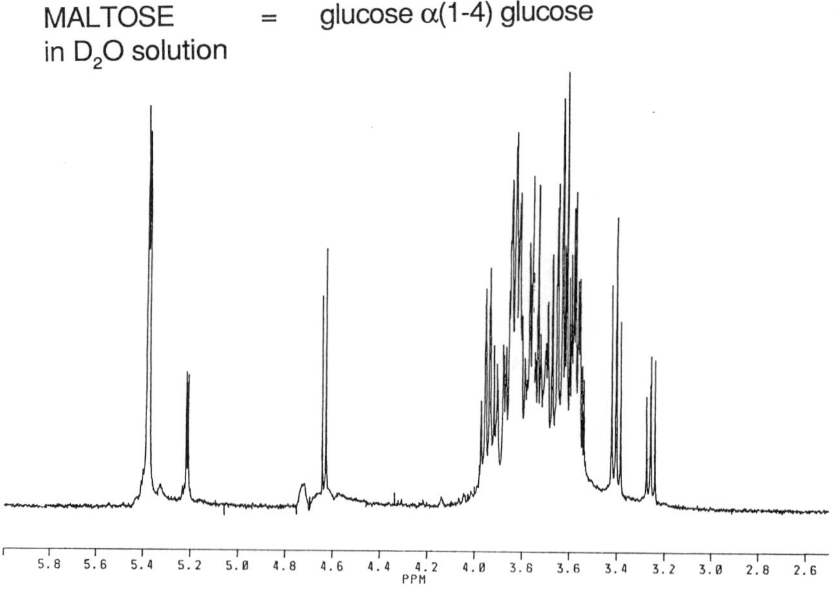

Fig. 3.29 Chemical shift regions of oligosaccharides. The entire chemical shift range is 3–6 p.p.m. The anomeric protons resonate in the 4.5–6 p.p.m. region. Note how crowded the 3–4 p.p.m. region is.

DQF COSY is much better than the magnitude or normal phase-sensitive versions. In simple systems most of the sugars can be followed; however, in the more common cases of interest, with 10 or more sugars, the resolution is not sufficient to identify many of the spin systems. The situation is improved by using RELAY or multiple RELAY spectra, taking advantage of the resolution in the anomeric region to extend connectivities further and further into the sugar. The number of RELAY steps that can be performed is limited by linewidth and by any local conformations in which small coupling constants occur. Recently, the RELAY has generally been replaced by TOCSY, since it is more efficient at getting coherence transferred from the anomeric proton out even to the H-6 and H-6′. For moderately complex systems a 1D version of TOCSY can be applied, taking the difference between spectra with and without inversion of a single anomeric proton resonance. After the mixing period the difference spectrum corresponds to a subspectrum, showing all resonances in the same sugar containing the inverted anomeric proton. TOCSY is clearly the most powerful method available for analysis of the sugar spin systems.

Many different sugars have the same basic topology as far as the coupling of spins goes. For example, glucose, mannose, and galactose are simply stereoisomers with the same composition. It is possible to use some information from coupling constants to distinguish these. Changing the stereochemistry at a particular site changes a proton from axial to equatorial on the six-member ring and, hence, there is a substantial change in the value of the coupling constant to neighbouring protons. The analysis of stereochemistry does not require accurate determination of the coupling constants, and sufficient resolution is often available even from slices through a TOCSY experiment to distinguish the two possibilities at each site.

In oligosaccharides considerable information can also be obtained from analysis of chemical shifts. Particular sugars have patterns of chemical shifts which are distinctive and are sensitive to sites of substitution. Unlike the case for proteins and nucleic acids, in which conformation-dependent contributions to the chemical shift are large compared with intrinsic differences, these shift patterns are primarily determined by the chemical bonding and, thus, are usually retained in different oligosaccharides. These can then be used in assigning protons to particular types of sugars.

3.5.2 Assignments

Two approaches have been used to obtain connectivities between sugars. First, NOESY (or ROESY) cross-peaks can be used to identify neighbours across the glycosidic bond. The success of this depends somewhat upon the local conformation, and for some geometries the protons are separated by too large a distance for the NOE to be seen. Unlike proteins, for which a number of alternative protons can be used to establish the sequential connectivity, there is no reliable alternative to the anomeric protons for polysaccharides. The total number of inter-residue NOEs is much smaller than for proteins, and many of the important protons are in crowded spectral regions. With polysaccharides of unknown, possibly branched, primary structure one must keep in mind that there may be several different sequential connectivities from the same residue. In addition, non-sequential NOEs also occur, which are difficult to identify when the primary structure is not known.

The second method is to use ^{1}H-^{13}C couplings across the glycosidic linkage. While these three-bond couplings are not very large, they can be identified in optimized HETCOR experiments (for example the constant delay version called COLOC). The correlations with directly bonded ^{1}H-^{13}C pairs must also be obtained. Unambiguous connections between residues can be obtained with both types of ^{1}H-^{13}C correlation. With TOCSY data, the assignments of the anomeric protons and the sites of other linkages can be extended throughout each sugar. The long-range heterocorrelated experiments also serve to assign some of the common sugar modifications, such as acetyl groups, which have no ^{1}H-^{1}H couplings and few if any NOEs. At natural abundance ^{13}C, the long-range correlation experiments are somewhat demanding in sensitivity but, with modern inverse detection methods, obtaining the needed data from moderate amounts of sample should become routine.

There have already been applications of 3D NMR to oligosaccharides. The ^{1}H COSY spectrum can be edited by using a 3D natural abundance ^{13}C HMQC-COSY experiment (Fesik *et al.* 1989). The ^{13}C resonances provide an additional chemical-shift axis, similar to the ^{15}N resonances used for peptide assignments. Additional assignment information can be obtained from a 3D NOESY-TOCSY experiment (Vuister *et al.* 1989).

3.6 Discussion

This section has tried to present both the ideas behind methods for systematic determination of resonance assignments for biomolecules and some of the details of how it is done for proteins, nucleic acids, and carbohydrates. In all systems and with all methods the basic requirements are to use couplings within residues to identify the basic chemical structure. Although it is not always possible to do a complete identification, reliable determination of the number of spins on the residue and the distribution of chemical shifts is sufficient to proceed with the analysis. The second crucial step is the determination of connectivities between residues. This is achieved with either inter-residue NOEs or heteronuclear couplings. In systems with good spectral dispersion (no degeneracies in chemical shift) these inter-residue connections are sufficient to complete assignments. In the more typical case, in which degeneracies are present, residue identities and sequential connections must be used together with the primary sequence to complete the assignments. An alternate approach is to use independent NOE paths to check for self-consistency. This latter approach is particularly attractive for automation of the process. To extend applications to larger and larger systems, further improvements in resolution are needed. This can be achieved either by 3D NMR (with or without uniform isotope labelling), or by selective incorporation of labelled residues. Both of these methods have already been applied successfully and have great promise for future work.

The logic used initially in proteins and then extended to other types of biopolymers can be applied to other systems as well. The only basic requirement is that there be a covalently linked chain which keeps some spins on neighbouring residues within the range of providing either a sequential NOE or a spin-spin coupling path.

References

Aue, W.P., Bartholdi, E., and Ernst, R.R. (1976). Two-dimensional spectroscopy. Application to nuclear magnetic resonance. *Journal of Chemical Physics* **54**, 2229–46.

Bax, A., and Davis, D.G. (1985a). MLEV-17-based two-dimensional homonuclear magnetization transfer spectroscopy. *Journal of Magnetic Resonance* **65**, 355–60.

Bax, A., and Davis, D.G. (1985b). Practical aspects of two-dimensional transverse NOE spectroscopy. *Journal of Magnetic Resonance* **63**, 207–13.

Bax, A., and Drobny, G. (1985). Optimization of two-dimensional homonuclear relayed coherence transfer NMR spectroscopy. *Journal of Magnetic Resonance* **61**, 306–20.

Billeter, M., Braun, W., and Wüthrich K. (1982). Sequential resonance assignments in protein ^{1}H nuclear magnetic resonance spectra, computation of sterically allowed proton-proton distances and statistical analysis of proton-proton distances in single crystal protein conformations. *Journal of Molecular Biology* **155**, 321–46.

Bothner-By, A.A., Stephens, R.L., Lee, J., Warren, C.D., and Jeanloz, R.W. (1984). Structure determination of a tetrasaccharide: transient nuclear Overhauser effects in the rotating frame. *Journal of the American Chemical Society* **106**, 811–13.

Braunschweiler, L., and Ernst, R.R. (1983). Coherence transfer by isotropic mixing: application to proton correlation spectroscopy. *Journal of Magnetic Resonance* **53**, 521–8.

Braunschweiler, L., Bodenhausen, G., and Ernst, R.R. (1983). Analysis of networks of coupled spins by multiple quantum NMR. *Molecular Physics* **48**, 535–60.

Brown, S.C., Weber, P.L., and Mueller, L. (1988). Toward complete ^{1}H NMR spectra in proteins. *Journal of Magnetic Resonance* **77**, 166–9.

Brush, C.K., Stone, M.P., and Harris, T.M. (1988). Selective reversible deuteration of oligodeoxynucleotides: simplification of two-dimensional nuclear Overhauser effect NMR spectral assignment of a non-self-complementary dodecamer duplex. *Biochemistry* **27**, 115–22.

Chazin, W.J., and Wüthrich, K. (1987). Optimization of homonuclear relayed coherence transfer experiments with proteins in H_2O solution. *Journal of Magnetic Resonance* **72**, 358–63.

Chou, S.H., Flynn, P., and Reid, B.R. (1989). Solid phase synthesis and high-resolution NMR studies of two synthetic double-helical RNA dodecamers r(CGCGAAUUCGCG) and r(CGCGUAUACGCG). *Biochemistry* **28**, 2422–34.

Davis, D.G., and Bax, A. (1985). Assignment of complex ^{1}H NMR spectra via two-dimensional homonuclear Hartmann-Hahn spectroscopy. *Journal of the American Chemical Society* **107**, 2820–1.

Derome, A.E. (1987). *Modern NMR techniques for chemistry research.* Pergamon Press, Oxford.

Drobny, G., Pines, A., Sinton, S., Weitekamp, D., and Wemmer, D. (1979). Fourier transform multiple quantum nuclear magnetic resonance. *Symposium of the Faraday Society* **13**, 49–55.

Dubs, A., Wagner, G., and Wüthrich, K. (1979). Individual assignments of amide proton resonances in the proton NMR spectrum of the basic pancreatic trypsin inhibitor. *Biochimica et Biophysica Acta* **577**, 177–94.

Eich, G., Bodenhausen, G., and Ernst, R.R. (1982). Exploring nuclear spin systems by relayed magnitization transfer. *Journal of the American Chemical Society* **104**, 3731–2.

Englander, S.W., and Wand, A.J. (1987). Main-chain-directed strategy for the assignment of H–1 NMR spectra of proteins. *Biochemistry* **26**, 5953–8.

Ernst, R.R., Bodenhausen, G., and Wokaun, A. (1987). *Principles of nuclear magnetic resonance in one and two dimensions.* Oxford University Press, New York.

Farrar, T.C., and Becker, E.D. (1971). *Pulse and Fourier transform NMR.* Academic Press, New York.

Feng, Y., Roder, H., Englander, S.W., Wand, A.J., and DiStefano, D.L. (1989). Proton resonance assignments of horse ferricytochrome c. *Biochemistry* **28**, 195–203.

Fesik, S.W., Gampe, R.T., Jr, and Zuiderweg, E.R.P. (1989). Heteronuclear three-dimensional NMR spectroscopy. Natural abundance ^{13}C chemical shift editing of ^{1}H-^{1}H COSY spectra. *Journal of the American Chemical Society* **111**, 770–2.

Frey, M.H., Leupin, W., Sorensen, O.W., Denny, W.A., Ernst, R.R., and Wüthrich, K. (1985). Sequence-specific assignment of the backbone ^{1}H- and ^{31}P-NMR lines in a short DNA duplex with homo- and heteronuclear correlated spectroscopy. *Biopolymers* **24**, 2371–80.

Gewirth, D.T., Abo, S.R., Leontis, N.B., and Moore, P.B. (1987). The secondary structure of 5S RNA: ^{1}H experiments on RNA molecules partially labelled with ^{15}N. *Biochemistry* **26**, 5213–20.

Griffey, R.H., Poulter, C.D., Yamaizumi, Z., Nishimura, S., and Hurd, R.E. (1982). ^{1}H NMR studies of ^{15}N-labeled *Escherichia coli* tRNA^{Met}. Use of $^1J_{1H-15N}$ couplings to identify imino resonances of uridine-related bases. *Journal of the American Chemical Society* **104**, 5810–11.

Gross, K.H., and Kalbitzer, H.R. (1988). Distribution of chemical shifts in ^{1}H nuclear magnetic resonance spectra of proteins. *Journal of Magnetic Resonance* **76**, 87–99.

Harris, R.K. (1983). *Nuclear magnetic resonance spectroscopy*. Ritman, London.

Hyberts, S.G., Maumlautrki, W., and Wagner, G. (1987). Stereospecific assignments of side-chain protons and characterization of torsion angles in eglin c. *European Journal of Biochemistry* **164**, 625–35.

Jeener, J. (1971). *Ampere International Summer School*. Basko polje, Yugoslavia.

Jeener, J., Meier, B.H., Bachmann, P., and Ernst, R.R. (1979). Investigation of exchange processes by two-dimensional NMR spectroscopy. *Journal of Chemical Physics* **71**, 4546–53.

Kumar, A., Ernst, R.R., and Wüthrich, K. (1980). A two-dimensional nuclear Overhauser enhancement (2D NOE) experiment for the elucidation of complete proton-proton cross-relaxation networks in biological macromolecules. *Biochemical and Biophysical Research Communications* **95**, 1–6.

LeMaster, D.M., and Richards, F.M. (1988). NMR sequential assignment of *Escherichia coli* thioredoxin utilizing random fractional deuteration. *Biochemistry* **27**, 142–50.

Macura, S., and Ernst, R.R. (1980). Elucidation of cross relaxation in liquids by two-dimensional NMR spectroscopy. *Molecular Physics* **41**, 95–117.

Marion, D., and Lancelot, G. (1984). Sequential assignment of the ^{1}H and ^{31}P resonances of the double stranded deoxynucleotide d(ATGCAT)$_2$ by 2D-NMR correlation spectroscopy. *Biochemical and Biophysical Research Communications* **124**, 774–83.

Maudsley, A.A., and Ernst, R.R. (1977). Indirect detection of magnetic resonances by heteronuclear two-dimensional spectroscopy. *Chemical Physics Letters* **50**, 368–72.

Müller, L. (1979). Sensitivity enhanced detection of weak nuclei using heteronuclear multiple quantum coherence. *Journal of the American Chemical Society* **101**, 4481–4.

Oh, B.H., Westler, W.M., Darba, P., and Markley, J.L. (1988). Protein carbon–13 spin systems by a single two dimensional nuclear magnetic resonance experiment. *Science* **240**, 908–11.

Otting, G., and Wüthrich, K. (1987). PreTOCSY, a new experiment for obtaining complete 2D ^{1}H NMR spectra of proteins in H_2O solution. *Journal of Magnetic Resonance* **75**, 546–9.

Otting, G., and Wüthrich, K. (1989). Studies of protein hydration in aqueous solution by direct NMR observation of individual protein bound water molecules. *Journal of the American Chemical Society* **111**, 1871–5.

Pardi, A., Wagner, G., and Wüthrich, K. (1983a). Protein conformation and proton nuclear magnetic resonance chemical shifts. *European Journal of Biochemistry* **137**, 445–54.

Pardi, A., Walker, R., Rapoport, H., Wider, G., and Wüthrich, K. (1983b). Sequential assignments for the ^{1}H and ^{31}P atoms in the backbone of oligonucleotides by two-dimensional nuclear magnetic resonance. *Journal of the American Chemical Society* **105**, 1652–3.

Rance, M., and Wright, P.E. (1986). Analysis of ^{1}H NMR spectra of proteins using multiple-quantum coherence. *Journal of Magnetic Resonance* **66**, 372–8.

Rance, M., Sorensen, O.W., Bodenhausen, G., Wagner, G., Ernst, R.R., and Wüthrich, K. (1983). Improved spectral resolution in COSY ^{1}H NMR spectra of proteins via double quantum filtering. *Biochemical and Biophysical Research Communications* **117**, 479–85.

Remerowski, M.L., Glaser, S.J., and Drobny, G.P. (1989). A theoreetical study of coherence transfer by isotropic mixing. Calculation of pulse sequence performance for systems of biological interest. *Molecular Physics* **68**, 1191–218.

Roy, S., and Redfield, A.G. (1981). Nuclear Overhauser effect study and assignment of D stem and reverse-Hoogsteen base pair proton resonances in yeast tRNA[Asp]. *Nucleic Acids Research* **9**, 7073–83.

States, D.J., Haberkorn, R.A., and Ruben, D.J. (1982). A two-dimensional nuclear Overhauser experiment with pure absorption phase in four quadrants. *Journal of Magnetic Resonance* **48**, 286–92.

Stockman, B.J., Reily, M.D., Westler, W.M., Ulrich, E.L., and Markley, J.L. (1989). Concerted two-dimensional NMR approaches to hydrogen-1, carbon-13, and nitrogen-15 resonance assignments in proteins. *Biochemistry* **28**, 230–6.

Szilagyi, L., and Jardetzky, O. (1989). α-Proton chemical shifts and secondary structure in proteins. *Journal of Magnetic Resonance* **83**, 441–9.

Vuister, G.W., de Waard, P., Boelens, R., Vliegenthart, J.F.G., and Kaptein, R. (1989). The use of 3D NMR in structural studies of oligosaccharides. *Journal of the American Chemical Society* **111**, 772–4.

Wand, A.J., and Nelson, S.J. (1988). Refinement and automation of the main chain directed assignments of ^{1}H spectra of proteins. *Transactions of the American Crystallographic Association* **24**, 131–44.

Weber, P.L., Morrison, R., and Hare, D. (1988). Determining stereo-specific ^{1}H nuclear magnetic resonance assignments from distance geometry calculations. *Journal of Molecular Biology* **204**, 483–7.

Wemmer, D.E., Kumar, N.V., Metrione, R.M., Lazdunski, M., Drobny, G., and Kallenbach, N.R. (1986). NMR analysis and sequence of toxin II from the sea anemone *Radianthus paumotensis*. *Biochemistry* **25**, 6842–9.

Wittekind, M., Reizer, J., Deutschen, J., Saier, M., and Klevit, R.E. (1989). Common structural changes accompany the functional inactivation of Hpr by seryl phosphorylation or serine to aspartate substitution. *Biochemistry* **28**, 9908–12.

Wüthrich, K. (1986). *NMR of proteins and nucleic acids*. Wiley, New York.

Wüthrich, K., Wider, G., Wagner, G., and Braun, W. (1982). Sequential resonance assignments as a basis for determination of spatial protein structures by high resolution proton nuclear magnetic resonance. *Journal of Molecular Biology* **155**, 311–19.

Wüthrich, K., Billeter, M., and Braun, W. (1984). Polypeptide secondary structure determination by nuclear magnetic resonance observation of short proton-proton distances. *Journal of Molecular Biology* **180**, 715–40.

Zuiderweg, E.R.P., Boelens, R., and Kaptein, R. (1985). Stereospecific assignments of ^{1}H-NMR methyl lines and conformation of valyl residues in the *lac* repressor headpiece. *Biopolymers* **24**, 601–11.

4

Modelling of related protein structures

Janet M. Thornton and Mark B. Swindells

4.1 Introduction

Understanding protein function requires detailed structural analysis of atomic resolution data which can be determined by X-ray crystallography (see Chapter 1) and NMR spectroscopy (Chapter 2). However, these methods are often slow and dependent on numerous factors for their success. For example, a crystallographic determination depends on the ability to grow well-ordered crystals, whereas NMR spectroscopy is only applicable to smaller proteins of up to 200 residues. In the absence of such definitive information a structure may often be predicted by detecting similarities between the new sequence and previously solved structures. Predictions that use related proteins of known structure are usually referred to as model-building strategies, the principles of which will be discussed in this chapter.

With the rapid growth of sequence databases, model building is now the subject of numerous investigations. There are currently more than 25 000 sequences available in the OWL composite sequence database (Bleasby and Wootton 1990), many of which can be clustered into homologous families. If at least one structure within a family is available, modelling the others becomes a realistic possibility. The subsequent design of mutagenesis experiments can benefit substantially from the models produced by these procedures.

Modelling is fundamentally based on an understanding of the relationship between protein sequence and structure. The fact that it is not yet possible to predict protein structure *ab initio* from sequence emphasizes our lack of understanding of the subject. However, recent progress in analysing and classifying protein structures, especially those in homologous families, has allowed the development of semi-automatic techniques for modelling related structures using a combination of knowledge-based and *ab initio* methods. Although molecular mechanics, dynamics, and distance geometry can contribute substantially to modelling, this chapter will concentrate on the application of knowledge derived from the Protein Data Bank (Bernstein *et al.* 1977) to this work.

4.2 The starting point

Knowledge-based model building (see Glossary) has a large range of applications. The simplest example predicts the effects of a point mutation on an experimentally derived structure and is often used to guide site-directed mutagenesis experiments. At the next level of complexity, multiple mutations, including insertions and deletions, may be modelled. As the number of alterations to the original structure increases, so does the complexity of the modelling process (see Fig. 4.1).

When modelling does not involve simple modifications to a known structure it is first necessary to identify a related structure through sequence comparisons. For instance, when two sequences of a reasonable length (for example, 50 residues) are identical except for one amino-acid substitution, we can be confident that they will have a similar fold. However, as the sequences diverge, the ability to distinguish significant alignments from coincidental similarities diminishes. In general, at least 20 per cent of the residues must be identical before we can be confident of similarity. Below this value structural similarities may also occur, even though we are currently unable to identify them. This situation is not uncommon. In the globins only two haem-binding residues are completely conserved, yet the gross structure is retained (Lesk and Chothia 1980).

The rest of this chapter will cover four areas, beginning with a review of data, derived from the Protein Data Bank, which forms the basis of our current knowledge. A brief summary will then be given of the first attempts to model structures, because these provide the basis for present methods. This will be followed by a detailed description of procedures currently used in modelling, as well as ways of assessing the accuracy of a model. Finally, we will discuss specific examples of model-building exercises, outlining the areas of success and failure.

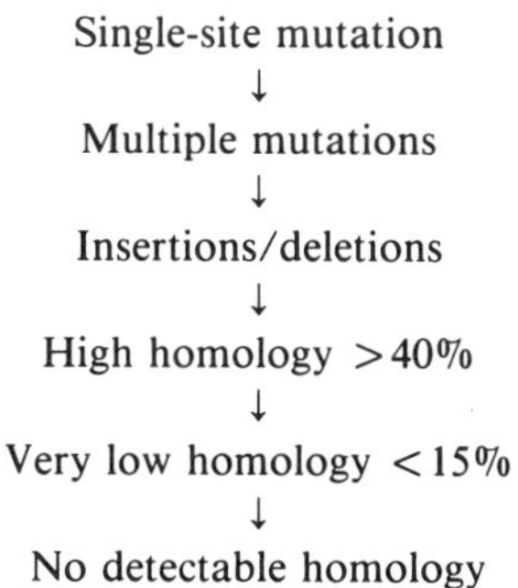

Fig. 4.1 The modelling problem — a spectrum of starting points. Ordering is with respect to increasing difficulty.

4.3 Comparison of homologous protein structures

The Protein Data Bank (Bernstein *et al.* 1977) now holds the coordinates for over 400 proteins, which can be grouped into more than 50 different families on the basis of sequence homology. These families range in size from only one member, in the case of interleukin-1β, up to 13 members in the globin family. Within families the identity ranges from 100 per cent down to less than 20 per cent. Many comparisons of homologous protein structures have been made, including the globins (Lesk and Chothia 1980; Bashford *et al.* 1987), the immunoglobulins (Lesk and Chothia 1982), the serine proteinases (Greer 1981; Read *et al.* 1984), and the aspartyl proteinases (Tang *et al.* 1978). Together these analyses have led to observations that now form the backbone of current modelling procedures. Detailed comparisons for individual families are given in the following chapters. However, with reference to modelling in general, the following important observations have been made.

Structures are conserved much more strongly than amino-acid sequence. This situation is observed both within homologous families, where sequence similarities can vary dramatically, and between certain families, whose protein folds are similar despite no detectable sequence similarity. A protein can be divided into structurally conserved regions (SCRs) and structurally variable regions (SVRs) (Greer 1981). The SCRs are often referred to as framework regions or, collectively, as the core (Fig. 4.2). Usually they com-

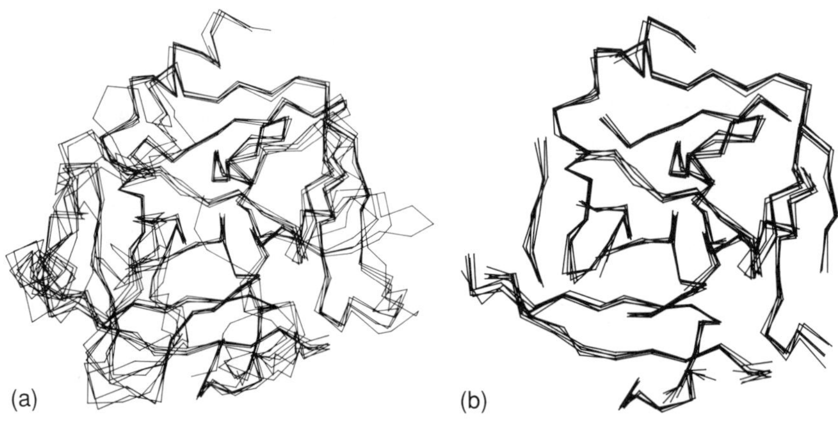

Fig. 4.2 Superposition of structures from the mammalian serine proteinase family. The structures, 3RP2, 2PTN, 2GCH, 1EST, and 2PKA, are least square fitted using the program MNYFIT (Sutcliffe *et al.* 1987a). (a) the resultant superposition of these structures, and (b) the structurally conserved regions (SCRs) from this alignment. (Figures taken with permission from Overington 1990.)

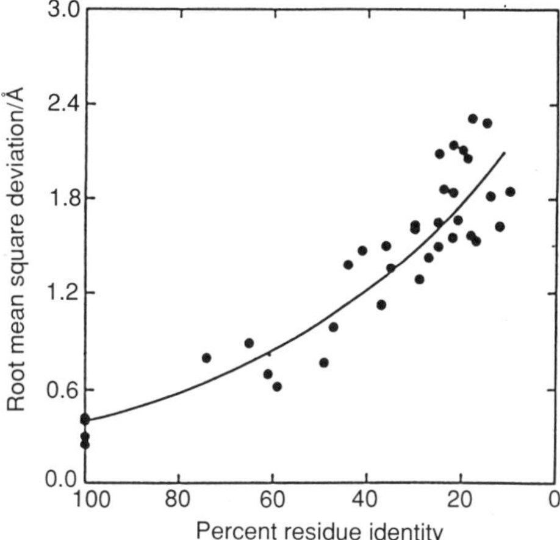

Fig. 4.3 The relationship of residue identity and RMS deviation of the backbone atoms from the common cores of 32 pairs of homologous proteins. (Taken from Chothia and Lesk 1986, with permission.)

prise the regions of regular secondary structure, such as α-helices and β-sheets, whereas the SVRs consist of the more variable connecting loops.

Comparisons of homologous structures show that there is a definite relationship between the percentage sequence identity in the core and the corresponding structural similarity measured. This is usually calculated as the root mean square (RMS) difference between the overlapping core regions (Chothia and Lesk 1986). As illustrated in Fig. 4.3, for structures with 60 per cent sequence identity the RMS overlap between core regions is about 0.8 Å, whereas at 20 per cent sequence identity the difference has risen to 1.8 Å. This relationship provides an empirical measure of the errors likely to occur when modelling on the basis of a related structure, although energy refinement may improve the final accuracy.

Insertions or deletions almost always occur in the SVRs and often lead to extensive rearrangements in conformation (Fig. 4.4). When they occur in a β-strand they are accommodated by a β-bulge which minimizes structural disruption. Within helices, alterations are confined to their termini in order to avoid altering the packing and register of the whole helix. The relationship between sequence identity and RMS overlap illustrates how the core packing changes as mutations are accommodated. Often the secondary structures are conserved but the packing between them is subtly altered. In helices, the axis may rotate or translate, whereas in sheets local variations propagate to give larger differences over their length. Such shifts are common and become larger as the proteins diverge. They are difficult to model empirically and so

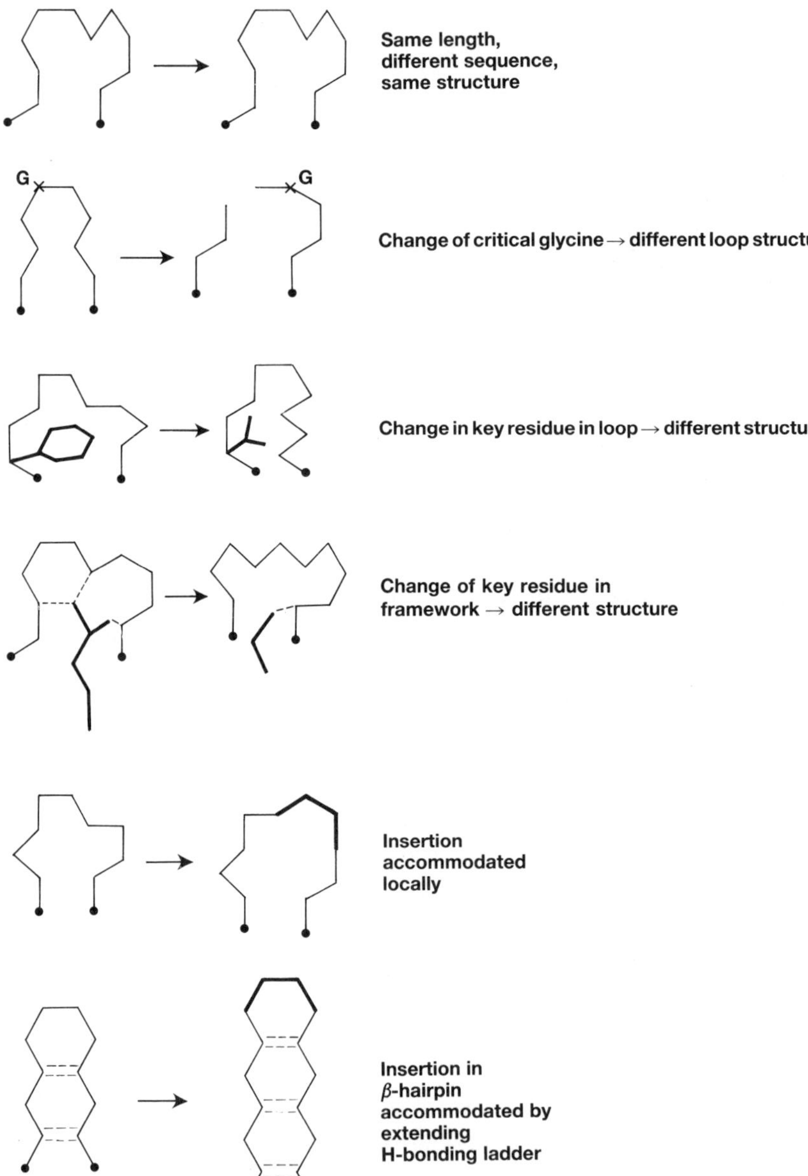

Fig. 4.4 A schematic representation of the typical conformational changes observed when modelling rearrangements in a loop structure.

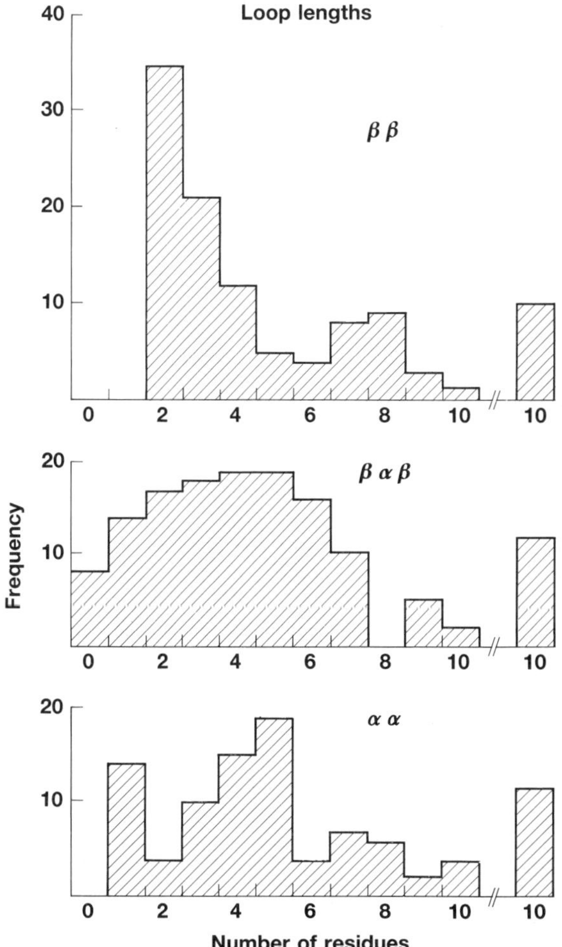

Fig. 4.5 The distributions of different loop lengths within three types of supersecondary structures: β-hairpins, $\beta\alpha\beta$-units, and α-loop-α motifs.

molecular mechanics and dynamics techniques must be used. However, even these have so far had limited success.

Loop regions are the most variable and therefore the most difficult to model. However, analyses have led to the following observations. Loops are generally short, with most having $\leqslant 5$ residues (Fig. 4.5). Such regions are tightly constrained by end-point geometries and consequently a limited set of preferred conformations are observed (Thornton *et al.* 1988). For example, β-hairpins can be divided into four common structural families.

Table 4.1 Canonical loop structures in three types of supersecondary motif

Type[1]	Conf[2]	Number[3]	Sequence[4]
$\beta\beta$-Hairpins			
2:2	$\alpha_L \gamma_L$	16	G/D/N/G
2:2	$\varepsilon\alpha_R$	10	GS/T
3:5	$\beta\alpha_R \gamma_R \gamma_L \beta$	10	βTTGβ
4:4	$\alpha_R \alpha_R \gamma_R \alpha_L$	5	αTTG
$\beta\alpha\beta$-Units			
$\alpha\beta$1	γ_L	7	*----G/D----*
$\alpha\beta$3	$\gamma_L \beta\alpha_R$	4	*----GA/H--*
$\beta\alpha$3	–	5	i--*-G*G-*G
$\beta\alpha$0	–	4	*iT/S-i-*
$\alpha\alpha$-Loops			
0:1	β	13	*--iPi-*
2:4	$\gamma_R \alpha_L \beta\beta$	11	*-----G*-i--*
3:5	$\beta\alpha_R \gamma_R$	3	*--i--Pi-i

[1] Classifies these motifs as β-hairpins, $\beta\alpha\beta$-units, and α-loop-α structures. The nomenclature describes the number of residues in a loop in terms of the hydrogen-bonding patterns observed (see Edwards *et al.* 1987 and Sibanda *et al.* 1989 for more details).
[2] Refers to the ϕ, ψ conformation of each residue in the loop, where α_L = left-handed α-helix; α_R = right-handed α-helix; β = β-strand; ε = residues with $+\phi$ and $-\psi$; γ refers to the bridge region at $\phi = +/-60$ and $\psi = 0$.
[3] The number of motifs observed in the Protein Data Bank in 1987.
[4] Sequence uses a one-letter code for the amino acids; *, hydrophobic; i, polar/hydrophylic; β, residue favouring the β-strand conformation; —indicates any type of residue;/shows alternatives.

These occur in proteins from different homologous families and probably represent particularly stable conformers. A compilation of canonical structures for short loops is given in Table 4.1 and a detailed classification for β-hairpins is shown in Fig. 4.6 (Sibanda *et al.* 1989).

Many loops include critical or key residues which are essential in determining loop conformation. Glycine frequently occurs in short loops, especially in β-turns which usually exploit the unique ability of this residue to adopt a positive ϕ angle. Substitutions at these locations would inevitably disrupt loop conformation. Prolines are also popular in loops and turns, where they effectively terminate the adjacent secondary structures. However, proline can only be considered essential when it is preceded by a *cis* peptide. This structure cannot be mimicked accurately by a *trans* peptide, and since prolines are unique in having a significant percentage of *cis* peptides preceding them (5.8 per cent of prolines in the Protein Data Bank, as shown by MacArthur and Thornton 1991), their mutation will almost certainly alter the conformation.

Other critical (key) residues determine loop conformation through specific side-chain interactions. These may be hydrophobic residues which pack against the core, or hydrophilic side-chains involved in hydrogen bonds. In either case, their mutation usually disrupts the loop conformation. Apart from these critical residues, loop conformations are surprisingly tolerant to mutations. There are several examples where the sequence of a loop is completely different in two homologous structures and yet the conformation is retained. Although end-point and backbone geometry is clearly important in determining short loop conformation, it is uncertain how relevant these restrictions are when loop length increases. In the larger loops, insertions and deletions are often accommodated locally without disrupting the rest of the loop structure (see Fig. 4.7).

A comparison of side-chain conformers in homologous structures has been made for several families of structures with sequence identities ranging from 16 to 60 per cent (Summers *et al.* 1987). The basic rule for determining the transfer of side-chain conformation from one structure to another involves maintaining maximal overlap of atom positions between the two structures. Summers *et al.* found that 60–97 per cent of γ-atoms in homologous structures were transferable when the residue was not mutated, and this value was only slightly less for the δ-atoms. For mutated residues the values were 60-89 per cent for γ-atoms and 56-84 per cent for δ-atoms. In general, the side-chain conformations are more conserved when the residue identity between the proteins is higher. This conservation partly reflects constraints imposed by the core's backbone conformation rather than specific side-chain interactions. For example, aromatic amino acids in helices have χ_1 values in the g^- or g^+ conformations, whereas the *trans* conformer is not observed due to steric hindrance with the previous turn of the helix (McGregor *et al.* 1987). In contrast, many exposed solvent-accessible side-chains are not as restrained. Indeed they are often mobile, even in the crystal structure. In the three structures of human interleukin-1β available in the Protein Data Bank, all are resolved to 2.0 Å, have R-factors of around 20 per cent, and are in the same space group, yet only 52 per cent of the accessible residues (76 per cent of all residues) have side-chains in the same energy well. This presumably reflects the poor quality of the electron density for the exposed side-chains.

Interactions between side-chains often exhibit preferred orientations, which can be detected by extracting pairs of side-chains from the data bank and superposing them as illustrated in Fig. 4.8. Side-chain pairs that can form hydrogen bonds show clustered distributions in three dimensions (e.g. arginine interacting with carbonyls or carboxylates). An atlas of all such side-chain interactions (400 in total) has been constructed from high-resolution data (Singh and Thornton 1990). Similarly, the distributions of atoms around the 20 side-chains also display distinct preferences. For example, oxygen atoms cluster around the edges of aromatic rings, whereas aliphatic carbons group around the face (Fig. 4.9). In contrast, interactions

SYSTEMATIC MODELLING OF β-HAIRPINS

⟵ REPLACEMENT ⟶

DELETION ↑

SET	DOUBLE H-BOND				SINGLE H-BOND			ALTERNATIVES
A	2:2 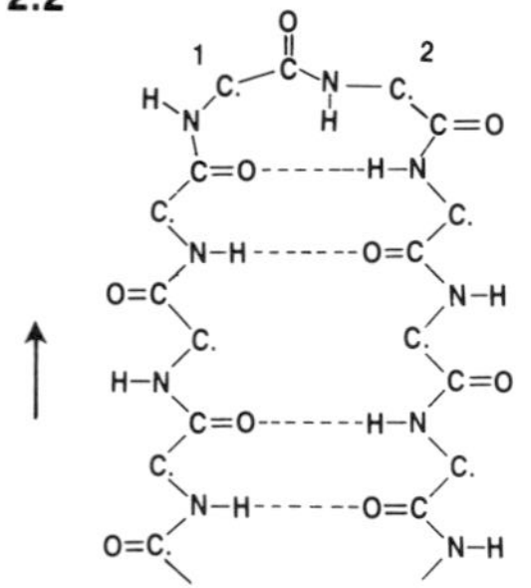	★ Type I' Gly —Asn—Gly— Asp α_L γ_L	★ Type II' —Gly—Ser— Thr ε γ_R	Type I —X—X— α_R α_R	2:4 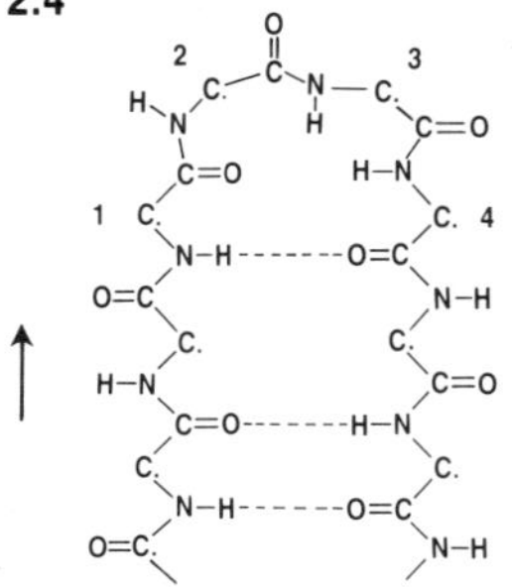	Unusual- Various		6:6 6:8 10:10 10:12
B +1	3:3 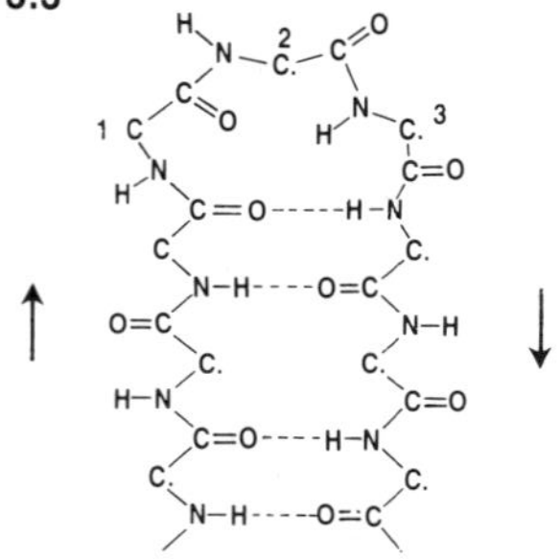	Rare - Various			3:5 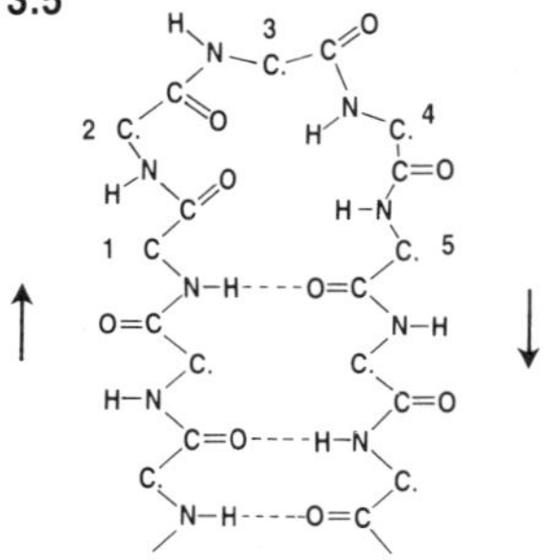	★ Type I [1–4] + G1 β-bulge —X—X—X—Gly—X— β α_R γ_R γ_L β	Various	7:7 7:9 11:11 11:13

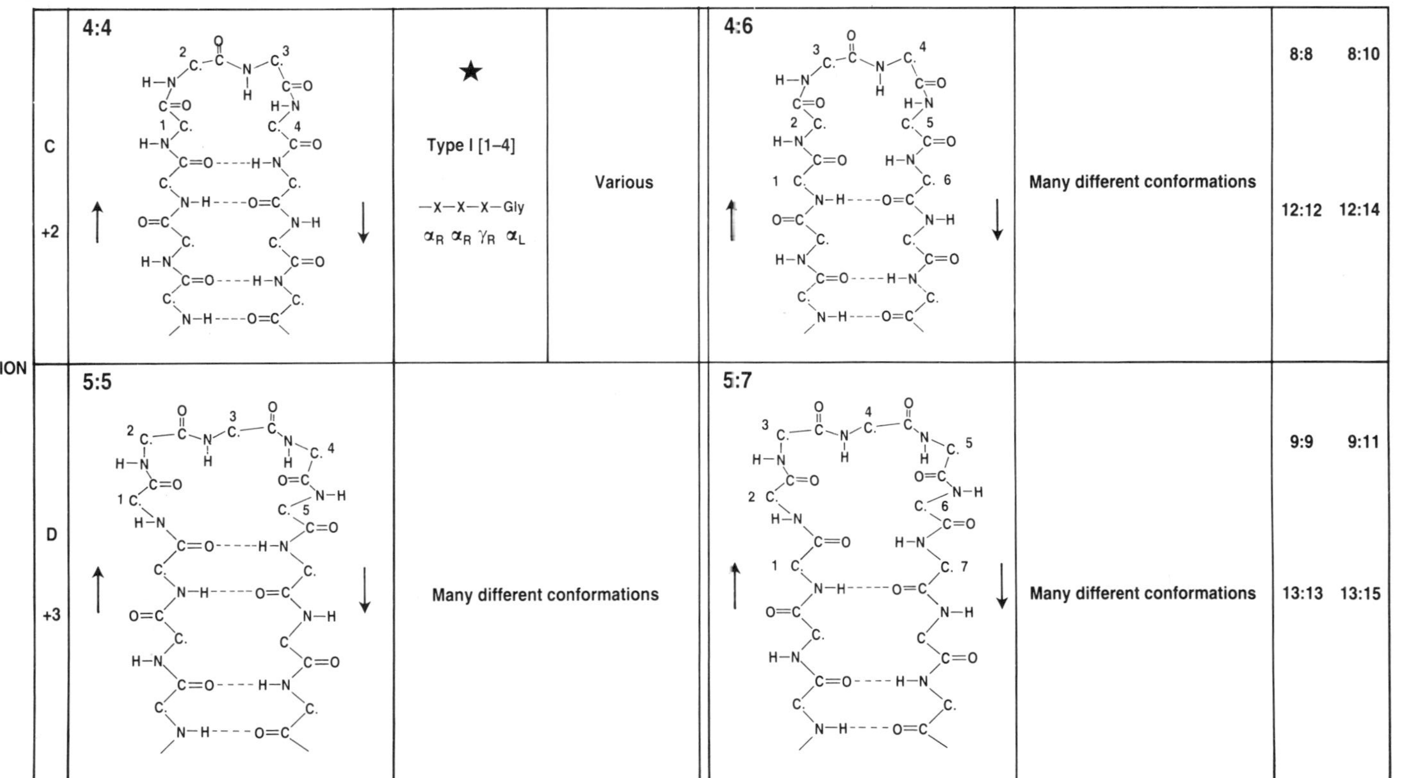

Fig. 4.6 A schematic classification of b-hairpin structures. Structures in the horizontal lines have identical numbers of residues in their loops but different conformations and sequences. These are useful in modelling residue substitutions. Structures in the columns represent loops in β-hairpin conformations with different numbers of residues, and these are used in modelling insertions and deletions.

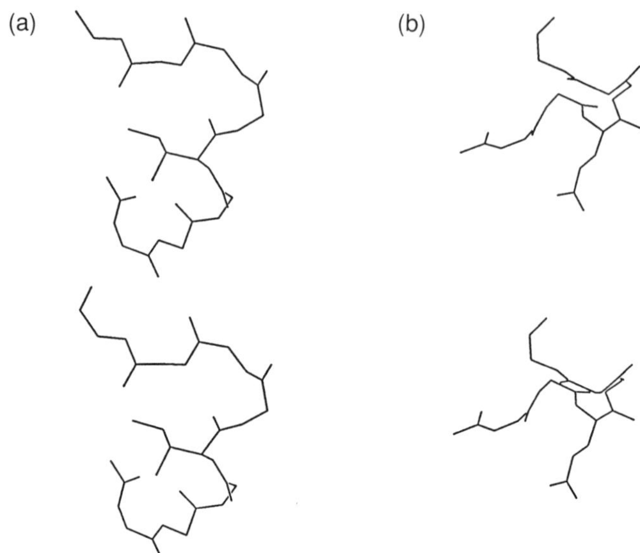

Fig. 4.7 (a) Main-chain trace of the loop containing the hydrophobic residue isoleucine 29 from the immunoglobulin structure 1REI. Ile 29 determines loop conformation by interacting with framework residues. (b) View of the loop between strands D1 and E1 in trypsin (2PTN). Glutamate 77 interacts with the framework around Asn 72.

between non-polar aliphatic residues have few preferences, and analyses of homologous structures show that these interactions frequently change as the percentage residue identity decreases.

4.4 First attempts to model by homology

The first detailed attempt to build a model on the basis of a homologous structure was made by Browne *et al.* (1969) who modelled the sequence of α-lactalbumin on to the structure of hen egg-white lysozyme, with which the sequence identity is about 40 per cent. Over the next 10 years modelling studies were carried out on several systems, including the serine proteinases (McLachlan and Shotton 1971) and the immunoglobulins (Padlan and Davies 1975). However, it was not until 1981 that a rigorous approach to model building was developed, which utilized the additional information available from multiple structures (Greer 1981). This work, which was based on comparisons of the mammalian serine proteinases, was to establish the principal steps in model building (Fig. 4.10).

A short overview of each stage will be given following the order established in Fig. 4.10.

(a)

(b)

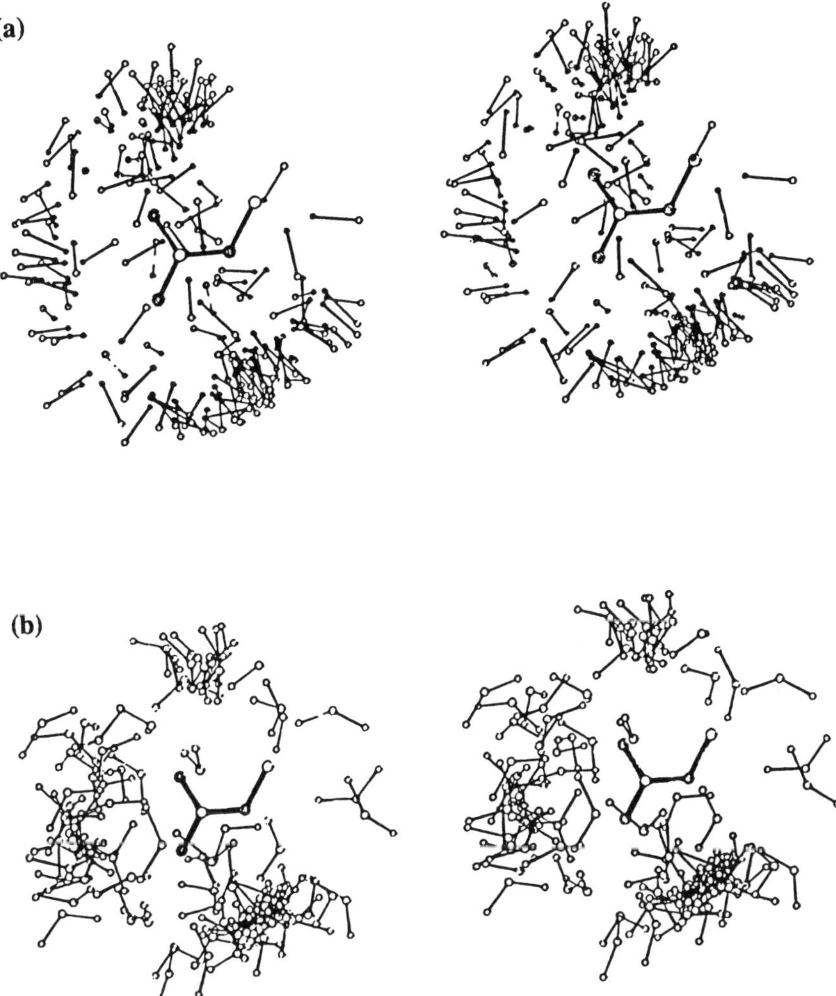

Fig. 4.8 Side-chain distributions derived from high-resolution protein structures. Interactions are extracted from Protein Data Bank coordinate files and the distributions are generated by superposing on the reference arginine (Singh and Thornton 1990). (a) Stereo-plot of transformed carbonyl groups around arginine. (b) Stereoplot of the transformed aspartate and glutamate carboxyl groups around arginine. The reference arginine consists of the CD, NE, CZ, NH1, and NH2 atoms.

Fig. 4.9 Stereo-plots of all transformed Tyr-oxygen pairs. Distributions were generated in a similar manner to those for Fig. 4.8, using tyrosine as the reference.

1. Align family of structures
   ↓
2. Align family of sequences
   ↓
3. Define structurally conserved regions
   ↓
4. Model SCRs of novel sequence
   ↓
5. Model the loops
   ↓
6. Append side-chains
   ↓
7. Evaluate and refine model using molecular mechanics and dynamics

Fig. 4.10 Flow diagram showing the major steps in modelling by homology.

4.5 Defining the structurally conserved regions (SCRs)

It is generally accepted that regions likely to be conserved between structures should be modelled first. When more than one structure is known for a homologous family, the SCRs can be defined objectively. Over the past decade numerous methods of increasing sophistication have been developed for this purpose. These procedures are all based on rotating coordinate sets with respect to a reference frame. One of the first methods to orient two coordinate sets was made by Rossmann and Argos (1976), using a computationally intensive search around all three Eulerian angles. The advantage of such a method was that no manual intervention was required during the process. As a result they were able to investigate numerous relationships between homologous proteins.

One obvious way to increase the speed of such procedures is to manually define a set of topologically equivalent residues and least squares fit their α-carbons. Greer (1981) utilized this technique and defined the framework as residues with C_α separations of <1 Å. In an attempt to reduce the dependence on manual definitions, Sutcliffe *et al.* (1987a) developed an approach that iteratively extends the framework from a minimum of three user-defined equivalences, until the RMS deviation of the overlapping regions exceeds a predetermined value. The method also attempted to superpose multiple structures without biasing the result towards a particular homologue. Bias usually occurs because one structure is fixed as the reference to which other homologues are fitted. By deriving an average framework for the homologous family through iteration of the fitting routine, Sutcliffe *et al.* were able to avoid this situation (see Fig. 4.2).

These procedures work well for close relatives when there are no shifts in secondary structure, but in distantly related family members the size of the framework shrinks significantly. This may be due to global shifts of helical axes as well as changes in β-sheet stacking and domain packing. To overcome this problem Nishikawa and Ooi (1974) and Sippl (1982) developed methods that used distance matrices to compare homologous tertiary structures. The logical progression from comparison is to use these matrices to align a pair of structures. Distance matrices are ideally suited to this problem as they allow structural information to be manipulated without considering proteins as rigid bodies. Based on this approach, Taylor and Orengo (1989) calculated a set of weights that indicate the likelihood of each residue in one protein having structural equivalence with residues in another. The weights are generated by considering similarities between all pairs of distances, the differences in their relative orientations, and their hydrogen-bonding patterns. Once the weights have been calculated, a table of comparisons is filled. The number of spaces in the table is equal to the product of the two sequence lengths, and each weight represents the likelihood of a residue pair having structural equivalence. Dynamic programming is then used to find the optimal path through these weights from which a satisfactory alignment can be achieved.

In a similar technique, Sali and Blundell (1990) describe a protein structure as a cumulative set of weights which go into a similar table. In this case far more weights are used, which can be conveniently divided into two groups. The first describes residue properties such as type, solvent accessibility, and secondary structure classification, whereas the second classifies residue relationships such as hydrogen bonds, hydrophobic clusters, and nearest neighbour distances. Property weights are relatively easy to calculate, whereas relationships are more complicated. For instance, it is difficult to identify topological equivalence from hydrogen-bonding patterns alone. Such weights are computed with a simulated annealing algorithm, where an energy term is used to assess the desirability of a relationship. With all of the weights suitably assigned, the structures are again aligned using dynamic programming.

4.6 Aligning sequences

With the sequences of known structure now optimally aligned, the related sequences of unknown structure must be added. Many automated methods are now available for aligning pairs and families of sequences. The method used depends on the amount of information available (see Taylor 1988 for a review). Wherever possible, structural information should be incorporated into the alignment because insertions and deletions rarely occur in the framework regions (Barton and Sternberg 1987).

The ability to align families of sequences is directly related to the degree of residue identity between the sequences. At 50 per cent or more this task is fairly trivial. However, multiple sequence alignment and profile fitting methods such as those described by Gribskov *et al.* (1987) become essential as the identities decrease. In a detailed study of the globins, Bashford *et al.* (1987) showed that there was sufficient information in a protein family to successfully distinguish their sequences from all non-globins. At homologies of less than 25 per cent, it becomes necessary to use template-based and structure prediction methods to locate equivalences. Consensus sequences or templates are specifically derived from proteins of known structure and therefore highlight regions of particular importance (Taylor 1986). This approach, which was successfully used to detect distant similarities between the aspartic and retroviral proteinase families, will be discussed in further detail below.

It has been suggested that alignments could also be improved through substitution matrices which specifically consider a residue's structural environment instead of the standard Dayhoff (1978) type matrix. Such matrices might investigate the substitution patterns of buried residues whose main-chain conformations are α-helical, Overington *et al.* (1990). However, such specific tables may lack statistical significance due to insufficient crystallographic data.

The major task of sequence alignments is to identify correctly the SCRs for the protein of unknown structure. Although numerous methods have been developed for aligning sequences, the alignments become less secure as sequence similarities decrease. Consequently, rebuilding and reiteration of the model become increasingly necessary at low homology as possible errors are detected.

4.7 Construction of structurally conserved regions

At the present time, SCR construction is still generally approached in the manner defined by Greer (1981), using sequentially similar SCRs from the homologous proteins to define the new core. While adhering to this general approach, two distinct methods have been developed. One method decides which protein of known structure is most similar to the unknown and then extracts all of the framework regions from this single structure, whereas the second approach extracts each SCR individually on the basis of residue identity. It is generally accepted that the first approach is more beneficial as it eliminates the problems associated with packing SCRs from separate starting structures. While this is satisfactory when the percentage identity is at least 40 per cent, the errors are larger when the residue identities are lower. Sali *et al.* (1990) have suggested that a suitable alternative could involve a distance matrix approach, where each distance in the model has an appropriate distribution attached. From this information a structure could be reconstructed in a similar manner to structures derived from NMR data, using a combination of conjugate gradient minimization and simulated annealing.

4.8 Loop modelling

In homologous structures, most of the changes in sequence and structure are restricted to the loop regions, therefore building loops accurately is one of the most demanding steps in producing a reliable model. As a result, a variety of methods have been developed which are often used in combination.

4.8.1 Loop extraction from the protein structure database, using end-point geometry

The end-point coordinates of the loop, as defined by the framework, are used to select all loops that could fit this framework from the database

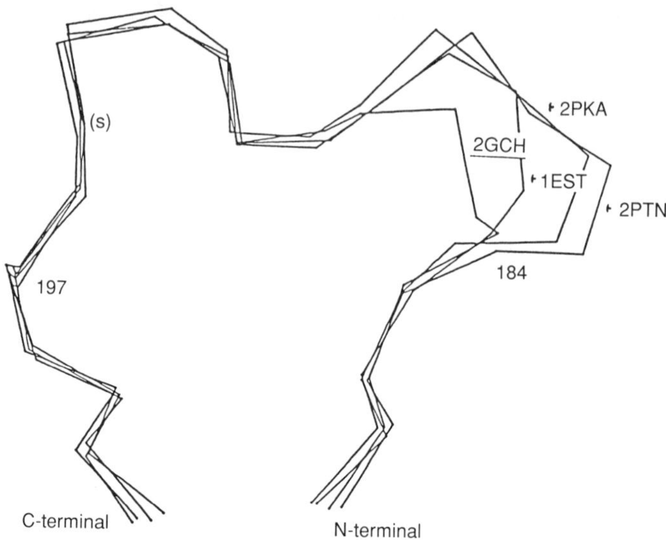

Fig. 4.11 Example of different loop conformations with similar end-point geometries. C_α traces are shown for the loop between strands I and J of the following serine proteinases: 2PKA (residues 184-297), 2PTN (184-197), 2GCH (185-197), and 1EST (185-197). The loops in 2PTN and 2PKA are both 15 residues long, 1EST has 14 residues, and 2GCH 13 residues. The effects of these insertions/deletions can be observed easily. (Taken with permission from Zvelebil 1987.)

of known structures (Fig. 4.11). The method of Jones and Thirup (1986) provides an elegant and fast search procedure. In this method a library of C_α separations for each protein is stored as an interatomic diagonal distance plot (Phillips 1970). Each element of the matrix (i,j) consists of the distances between $C_{\alpha i}$ and $C_{\alpha j}$. If two fragments have the same structure they will have the same matrix, although the reverse is not true. For loop modelling, the end-point C_α separations of the loop are used to search this library for matches of a given segment length. Claessons *et al.* (1989) tested the ability of such search methods to model insertions and deletions of residues. They found that crude selection criteria based solely on fragment length and geometric fit yielded realistic backbones in about two-thirds of the test cases. Given the importance of critical residues, it is often useful to impose additional sequence constraints on the loop searches. Glycines, as well as buried residues which participate in critical hydrogen bonds, should also be conserved as they are often important in determining loop structure. These tests have shown that such sequence constraints improve the agreement between extracted and observed loops.

4.8.2 Global searching

When no satisfactory loop conformers are detected in the database, alternative strategies must be adopted. In global searching ϕ, ψ space is scanned for all suitable loop conformations. Members of this reduced set are then tested for their ability to undergo satisfactory loop closure (Moult and James 1986). Due to the computational complexity of such a search, this approach is only really feasible for shorter loops. A variation on this approach uses the method of Go and Scheraga (1970) to close the loop. Given two end-point constraints between which there are three residues, this method will solve equations to find suitable ϕ, ψ values for the linking tripeptide. Thus, for a loop of n residues the conformational search can be restricted to n-3 residues, after which suitable tripeptide linkers are calculated to close the loop. This has two advantages. First, it significantly reduces the number of conformations to be searched; secondly, it closes the loop with good geometry.

4.8.3 Energy minimization/molecular dynamics

It should be possible to build in one loop structure and then introduce variations using molecular dynamics or simulated annealing. In this way a variety of conformational states can be explored and the lowest energy conformation chosen. These methods are used primarily to differentiate between and refine loop conformers generated from database and global search techniques.

Although the correct loop conformer is often present in a selection of structures, the absence of a solvation term in the energy calculation often precludes its selection. However, the addition of a solvation term significantly complicates the calculation, with the result that several groups have considered accessibility as an alternative screen. For example, to distinguish between two conformers with similar energies, $(\Delta E \approx 8.4 \text{ kJ mol}^{-1})$, the one whose hydrophobic residues have the lowest solvent accessibility is chosen (Bruccoleri *et al.* 1988). This selection procedure assumes that the conformation with the lowest energy will be the most closely packed. Similarly, Schiffer *et al.* (1990) show that solvation is crucial in predicting the conformation of accessible side-chain residues. Without this term the side-chains tend to flatten against the protein during minimization by creating intramolecular hydrogen bonds rather than the protein-water hydrogen bonds observed in crystal structures. In their prediction of rat trypsin built using the bovine homologue, Schiffer *et al.* show that nearly all of the external polar residues have lower solvation energies in the crystal structures than in their predicted model. To evaluate solvation energy they use Eisenberg and McLachlan's approach (1986) whereby each atom type is assigned an experimentally derived atomic solvation parameter. For a given conformer the accessibility of each atom is calculated to find the fraction of exposed surface and is multiplied by its solvation parameter in order to evaluate its contribution to the total solvation energy. Similarly, Martin *et al.* (1989) select loop conformers on the basis of a much simplified energy term. Having chosen the five lowest energy loops, they select the final structure on the basis of minimum solvent accessibility of hydrophobic atoms. Ideally, the solvation term should be included in the original energy evaluations, but the form for such a potential function is very primitive and accessibility calculations are still computationally expensive.

4.9 Side-chains

Currently, most modelling approaches involve delineating the main-chain conformation and then appending the side-chains. Several different methods have been developed, based on a similar rationale to those applied in loop modelling. It is known that side-chain conformation is often conserved between homologous residues even when a replacement has occurred (Summers *et al.* 1987). Where possible, therefore, χ values are taken from the parent structure(s) and used to model the substitute. Figure 4.12 shows an example of such a mutation within the globin family, where lysine is replaced with glutamine. Other approaches consist of conformational searches and energy evaluations in χ space. These methods involve grid sampling χ values, say every $10°$, and calculating the energy at each point. With the largest side-chains, such as arginine with five χ angles, search time becomes prohibitive, involving 36^5 energy evaluations. One alternative is to use a wider grid, for instance $30°$ steps, and to explore areas of conformational space using energy minimization and dynamics.

When a homologous protein does not contain all of the necessary information to enable side-chain replacement, the knowledge-based approach can be

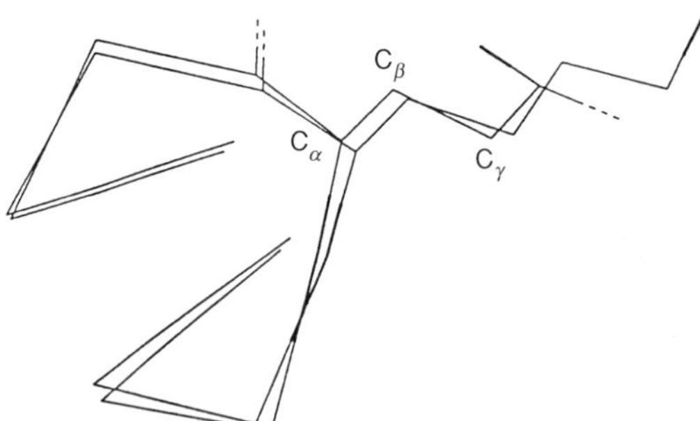

Fig. 4.12 Example of side-chain χ_1 angle conservation when mutations occur between homologous proteins. The structurally equivalent residues, lysine 56 from 3HHB and glutamine 61, are shown superimposed on their C_α atoms. The similar location of the C_γ atoms is clearly seen, indicating the conservation of χ_1.

extended to derive statistical conformational preferences for each residue type (e.g. Janin *et al.* 1978; McGregor *et al.* 1987). Such information would be needed, for instance, if an arginine residue were to replace alanine, because alanine does not contain any χ angle information. Using an analysis of homologous structures, Sutcliffe *et al.* (1987b) have taken this a step further by defining the most probable side-chain conformations for each residue type in the three different secondary structures: α-helix, β-sheet, coil. From these analyses they have derived a set of rules to model side-chain substitutions.

Comparisons of modelled and predicted side-chain χ angles can often be misleading, because totally different χ angles can often correspond to equivalent conformations. For instance, with phenylalanine there is perfect overlap between two conformers for which χ_2 differs by 180°. Therefore, Summers *et al.* (1987) and Sutcliffe *et al.* (1987b) use the concept of maximizing the atomic occupancies to compare conformers. Similarly, Schiffer *et al.* (1990) use a volume error overlap integral in three dimensions. To evaluate the volume overlap they superpose the modelled and observed structures and sum over all points on a 0.1 Å grid that fall within the overlap region for a given residue.

Summers and Karplus (1989) present an excellent and detailed procedure for rule-based side-chain modelling. They test it by modelling the side-chains of the C-terminal lobe of rhizopuspepsin onto the observed main-chain structure, but start with side-chain information from penicillopepsin with which it has 35 per cent residue identity. Their method involves four stages. In the first stage, all atom coordinates for which information can be derived from the template structure are generated. In the second and third stages, rigid rotation samples χ values in 10° increments and evaluates their energies while holding the rest of the protein fixed. This adjusts the side-chain conformation of those residues defined in the first stage and constructs new conformations for those where no structural information is available from the homologous structure. The fourth stage involves extensive model refinement and evaluation, involving calculation of self-energy, interaction energy, accessibility, hydrogen bonding, and energy minimization. After the refinement on the test structure was completed, 92 per cent of the χ_1 and 81 per cent of the χ_2 angles were correctly positioned, and disagreement was generally restricted to highly accessible side-chains. This method, which combines template knowledge with energetic considerations, is to be recommended.

A novel method for predicting side-chain conformations given the backbone structure but no knowledge of side-chain orientations is soon to be published (Lee and Subbiah, private communication). This rapid and completely automated method uses simulated annealing to optimize side-chain packing, considering only van der Waals interactions. Over nine proteins, the program gave RMS deviations of 1.77 Å from the native structure, dropping to 1.25 Å for buried core residues. This method is clearly an important advance and will be especially relevant for modelling the side-chain conformations of proteins based on relatives with very low residue identities. The results suggest that given the backbone conformation, van der Waals forces are the primary determinant of packing in the core. In reality, side-chain packing will help to determine main-chain conformation and so some iteration may be required for modelling. This procedure may also assist the interpretation of low-resolution electron density maps.

4.10 Energy refinement

All methods require energy refinement, not only to regularize the covalent geometry of the model but also to relieve local non-covalent bad contacts and optimize interaction energies. Conventional energy minimization techniques, which explore local minima, can help to regularize main-chain bond length and angle distortions that occur when melding loops onto the framework. More recently, molecular dynamics algorithms have been used to explore wider areas of conformational space.

Energy evaluations usually involve summing the contributions from terms representing bond lengths, bond angles, and torsional angles, as well as non-covalent terms consisting of van der Waals contributions and electro-statics. Since there is no reliable method for evaluating the contribution of solvation, this effect is usually ignored (see below). Minimization techniques have evolved over many years, yielding standard methods such as steepest descent and conjugate gradient minimization. Both of these methods have a very small radius of convergence which produces only small changes in structure (usually less than 0.5 Å). For larger shifts, such as those observed in distantly related structures, other methods, such as simulated annealing, are required. In this method the protein is first heated to a high temperature (up to 10 000 K) which enables it to move freely across energy barriers. The temperature is then systematically reduced to anneal the structure into its lowest energy state. Application of this novel technique to modelling homologous structures has so far been limited, but it provides the possibility of predicting the differences between homologues *ab initio*.

Molecular dynamics can also be used to calculate the free energy differences between similar structures (see Karplus and Petsko 1990 for a review). Such calculations on single-site-directed mutants, for which experimental free energies are readily available, provide valuable tests for the methodology and the potential functions used.

Ideally, modelling requires an automated method which will predict the effects of mutations in both the loop and core regions. Minimization alone would never achieve this, since the energy surface is convoluted with multiple local minima. Although simulated annealing may greatly assist in this area, a successful method will also require the development of accurate potential functions which include solvation effects. This will be an area of intense research activity in the next few years.

4.11 Establishing the accuracy of a model

Once a model has been built it is important to assess its quality. A rapid means of assessing the general accuracy is shown in Fig. 4.1, which gives a measure of the likely discrepancy between C_α core atoms for a given percentage sequence identity. At a more sophisticated level one would expect the correct conformation of a protein to have the lowest energy among a choice of alternative conformers. By selecting two proteins of identical length from different structural classes (e.g. all α and all β) and swapping their sequences, one should be able to distinguish the correct conformations using energy calculations.

Novotny *et al.* (1988) tried this using the helical protein, hemerythrin, and the immunoglobulin variable light-chain domain from mouse myeloma cells, which is composed of β-strands. After generating incorrect models by swapping their sequences, the structures were energy minimized in order to remove side-chain overlaps. Unfortunately, after minimization they found that the native and incorrect conformations for each fold had similar energies. Recently, another attempt to address the same problem was made by Hendlich *et al.* (1990), by calculating the energies using potentials derived from crystallographic data. The method first derives potentials for the atomic interactions of all 20×20 amino-acid pairs as a function of sequential separation between the residues. For simplicity, side-chain interactions are limited to the C_β atoms of each residue, although there is no reason why this should not be extended to all atoms. These potentials are then reapplied to the test protein in order to calculate its energy. Using this approach they were able to distinguish successfully the native conformations in a similar experiment to Novotny's. With suitable modifications this method could easily be applied to any newly constructed model.

An alternative critical assessment involves the comparison of models whose structures are subsequently solved by X-ray crystallography. Two such publications confirm that regions of higher sequence conservation are more likely to be correctly modelled. Greer (1990) has addressed this problem by comparing serine proteinase models from his 1981 paper with their experimental structures. Generally, he found that modelling errors were confined to the loop regions. With the increased number of known structures, a new alignment of 35 proteinase sequences was proposed. In another paper (Weber 1990) two independently derived models of human immunodeficiency virus (HIV) proteinase were compared with the subsequently derived crystal structure. The analysis clearly demonstrated the advantage of basing a model on the most sequentially homologous structure.

4.12 Examples of protein modelling

Numerous investigations of significant therapeutical importance have already been assisted by these modelling techniques. In this section we will describe two specific applications. The first shows how loop modelling procedures are used to construct antibody hypervariable regions, whereas the second reviews work that led to a proposed structure of HIV proteinase.

4.12.1 Modelling antibody hypervariable regions

The hypervariable regions of the antibody molecule comprise six loops, which together form the antibody combining site used to recognize the antigen. Its biological importance has led many groups to model specific antibody sequences using different approaches. Ultimately the aim is to design an antibody of predetermined specificity. The immunoglobulin fold is also found in many functionally different molecules of the immune system, including receptors. As the framework regions of antibodies are similar, the major challenge is to model the loops correctly. Here four attempts to model the combining site are described.

4.12.1.1 Knowledge-based approach   Chothia, Lesk, and colleagues have made a detailed study and comparison of the immunoglobulin hypervariable regions (Chothia *et al.* 1986, 1989; Chothia and Lesk 1987). By looking at each of the six hypervariable loops (labelled L1, L2, L3 from the light-chain hypervariable loops, and H1, H2, H3 from the heavy chain) in six immunoglobulins of known structure, they propose a canonical structure hypothesis in which antibodies have only a few main-chain conformations or canonical structures for each hypervariable region. They suggest that most sequence variations only modify the interacting surface provided by the side-chains on a canonical main-chain structure. However, at a few specific positions, sequence changes cause the main chain to switch to a different canonical conformation. Thus, the L2 and H1 regions exhibit only one canonical structure, the L3 region has three distinct main-chain conformations, whereas L1 and H2 can adopt four canonical structures. The sixth region, H3, is a long loop with many alternative conformations, which cannot be modelled using this method. In their paper, Chothia *et al.* (1989) test the canonical structure model by using it to predict the structures of four immunoglobulin combining sites. Their method involves studying the sequence of the loop to determine whether it has the same length and set of residues responsible for a known conformation. This analysis indicated that a knowledge-based prediction could be made for 19 out of the 24 hypervariable regions in the test set. Four H3 regions and one L3 loop were too long to be modelled using this approach. Few details concerning side-chain conformations or loop-melding procedures are given. However, the prediction of local main-chain conformation was very good. RMS differences in atomic positions of main-chain atoms after optimal super-position ranged from 0.3 to 1.4 Å, with 16 out of 19 loops having RMS differences of less than 1 Å. The prediction of loop orientation relative to the framework was not so good, with C_α positions after superposition of the framework differing by 0.7–3.0 Å. For general applicability this approach requires automation involving the identification of critical residues, extraction of loops from the protein database with priority given to the immunoglobulins, as well as automated melding, building, and refinement procedures.

4.12.1.2 Conformational search approach Bruccoleri *et al.* (1988) used their conformational search program, CONGEN (Bruccoleri and Karplus 1987), to reconstruct the six hypervariable loops in McPC603. These six loops were deleted from the structure and reconstructed in order of increasing length, which for this antibody is: L2-H1-L3-H2-H3-L1. The three shortest loops, which are closer to the framework and not interacting with each other, provide a basis on which to construct the longer loops. For each loop the CONGEN program generates a set of conformations by sampling ϕ, ψ space using a selectable angular grid. The loops are made to satisfy the end-point constraints of the framework and closure is achieved using a modified version of the Go and Scheraga algorithm (1970). The selected conformers must have acceptable potential energies, which depend critically on loop-loop interactions and therefore reflect the order of loop reconstruction. All conformations were assessed by their empirical potential energies. If the energies were close (within 8.4 kJ mol^{-1}), then the loop with the smallest solvent exposure was chosen. The longer loops (H2 and L1) were constructed in parts by searching over small sections and using low-energy conformers as starting points for the next search. The model of the McPC603 has 2.4 Å total and 1.7 Å backbone RMS shifts from the X-ray structure. This method appears less accurate than the knowledge-based approach and is computer intensive. Often the best agreements with X-ray structures were not always the conformers with the lowest energies.

4.12.1.3 Molecular dynamics approach Fine *et al.* (1986) predicted the conformations of four loops in the antibody combining site of McPC603 by generating a large number of random loop backbone structures, followed by either minimization or molecular dynamics. They prepared libraries of randomly generated main chain plus C_β loop conformations, which were screened to eliminate van der Waals overlaps within the loop or with the rest of the molecule. These loops were then energy minimized or subjected to molecular dynamics, while the rest of the molecule was held fixed. The lowest energy conformers were examined by generating possible side-chains. They performed two types of calculation: first, considering a loop *in situ* with the rest of the molecule present and, secondly, by modelling the loop after deleting all of the hypervariable loops in the structure. The programs DISCOVER (Biosym Corp., San Diego, USA) and CHARMm (Brooks *et al.* 1983) were used to perform the energy minimizations. Randomly generated ϕ, ψ values were iteratively adjusted to produce the fixed end-point geometry required by the framework with minimal conformational perturbation to the loop.

Their results showed that the two shortest loops (H1 and L2) could be modelled using only alanines, suggesting that these loops fold independently of sequence variation (cf. only one canonical structure was found by Chothia *et al.* 1989). They found that L3 showed significant conformational variability and no single structure could be chosen. The fourth loop, H3, required side-chain interactions to fix the conformation into that observed in the crystal structure (cf. Chothia *et al.* could not identify canonical structures for this loop). This study was significant because it emphasized the differences between loops, including their environment. As observed in crystal structures some loops are well defined, whereas others remain rather flexible and do not adopt a single conformer.

4.12.1.4 Combined knowledge-based and random loop generation approach   The consensus opinion is that the most accurately modelled structures will be made by combining molecular mechanics, dynamics, and distance geometry with knowledge-based techniques. Such an approach was applied to the hypervariable loops of antibodies by Martin *et al.* (1989). This method involves the following main steps.

1.   Short loops ($\leqslant 5$ residues) are modelled by a global ϕ, ψ grid search using CONGEN (Bruccoleri and Karplus 1987).

2.   Longer loops are found by searching for suitable conformations in all Protein Data Bank proteins. The N- and C-terminal residues are used to search the database of C_α separations created from all proteins in the Protein Data Bank. Two searches are applied to each loop, starting from the N- and C-terminal C_α atoms. The results of both searches are then merged to extract loops that satisfy both constraints. If the loop is of eight or more residues, the middle section of each loop is deleted and reconstructed using CONGEN, which generates multiple conformations.

3. Loops are melded onto the framework by overlapping their N- and C-termini with their respective take-off points as closely as possible. The loop is then rotated so that the plane defined by its two termini and the centre

of mass of its backbone are co-planar with that similarly defined by the parent loop.

4. The side-chains are appended using parent χ angles where available and conformational searches in the remaining regions.

5. The loops are refined using energy minimization and the 'best' candidates are selected, based on an energy function incorporating solvent accessibility.

This approach is fully automated and has been used to model, *in situ*, two complete antibody combining sites (HyHel-5 in the presence of the antigen lysozyme and Gloop 2). All 12 loops were modelled; six on the basis of known antibody loop structures, three using non-antibody loops, and three were generated by conformational search techniques. Only three loops were not constructed, at least in part, by CONGEN. A comparison of the predicted and observed structures shows RMS superpositions for backbone atoms ranging from 0.9 to 1.5 Å (average 1.0 Å) and all atoms, including side-chains, from 1.0 to 2.9 Å (average 2.0 Å). Unfortunately, no attempt was made to systematically assess the effect of using CONGEN for the reconstruction.

4.12.2 Modelling a very distantly related protein

When modelling distantly related proteins (<20 per cent sequence identity), major problems are encountered when aligning the sequences. An excellent example was the modelling of the HIV proteinase, a distant relative of the aspartyl proteinase family, by Pearl and Taylor (1987).

Many retroviruses express a proteinase that is vital for processing the polyprotein into separate core proteins. It had been observed previously that the sequence Asp-Thr-Gly, which forms the active site of aspartyl proteinases, was also conserved in the retroviral proteinases, suggesting a possible relationship. However, the retroviral proteinases are less than 130 amino acids long, compared with the two-domain aspartyl proteinases, which have more than 300 residues. These two domains show a distant relationship in sequence and twofold structural symmetry, which suggested possible gene duplication and fusion during evolution (Tang *et al.* 1978). Therefore it seemed possible that the viral proteinase sequence might correspond to a single domain of an aspartyl proteinase and function in a dimeric form. Pearl and Taylor (1987) examined the sequences and structures of both families and used pattern matching, structure prediction, and molecular modelling techniques to propose a three-dimensional structure for the dimer. Subsequent X-ray crystallographic studies have shown that the model has the correct topology. For the monomer the difference between the model and the real C_α coordinates is 2.61 Å for 87 common C_α atoms (Weber 1990).

In the first step of the modelling procedure a structural alignment of the

aspartyl proteinase family was made to identify SCRs in the known structures of penicillopepsin and endothiapepsin. Retroviral proteinase sequences
were then aligned in a separate group, using dynamic programming.
Afterwards consensus templates were separately generated for the aspartyl
and retroviral proteinases and refined until each template only detected its
own family members in a search of the entire sequence data bank. With both
templates constructed, an alignment of both families was made, based on
matching patterns of higher residue conservation in both templates. This was
enhanced by predictions of secondary structure, which were used to identify
possible SCRs in the retroviral proteinases. These methods allowed a rough
correspondence to be established in regions where no definite patterns were
found. From this starting point the joint template was refined as before.

The joint template resulting from this alignment included the active site
and hydrophobic residues in the β-sheet core. Only three residues were
totally conserved: the functional aspartate and glycine, together with
another glycine about 100 residues distant sequentially but nearby in three
dimensions. Having made the alignment, the core was built and the loops
appended using procedures similar to those described earlier. The two active-
site loops were essentially conserved but other loops were shorter, making
the retroviral proteinase a parsimonious version of the aspartyl proteinase
domains. The only topological change involved the deletion of one strand,
which changed the C-terminal hairpin into a single β-strand. Using the
active-site twofold axis from endothiapepsin (Pearl and Blundell 1984),
Pearl and Taylor went on to propose a structure for the dimer. The subsequent solution of the retroviral proteinase structure showed that the
prediction was remarkably good considering the lack of sequence homology.

The availability of much closer relatives to the HIV proteinase inevitably
leads to a more accurate model. This was clearly shown by Weber (1990) who
based his model on the recently solved Rous sarcoma virus (RSV) proteinase
(Miller *et al.* 1989; Wlodawer *et al.* 1989). However, the alignment then
becomes a trivial problem due to the similarity between the retroviral
proteinase sequences. Such a model should not be critically compared with
another derived from sequences with only three residue identities.

4.13 Future directions

This review has concentrated on the standard approaches to modelling by
homology. Future developments will probably involve greater use of
distance geometry to manipulate structural data, allowing rapid structural
comparisons and database searches (Jones and Thirup 1986). Recently, an
alternative to distance geometry has been developed (Rackovsky 1990) which
has the added advantage that chirality of the structure is retained. Also, it

has been suggested that neural networks could be applied to distance matrices in order to derive the C_α trace of an unknown protein. Although this approach has only been tested by modelling a highly homologous (74 per cent residue identity) structure, it emphasizes the flexibility of such networks.

Many newly solved structures reveal previously observed topologies despite having almost undetectable sequence similarities. For example, the structure of human interleukin-1β closely resembles *Erythrina* trypsin inhibitor (Onesti *et al.* 1991), even though only six identities exist at topologically equivalent positions out of a total 153 residues. Such low homology cannot currently be recognized by standard sequence alignment programs, although many groups are working towards this goal. The possibility of a significant fraction of the 25 000 sequences currently available in the data banks adopting already known topologies highlights the importance of developing reliable modelling packages. In addition, similar domain structures are even more frequent. If there are, in fact, only a limited number of topologies available to a protein, this would significantly assist the prediction of structure from sequence. In our opinion, improvements in structure prediction will only be achieved by consideration of tertiary structure topology and packing. Unfortunately, current methods of predicting topology have been somewhat disappointing. However, this area is currently receiving a lot of attention and it would not be unreasonable to expect significant progress in the 1990s. Once the human genome project comes on line such prediction and modelling strategies will become vital.

References

Barton, J.G. and Sternberg, M.J.E. (1987). Evaluation and improvements in the automatic alignment of protein sequences. *Journal of Molecular Biology* **198**, 333–7.

Bashford, D., Chothia, C., and Lesk, A.M. (1987). Determinants of protein fold—unique features of the globin amino acid sequences. *Journal of Molecular Biology* **196**, 199–216.

Bernstein, F.C., *et al.* (1977). The protein databank: A computer-base archival file for macromolecular structures. *Journal of Molecular Biology* **112**, 535–42.

Bleasby, A.J. and Wooton, J.C. (1990). Construction of a validated, non-redundant composite protein sequence database. *Protein Engineering* **3**, (3), 153–9.

Brooks, B.R., Bruccoleri, R.E., Olafson, B.D., States, D.J., Swaminathan, S., and Karplus, M. (1983). CHARMm: A program for macromolecular energy minimisation and dynamics calculations. *Journal of Computational Chemistry* **4**, 187–217.

Browne, W.J., Norton, A.C.T., Phillips, D.C., Brew, K., Vanamann, T.C., and Hill, R. L. (1969). A possible three dimensional structure of bovine α lactalbumin based on that of hen egg white lysozyme. *Journal of Molecular Biology* **42**, 65-6.

Bruccoleri, R.E. and Karplus, M. (1987). Prediction of the folding of short polypeptide units by uniform conformational sampling. *Biopolymers* **26**, 137-68.

Bruccoleri, R.E., Haber, E., and Novotny, J. (1988). Structure of antibody hypervariable loop reproduced by a conformational search algorithm. *Nature* **335**, 564-8.

Chothia, C. and Lesk, A.M. (1986). The relation between the divergence of sequence and structure in proteins. *EMBO Journal* **5**, (4), 823-6.

Chothia, C. and Lesk, A.M. (1987). Canonical structures for the hypervariable regions of immunoglobulins. *Journal of Molecular Biology* **196**, 901-17.

Chothia, C., *et al.* (1986). The predicted structure of immunoglobulin D1.3 and its comparison with the crystal structure. *Science* **233**, 755-8.

Chothia, C., *et al.* (1989). Conformations of immunoglobulin hypervariable regions. *Nature* **342**, 877-83.

Claessons, M., Van Cutsen, E., Lasters, I., and Wodak, S. (1989). Modelling the polypeptide backbone with spare parts from known structures. *Protein Engineering* **2**, (5), 335-45.

Dayhoff, M.O. (1978). Atlas of protein sequence and structure. *National Biomedical Research Foundation* **5**, (3), 345-58.

Edwards, M.S., Sternberg, M.J.E., and Thornton, J.M. (1987). Structural and sequence patterns in the loops of $\beta\alpha\beta$ units. *Protein Engineering* **1**, (3), 173-81.

Eisenberg, D. and McLachlan, A. (1986). Solvation energy in protein folding and binding. *Nature* **319**, 199-203.

Fine, R.M., Wang, H., Shenkin, P.S., Yarmush, D.L., and Levinthal, C. (1986). Predicting antibody hypervariable loop conformations II: minimisation and molecular dynamics studies of MCP603 from many randomly generated loop conformations. *Proteins* **1**, 342-62.

Go, N. and Scheraga, H.A. (1970). Ring closure and local conformational deformation of chain molecules. *Macromolecules* **3**, 178-87.

Greer, J. (1981). Comparative model-building of the mammalian serine proteases. *Journal of Molecular Biology* **153**, 1027-42.

Greer, J. (1990). Comparative modelling methods—application to the family of mammalian serine proteases. *Proteins* **7**, (4), 317-34.

Gribskov, M., McLachlan, A.D., and Eisenberg, D. (1987). Profile analysis: Detection of distantly related proteins. *Proceedings of the National Academy of Sciences USA* **84**, 4355-8.

Hendlich, M., *et al.* (1990). Identification of native folds amongst a large number of incorrect models. *Journal of Molecular Biology* **216**, 167–80.

Janin, J., Wodak, S., Levitt, M., and Maigret, B. (1978). Conformation of amino acid side chains in proteins. *Journal of Molecular Biology* **125**, 357–86.

Jones, T.A. and Thirup, S. (1986). Using known substructures in protein model building and crystallography. *EMBO Journal* **5**, (4), 819–22.

Karplus, M. and Petsko, G.A. (1990). Molecular dynamics simulations in biology. *Nature* **347**, 631–9.

Lesk, A.M. and Chothia, C. (1980). How different amino acid sequences determine similar protein structures: the structure and evolutionary dynamics of the globins. *Journal of Molecular Biology* **136**, 225–70.

Lesk, A.M. and Chothia, C. (1982). Evolution of proteins formed by β sheets. The core of immunoglobulin domains. *Journal of Molecular Biology* **160**, 325–42.

MacArthur, M.W. and Thornton, J. M. (1991). Influence of proline residues on protein conformation. *Journal of Molecular Biology*, in press.

McGregor, M., Islam, S., and Sternberg, M.J.E. (1987). Analysis of the relationship between side-chain conformation and secondary structure in globular proteins. *Journal of Molecular Biology* **198**, 295–310.

McLachlan, A.D. and Shotton, D.M. (1971). Structural similarities between α-lytic protease of myxobacter 945 and elastase. *Nature* **229**, 202–5.

Martin, A.C.R., Cheetham, J.C., and Rees, A.R. (1989). Modelling antibody hypervariable loops: a combined algorithm. *Proceedings of the National Academy of Sciences USA* **86**, 9268–72.

Miller, M., Jaskolski, M., Mohana Rao, J.K., Leis, J., and Wlodawer, A. (1989). Crystal structure of a retroviral protease proves relationship to aspartic protease family. *Nature* **337**, 576–9.

Moult, J. and James, M.N.G. (1986). An algorithm for determining the conformation of polypeptide segments in proteins by systematic search. *Proteins* **1**, 146–63.

Nishikawa, K. and Ooi, T. (1974). Comparison of homologous tertiary structures of proteins. *Journal of Molecular Biology* **43**, 351–74.

Novotny, J., Rashin, A.A., and Bruccoleri, R.E. (1988). Criteria that discriminate between native proteins and incorrectly folded models. *Proteins* **4**, 19–30.

Onesti, S., Brick, P., and Blow D.M. (1991). Crystal structure of a Kunitz-type trypsin inhibitor from *Erythrina caffra* seeds at 2.5 Å resolution. *Journal of Molecular Biology* **217**, 153–76.

Overington, J. P. (1990). *Knowledge-Based Protein Modelling*, Unpublished Ph.D. thesis. University of London.

Overington, J.P., Johnson, M.S., Sali, A.S., and Blundell, T.L. (1990). Tertiary structural constraints on protein evolutionary diversity:

templates, key residues and structure prediction. *Proceedings of the Royal Society of London* **241**, 132–45.

Padlan, E.A. and Davies, D.R. (1975). Variability of three dimensional structure in immunoglobulins. *Proceedings of the National Academy of Sciences USA* **72**, 819–23.

Pearl, L.H. and Blundell, T.L. (1984). The active site of aspartic proteinases. *FEBS Letters* **174**, (1), 96–101.

Pearl, L.H. and Taylor, W.R. (1987). A structural model for the retroviral proteases. *Nature* **329**, 351–4.

Phillips, D.C.P. (1970). *British biochemistry past and present*, (Goodwin edn), pp. 11–28. Academic Press, London.

Rackovsky, S. (1990). Quantitive analysis of known protein X-ray structures I. Methods and short length-scale-results. *Proteins* **7**, 378–402.

Read, R.J., Brayer, G.D., Jurasek, L., and James, M.N.G. (1984). A critical evaluation of comparative model building of *Streptomyces griseus* trypsin. *Biochemistry* **23**, 6570–5.

Rossmann, M.G. and Argos, P. (1976). Exploring structural homology of proteins. *Journal of Molecular Biology* **105**, 75–96.

Sali, A. and Blundell, T.L. (1990). Definition of topological equivalence in protein structures – a procedure involving comparison of properties and relationships through simulated annealing and dynamic programming. *Journal of Molecular Biology* **212**, (2), 403–28.

Sali, A., Overington, J.P., Johnson, M.S., and Blundell, T.L. (1990). From comparisons of protein sequences and structures to protein modelling and design. *Trends Biochemical Sciences* **15**, (6), 2235–40.

Schiffer, C.A., Caldwell, J.W., Kollman, P.A., and Stroud, R.M. (1990). Prediction of homologous structures based on conformational searches and energetics. *Proteins* **8**, (1), 30–43.

Sibanda, B.L., Blundell, T.L., and Thornton, J.M. (1989). Conformation of β-hairpins in protein structures. *Journal of Molecular Biology* **206**, 759–77.

Singh, J. and Thornton, J.M. (1990). SIRIUS – An automated method for the analysis of the preferred packing arrangements between protein groups. *Journal of Molecular Biology* **211**, 595–615.

Sippl, M.J. (1982). On the problem of comparing protein structures. *Journal of Molecular Biology* **156**, 359–88.

Summers, N.L. and Karplus, M. (1989). Construction of side chains in homology modelling. Application to the C terminal lobe of rhizopus-pepsin. *Journal of Molecular Biology* **210**, (4), 785–811.

Summers, N.L., Carlson, W.D., and Karplus, M. (1987). Analysis of side chain orientations in homologous proteins. *Journal of Molecular Biology* **196**, 175–98.

Sutcliffe, M.J., Haneef, I., Carney, D., and Blundell, T.L. (1987a).

Knowledge based modelling of homologous proteins, part 1: three dimensional frameworks derived from the simultaneous superposition of multiple structures. *Protein Engineering* **1**, 377–84.

Sutcliffe, M.J., Hayes, F.R.F., and Blundell, T.L. (1987b). Knowledge based modelling of homologous proteins, part 2: rules for the conformations of substituted side chains. *Protein Engineering* **1**, 385–92.

Tang, J., James, M.N.G., Hsu, I.N., Jenkins, J.A., and Blundell, T.L. (1978). Structural evidence for gene duplication in the solution of acid proteases. *Nature* **271**, 618–21.

Taylor, W.R. (1986). Identification of protein sequence homology by consensus sequence alignment. *Journal of Molecular Biology* **188**, 233–58.

Taylor, W.R. (1988). Pattern matching methods in protein sequence comparison and structure prediction. *Protein Engineering* **2**, 77–86.

Taylor, W.R. and Orengo, C.A. (1989). Protein structure alignment. *Journal of Molecular Biology* **208**, 1–22.

Thornton, J.M., Sibanda, B.L., Edwards, M.S., and Barlow, D.J. (1988). Analysis design and modification of loop regions in proteins. *BioEssays* **8**, (2), 63–9.

Weber, I.T. (1990). Evaluation of homology modelling of HIV protease. *Proteins* **7**, (2), 172–84.

Wlodawer, A., *et al.* (1989). Conserved folding in retroviral proteases: crystal structure of synthetic HIV-1 protease. *Science* **245**, 616–21.

Zvelebil, M.J.J.M. (1987). *Analysis and Prediction of Protein Structure and Function*, Unpublished Ph.D. thesis. University of London.

5

Protein motions: structural and functional aspects

Scott F. Sneddon and Charles L. Brooks III

5.1 Overview of protein motion

The ability of proteins to operate as catalysts clearly depends on subtle details of their structure. The overall shape and disposition of functional groups within an active site is what gives proteins some of their special properties: high substrate specificity, transition state stabilization, and the ability to transmit information to other units in a protein complex (allostery), to name but a few. As experimental and theoretical investigations have uncovered information about the structural component of protein function, so too have they shown that proteins exhibit a rich variety of motions (Gurd and Rothgeb 1979; Huber and Bennet 1983; Frauenfelder and Gratton 1986). If one considers just a few of the tasks proteins must perform in a living organism, it becomes apparent that motion, as well as structure, is an essential component of protein function. Receptor proteins lie partially buried in the cell membrane, and, on binding their target molecule, transmit this information into the cell by changing conformation, allowing the cell to respond to the external stimulus. Transport proteins (the oxygen transport protein haemoglobin, for example) bind ligands within their interior, releasing the ligand when conditions within the cell demand. The allosteric effects in haemoglobin involve motion of the subunits, while the entry and exit of the ligand requires an environmentally induced local conformational change (see Section 5.3.3). Many proteins are synthesized in a part of the cell distant from where they will perform their function. Transport of the protein from its point of manufacture to its point of use may require crossing membranes inside the cell. The folded protein is often too large to cross the membrane, and so must first unfold, traverse the membrane as a linear sequence, and refold on the other side (see Section 5.2.4). Catalytic proteins (enzymes) must often provide a very specific, non-aqueous environment for catalysis. Several known proteins completely engulf their substrate, perform the chemical reaction within the enclosed active site, then open to release the products. In considering just these few examples we see that, in addition to the importance of their structure,

Table 5.1 Time-scales and amplitudes of protein motions

Time-scale[a]	Amplitude[a]	Description
Short time-scale 10^{15}–10^{12} s^{-1}	Small amplitude 0.001–0.1 Å	Bond stretching; angle bending; constrained dihedral motion
Medium time-scale 10^{12}–10^{9} s^{-1}	Medium amplitude 0.1–10 Å	Unhindered surface side-chain motion; unhindered surface loop motion; collective motions
Long time-scale 10^{9}–10^{6} s^{-1}	Large amplitude 1-100 Å	Folding in small peptides; ring flips in the protein interior; helix–coil transitions
Very long time-scale 10^{6}–10 s^{-1}	Very large amplitude 10–100 Å	Protein folding

[a] The values given indicate orders of magnitude.

proteins exhibit a variety of motions as they operate within the cell.

In this chapter we discuss some of the theoretical and experimental evidence for protein motions on a variety of time and amplitude scales. Table 5.1 provides an outline of our catagorization of motions into different time and amplitude scales; this table gives the general outline of the rest of the chapter. We start by outlining the types of structural changes that occur as a result of motions of increasing amplitude (and time-scale). In each subsection specific examples are given, along with the experimental or theoretical techniques used in their investigation. Finally, we give examples of how these same kinds of motions manifest themselves in protein function (Section 5.3). Here we focus on how proteins move as a result of substrate binding, performing catalysis, and binding and releasing ligands.

5.2 Structural aspects of protein motions

The structural changes that take place as a result of protein motions can be related to structure through consideration of the so called 'degrees of freedom'. To illustrate the nature of motions arising from different degrees of freedom, consider the simple example of the butane molecule, in Fig. 5.1.

Focusing only on the carbon atoms, butane has four heavy atoms, three bonds between them ($r1, r2$, and $r3$), two bond angles ($\theta1$ and $\theta2$), and one torsional (or dihedral) angle ($\phi1$). At physiological temperature butane exhibits motion of all of these degrees of freedom. Each degree of freedom has its own intrinsic force constant, k, which determines the energy cost of deforming the molecule along this degree of freedom (Fig. 5.2). Generally, the bonds have the largest force constants, the angle force constants are

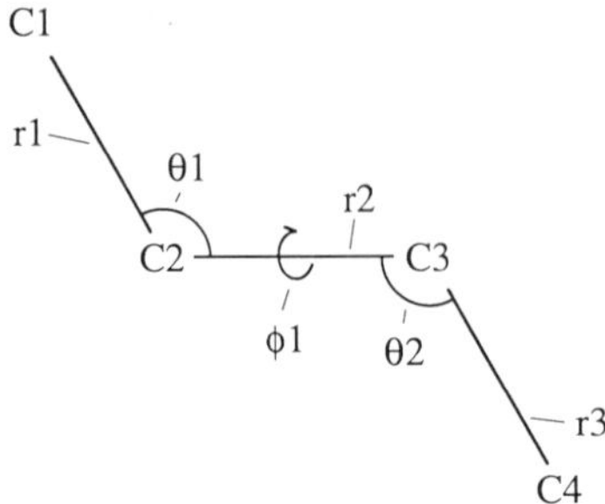

Fig. 5.1 Diagram of the degrees of freedom in butane. There are three bonds ($r1, r2, r3$), two bond angles ($\theta1$ and $\theta2$) and a single dihedral or torsion angle $\phi1$.

Bond Stretching: $E = 1/2k(r - r_0)^2$, $k \sim$ 300-500 kcal/angstrom^2

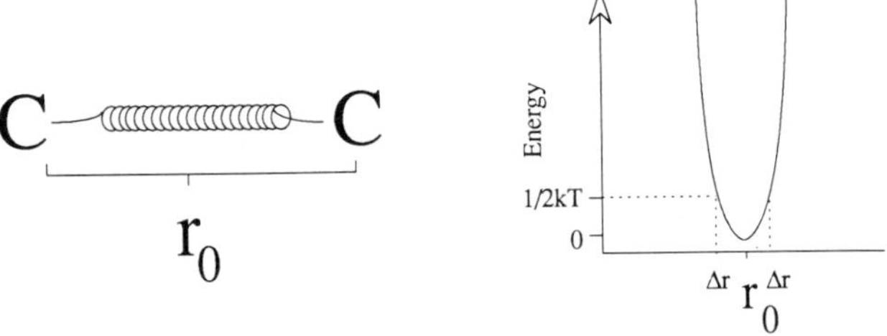

Angle Bending: $E = 1/2k(\theta - \theta_0)^2$, $k \sim$ 30-50 kcal/rad^2

Fig. 5.2 Representation of the energy for bond stretching and angle bending. The force constant k is much larger for bond stretching than for angle bending and therefore angles make larger excursions at a given temperature, T (the RMS fluctuations will be to energies of $+\frac{1}{2}kT$ and $-\frac{1}{2}kT$.

smaller, and the torsional force constants are smaller still; because of this difference in force constant the bonds and angles are referred to as 'hard' degrees of freedom, and the torsion angle is called a 'soft' degree of freedom. The ramification of these differences in force constant is that, at a given temperature, the 'hard' degrees of freedom (bonds and angles) will show the smallest fluctuations, and fluctuations in 'soft' degrees of freedom will be of much larger amplitude (Figs 5.2 and 5.3). Bond and angle motions, while smaller in amplitude than torsional motions, can play an important role in facilitating these larger scale motions. The bonds and angles can stretch slightly to greatly reduce the steric repulsion between atoms one and four as they swing past each other during a $\phi 1$ torsional transition (Fig. 5.3). In this way motions of smaller amplitude (and shorter time-scale) can combine to permit much larger amplitude motions (we will see examples of this below).

Dihedral Motion: $E = k[1+\cos(n\phi-\delta)]$, $k \sim .5\text{-}1.5$ kcal

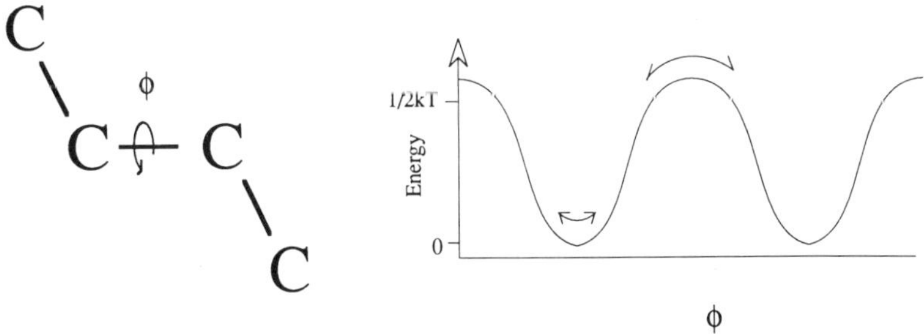

Atomic Repulsion and Dispersion Attraction: $E = 4\varepsilon[(\sigma/r)^{12} - (\sigma/r)^{6}]$,
$\varepsilon = 0.1 - 0.3$ kcal

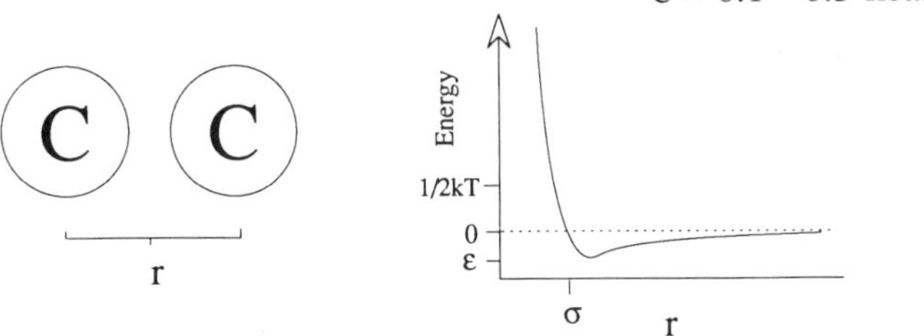

Fig. 5.3 (a) Representation of the energy cost of dihedral rotation and non-bonded atom-atom interaction. The barrier to dihedral rotation is generally on the order of kT (unless there is steric hindrance to rotation) and transitions between minima occur rapidly for the free molecule (see Section 5.2.2.1). (b) The shape of the interaction energy curve for two uncharged atoms, showing the small energy well at contact, and the rapid rise in energy as the atoms get closer than $r = \sigma$.

Protein structure is obviously far more complex than butane; however, there are analogous hard and soft degrees of freedom here as well. For our purposes we will simply discuss the key hard and soft degrees of freedom in proteins. The hard degrees of freedom are present in the covalent bonding along the polypeptide backbone, and within a given side-chain (along with the associated bond angles). The soft degrees of freedom include the two unhindered backbone dihedral angles ϕ and ψ, and side-chain torsional angles χ^n). The folded structure of proteins introduces new hard and soft degrees of freedom not encountered in the simple butane example. Proteins have a densely packed interior, and therefore atom-atom van der Waals repulsions (Fig. 5.3) can hinder motion of side-chain dihedral angles. Proteins also contain a large number of intramolecular hydrogen bonds (especially in secondary structures). These hydrogen bonds, while quite stable, have smaller force constants than covalent bonds, and therefore exhibit larger fluctuations (Fig. 5.4). Finally, many proteins are organized

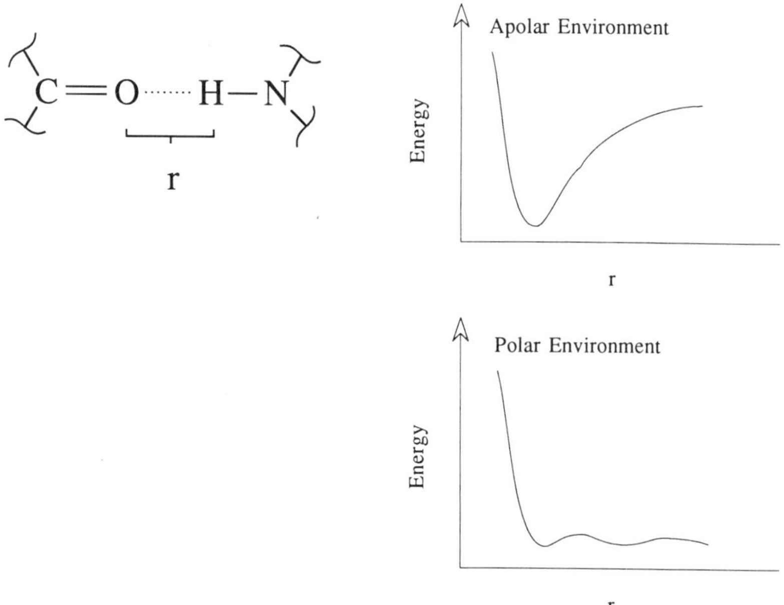

Fig. 5.4 Hydrogen bond energy in polar and apolar environments. The top figure shows the deep energy minimum when there are no other molecules in the vicinity to compete for the hydrogen bond. The bottom curve shows that when there are other polar molecules to compete for the hydrogen bond (solvent water in this case) that the stability of the hydrogen bond is greatly reduced (results from Sneddon *et al.* 1989).

into structural domains, and there can be very large-scale motions of these domains relative to one another.

In the sections that follow we will consider motions occurring over many orders of magnitude in both time-scale and amplitude. We will see that larger scale motions are facilitated by motions on smaller time-scales. The rate of motion is controlled by the effective force constant for fluctuations within a given conformation, and by the energy barriers between conformations for transitions between structures.

5.2.1 Short time-scale, small amplitude motions

The types of motions taking place on short time-scales, on the order of femtoseconds (10^{-15} s) to picoseconds (10^{-12} s), are governed by fluctuations in the so-called 'hard degrees of freedom': covalent bond stretching, angle bending, and constrained dihedral motions. Vibrational spectroscopy yields bond stretching energies on the order of a few thousand wavenumbers. This energy relates to an equivalent frequency of motion on the femtosecond time-scale. At physiological temperatures, fluctuations in bond lengths are not expected to be much greater than a few hundreths of an angstrom. Angle bending motions have smaller effective force constants, so these occur on slightly longer time-scales of femtoseconds to tens of femtoseconds. Angle bending motions will move atoms separated by two bonds by tenths of an angstrom. Combinations of bond stretching and angle bending motions play an important role in allowing larger amplitude motions. As the first and fourth atoms forming a dihedral angle move past each other during a transition, the bonds and angles can open slightly to reduce van der Waals repulsions (Fig. 5.3).

Another short time-scale, small amplitude motion is sterically hindered dihedral motion, which occurs on a time-scale of hundreds of femtoseconds. These motions occur on a short time-scale not because of an intrinsically large dihedral angle force constant, but because of the large *effective* force constant presented by close packing in the protein interior. The van der Waals repulsion between atoms increases rapidly (as approximately $1/(r_0\text{-}r)^{12}$, Fig. 5.4) and this interaction results in a large effective force constant for dihedral fluctuations when atoms are densely packed. Dihedral transitions do occur in the protein interior, but these are facilitated by the collective motion of a number of groups, and take place on a much longer time-scale (discussed in a later section) (see Section 5.2.3.2). An example of restricted side-chain motion in the protein interior is found in the fluorescence behaviour of ribonuclease T_1 (RNase T_1). The limiting anisotropy of tryptophan fluorescence in RNase T_1 is greater than 0.9, indicating that this side-chain is largely immobilized by the packing of surrounding residues (Lakowicz *et al.* 1983) Axelsen and co-workers have studied the microscopic nature of tryptophan motion in RNase T_1 by calculating the correlation times for motion of this tryptophan in a molecular dynamics simulation. They found that the computed anisotropy decayed to its limiting value (0.96 in their calculation) within 1 ps (Axelsen *et al.* 1988). In their simulations they found angular fluctuations of $\pm 17°$ occurring on the picosecond time-scale (Axelsen and Prendergast 1989). Chen and co-workers performed a similar calculation of the time-scales of motion for two tryptophan residues in apo-azurin (Chen *et al.* 1988). The tryptophan fluorescence in this protein indicates that the two tryptophans move at different rates, one on the picosecond time-scale, and the other on a much longer time-scale. While the simulation could not be run long enough to accurately probe the long time-scale motion, they found similar picosecond decay behaviour for tryptophan 48, which is in the interior (angular fluctuations were $\pm 10°$ degrees in their study).

The experimental techniques available to probe these motions in small molecules and in proteins include infra-red absorption and Raman spectroscopy, which probe an energy range between 200 cm^{-1} and 4000 cm^{-1}. Time-resolved fluorescence techniques are now reaching into the subpicosecond time domain, and have the potential of providing important structural information for the verification of the theoretical methods currently being applied to study picosecond motions (Levy and Szabo 1982; Ichiye and Karplus 1983; Ruggiero *et al.* 1990). NMR relaxation methods are also being employed in the study of picosecond motions in proteins, these will be discussed in more detail in the next section.

5.2.2 *Medium time-scale, medium amplitude motions*

Motions on the time-scale of 10^{12}–10^{9} s^{-1} have been difficult to characterize experimentally. X-ray structures provide some evidence for the magnitude expected of such motions, but time-scale information is lost. NMR probes also provide evidence for the amplitude of these motions, but again, the time-frame of the experiment is too long to give details of the time evolution of these motions. Other methods, low frequency Raman and time-of-flight neutron scattering do probe these time-scales, but have not provided detailed structural information in proteins because of the difficulty of assigning features in what is usually a very broad spectrum. Recent advances in time-resolved fluorescence and NMR relaxation methods have enabled more detailed studies of picosecond to nanosecond motions. Time-resolved fluorescence (and fluorescence anisotropy) measures decay from the singlet excited state of either tyrosine or tryptophan. The rate of decay depends both on the intrinsic decay rate, and on environmentally induced decay processes (Beecham and Brand 1985; Alcala *et al.* 1987). The analysis of fluorescence decays has provided information about structural changes taking place on time-scales ranging from picoseconds to nanoseconds (Lakowicz and Weber 1973; Lakowicz and Maliwal 1983; Chen *et al.* 1987). NMR relaxation methods are a powerful means of obtaining detailed information about protein motions occurring on time-scales from picoseconds to nanoseconds. The relaxation process is sensitive to the fluctuating magnetic fields produced by the motions of surrounding nuclei and the decay rate is related to the angular correlations of motion between atoms (Lipari and Szabo 1982a; 1982b; Levy 1985; Heatley 1986; Kitamaru 1986; London 1989). The relaxation of ^{13}C and ^{15}N-labelled protein atoms by their directly bonded protons has been interpreted in terms of motions occurring more rapidly than the overall tumbling of the protein (times less than 10 ns) (Levy *et al.* 1985; McCain *et al.* 1988; Weaver *et al.* 1988; Kay *et al.* 1989). Proton-proton relaxation is also being used to probe motions on this time-scale (Olejniczak *et al.* 1981; Torda and Norton 1989), though the interpretation of these experiments is made more difficult by the many proton-proton interactions that can contribute to relaxation.

Motion on this time-scale and amplitude range have been extensively characterized by theoretical methods. The fact that molecular dynamics calculations of several picosecond duration have reproduced both the magnitude of and variations in the RMS fluctuations seen in crystal structures supports the assertion that simulations are probing the major components of the motion (Levy *et al.* 1985; Brooks *et al.* 1988). Order parameters derived from NMR relaxation experiments have also been correlated with those from dynamics simulations, indicating that the simulations are reproducing picosecond motions (Lipari *et al.* 1982; Levy *et al.* 1985). Further indirect evidence is provided from normal mode calculations, which suggest that the majority of the RMS fluctuation in atomic position arises from low-frequency modes occurring on a time-scale of the order of 1–10 ps (Levy and Karplus 1979; Levy *et al.* 1982; McCammon and Harvey 1987). This work suggests that molecular dynamics simulations should be able to provide a link between the high-frequency regime studied by the methods discussed earlier, and the low-frequency motions, which can be studied directly by NMR methods. Picosecond resolved fluorescence spectroscopy and NMR relaxation spectroscopy will provide important verification of the predictions of these simulations. Molecular dynamics simulations are also an effective means of suggesting detailed motional models required to fit data from NMR relaxation and fluorescence decay experiments.

To give a qualitative feel for the kinds of motions proteins undergo on the picosecond to nanosecond time-scale, we will discuss results from four computer simulations on model systems. In the first system we characterize the motions of an unhindered surface side-chain in a nanosecond-long simulation of a blocked tyrosine monopeptide. Next, we will turn to unhindered loop motions in a simulation of a peptide fragment with constrained ends. In the final two systems we will explore the collective motions of secondary structures via normal mode calculations on a model α-helix and an antiparallel b-sheet. All simulations were performed with the molecular mechanics and dynamics program CHARMM and its associated potential function (Brooks *et al.* 1983).

5.2.2.1 Unhindered surface side-chain motion   Side-chains on the protein surface have greater freedom of rotation about the side-chain dihedral angles χ because they are not subject to the close packing interactions found in the protein interior. Surface side-chains can thus be expected to undergo dihedral transitions more often, as well as to show larger fluctuations within any given dihedral well. Surface side-chains are, however, subject to frictional forces from solvent. To demonstrate the qualitative features of unhindered surface side-chain motions (Section 5.3.1.1), we have performed a simulation of a single tyrosine residue in a gas-phase dielectric constant of 80 (to provide a representation of electrostatic screening by solvent), and using Langevin dynamics (to account for the frictional forces imposed by solvent). The dynamics time-step was 1.5 fs, and 1.5 ns of dynamics were run at a temperature of 300K. The characteristic motions upon which we will

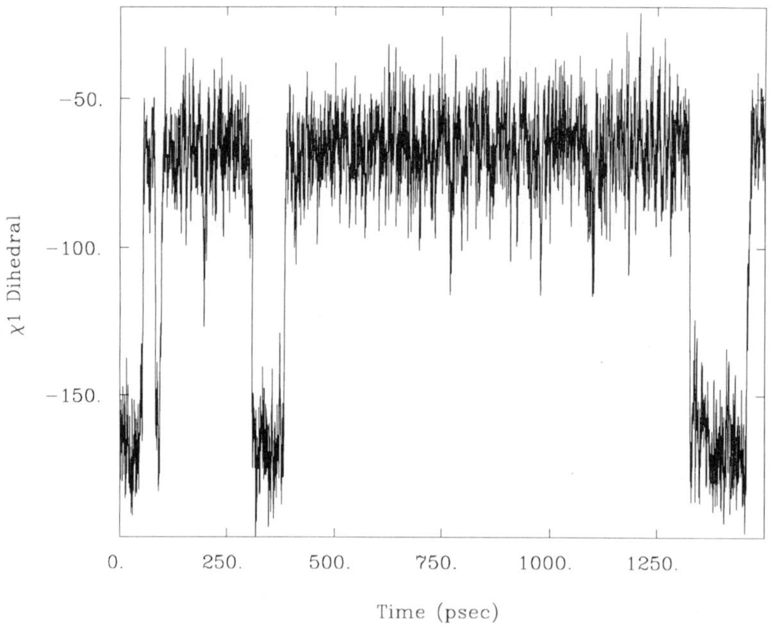

Fig. 5.5 Side-chain χ^{1} dihedral angle time-series from the simulation of isolated tyrosine.

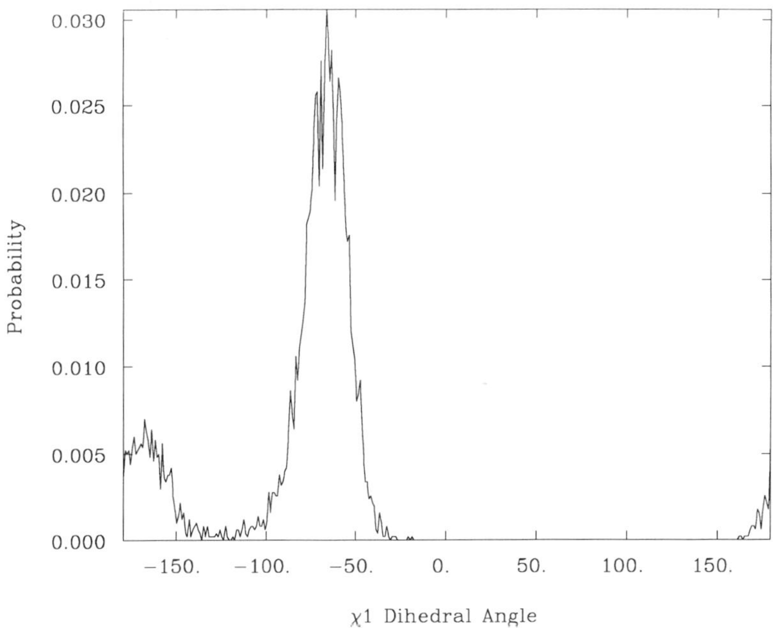

Fig. 5.6 Side-chain χ^1 dihedral angle normalized probability distribution.

focus are those within a particular dihedral minimum, and transitions between minima.

Figure 5.5 shows the time-series for the χ^1 angle (defined as the dihedral between atoms N-Cα-Cβ-Cγ), Fig. 5.6 shows the associated dihedral probabilities from the simulation. There are two primary conformations for χ^1, one at $\approx -160°$ (Fig. 5.6) and a more populated conformation at $-60°$. For the range of times between 500 ps and 1250 ps the average is $-66°$ and the RMS fluctuations are 13°. For times between ≈ 310 ps and 380 ps the average dihedral is $-167°$ with an RMS fluctuation of 13°. The less favourable $\chi^1 = 60°$ conformation was not sampled during the simulation. For rotation around the χ^2 dihedral (defined between atoms Cα-Cβ-Cγ-Cδ1) there are two distributions at $-90°$ and $+90°$ with several transitions between minima in the course of the simulation (≈ 7–8 per ns, Fig. 5.7). Within a single minimum (taken from times between 450 ps and 850 ps) the RMS fluctuation in χ^2 is 20° (Fig. 5.8). The remaining side-chain dihedral (involving atoms Cδ1, Cζ, O-η, and H-η) interconverts between minima at 0° and 180° on a time-scale of roughly 15 transitions ns^{-1} (Fig. 5.9). Within each minima (taken from times between 0 and 207 ps) the RMS fluctuation is 21° (Fig. 5.10). The simulation shows increased motion of atoms further from the main chain, a feature noted previously in small

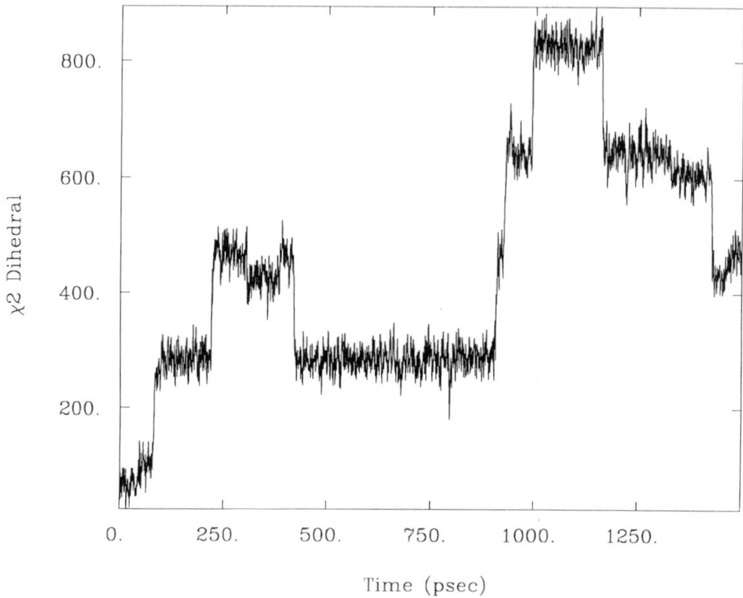

Fig. 5.7 Side-chain χ^2 dihedral angle time-series from Tyr side-chain simulation. The time-series is plotted as a cumulative value of the dihedral (not kept in the range -180 to 180) to show successive clockwise and counterclockwise transitions.

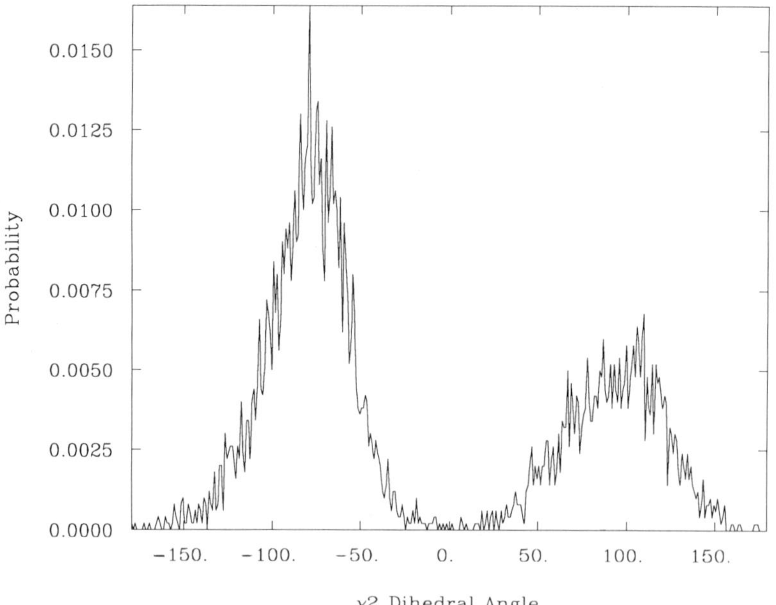

Fig. 5.8 The tyrosine χ^2 dihedral probability.

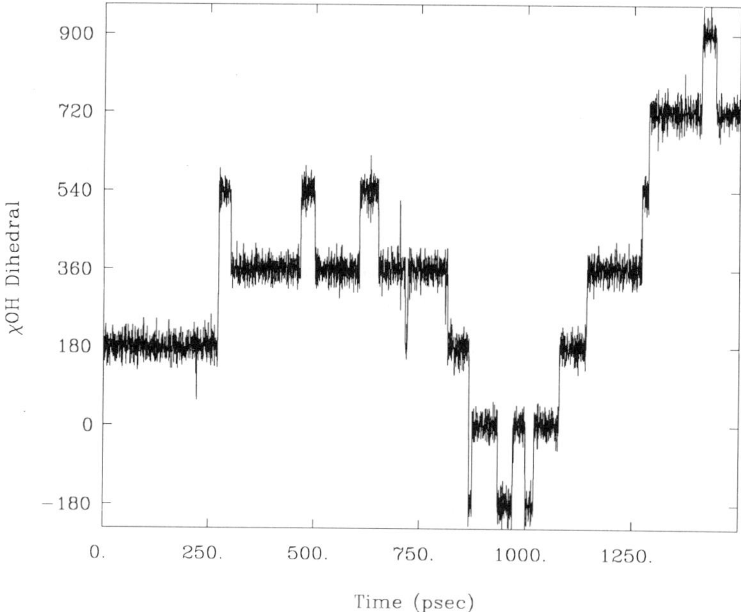

Fig. 5.9 Tyrosine hydroxyl dihedral angle time-series, plotted as a cumulative value of the dihedral (as in Fig. 5.7).

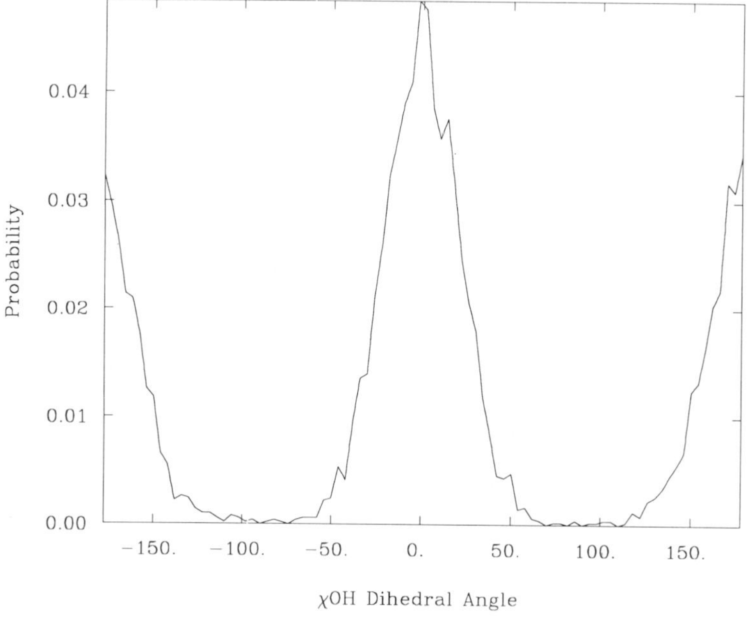

Fig. 5.10 The tyrosine χ^{OH} dihedral normalized probability.

molecule studies (Kitamaru 1986) and protein simulations (Post *et al.* 1989). The time-scale for transitions between conformations also decreases for these atoms, indicating that the potential barriers separating stable conformations are lower in energy. The features of the dihedral fluctuations presented here, from a gas-phase simulation with frictional forces, are in qualitative agreement with those from a much more extensive calculation carried out on a short peptide in water (Tobias *et al.* 1990). In this more rigorous calculation, which was propagated for 2.2 ns in a realistic solvent model, the dihedral minima were the same, as were the number and frequency of transitions between minima. Thus, we find that surface side-chain rotation involves localized oscillations within a minimum energy well, and larger transitions between wells. Furthermore, we see that the nature of these localized oscillations is similar between various minimum energy wells. Fluorescence depolarization measurements of tyrosine in short peptides have shown rotational correlation times for motion within the Tyr side-chain of ≈ 35 ps for tyrosine and blocked tyrosine, and from 80–100 ps for Tyr-Tyr and Leu-Tyr-Leu peptides (Lakowicz and Maliwal 1983). The time-scale for χ^1 rotation (≈ 25 ps) and χ^2 (≈ 15 ps) are in qualitative agreement with these experiments.

5.2.2.2 Unhindered surface loop motions   In many crystal structures some segment of the polypeptide is not clearly resolved, leading to the conclusion that this segment is highly mobile or disordered. Disorder in loop regions can also be inferred from NMR studies, either from narrow line widths or from the observations of NOEs consistent with multiple structures (see Chapter 3.) Some proteins also have surface loops that change conformation dramatically when substrate is bound (see Section 5.3.1.2). To gain a qualitative understanding of the types of structures sampled by motion of unhindered surface loops, as well as the rate of interconversion between conformations, we have performed a simulation similar to that described above for side-chain motion. The system consisted of a blocked alanine octapeptide simulated in a gas-phase dielectric of 80, using Langevin dynamics. The initial conformation of the peptide was generated by placing the first three residues in extended β-configuration, the fourth residue in α-configuration, and the remaining four residues in β-conformation, giving the L-shaped structure shown in Fig. 5.11. Dynamics were then propagated for 1.5 ns with a time-step of 1.5 fs and at a temperature of 300K. The amide nitrogen and carbonyl carbon between the blocking group and the first residue, and the corresponding atoms of the final residue and its blocking group, were held fixed to model the attachment of the loop to rigid secondary structure at the protein surface.

The loop structure quickly relaxed from its L-shaped starting structure and sampled a range of conformations during the 1.5 ns simulation. The

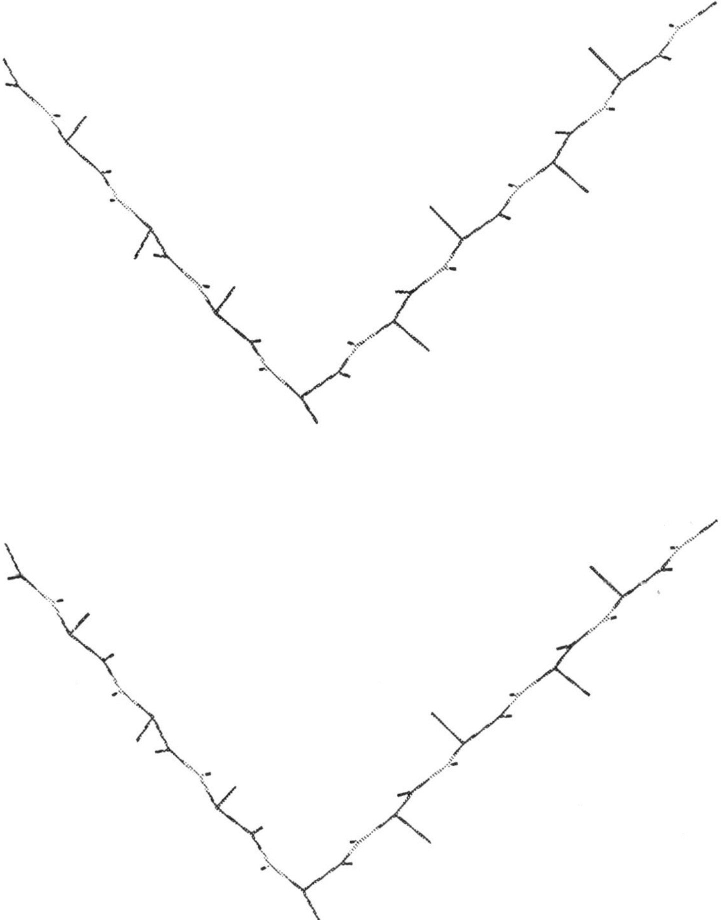

Fig. 5.11 Starting structure for unhindered loop simulation.

average structure from the trajectory (after a few cycles of minimization to relieve bad contacts) is shown in Fig. 5.12, and demonstrates that during the course of the simulation the loop spent roughly equal amounts of time on all sides of the line joining the loop ends. As would be expected, fluctuations about this average structure result in large excursions for groups near the loop centre, and smaller excursions closer to the fixed loop ends; the RMS fluctuations from the average structure, over the course of the entire trajectory, were 0.5 Å for residue 2, 1.9 Å for residue 3, 2.5 Å for residue 4, and 3.7 Å for residue 5 (these values were all taken from

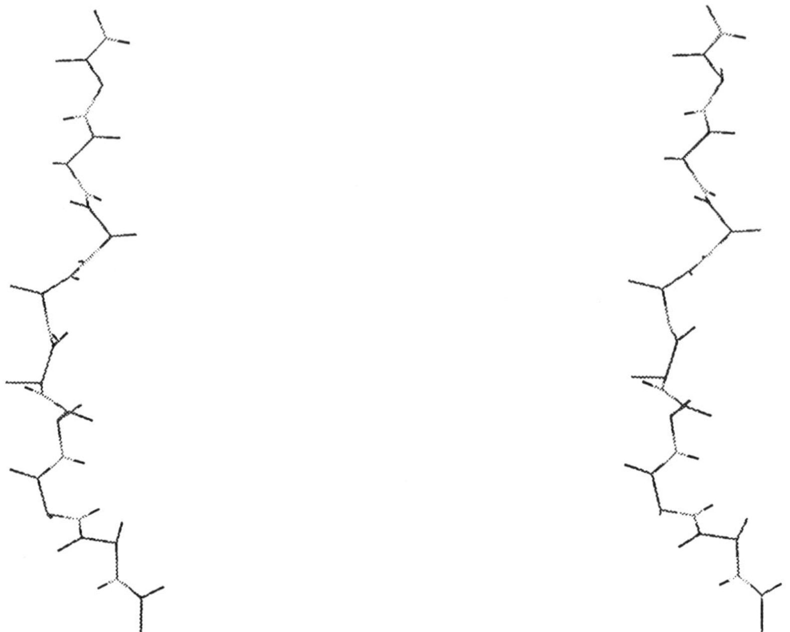

Fig. 5.12 Average structure (after minimization, see text) from unhindered loop trajectory.

$(\langle(\langle r_{(C\alpha)}\rangle - r_{i(C\alpha)})^2\rangle)^{1/2}$, where i denotes the time-step of the simulation). Figure 5.13 shows the time-series of $\langle r_{(C\alpha)}\rangle - r_{i(C\alpha)}$ for residue 5 (5-Cα). In the early stages of the simulation deviations from the average structure were quite large because the loop was still in the L-shaped starting structure. In the first 200 ps the loop oscillated about its initial position and the α-carbon of residue 5 remained at distances of between 4.5 and 7.0 Å from its average position. At approximately 250 ps the chain underwent a transition that brought 5-Cα close to its average position. During the remainder of the simulation the loop sampled configurations with residue 5 on all sides of the line joining the loop ends, with this residue approaching its average position at 250 ps, 650 ps, 825 ps, 875 ps, 1.0 ns, and 1.2 ns. Between these times 5-Cα sampled distances between 4 and 6 Å from its position in the average structure.

This trajectory showed five major changes in loop geometry occurring on time-scales of between 50 and 400 ps, with smaller fluctuations on the 5 to 20 ps time-scale occurring between major transitions. Our model included conformational restriction at the loop ends, dielectric screening of backbone-backbone interactions, and solvent frictional forces. Clearly, in a

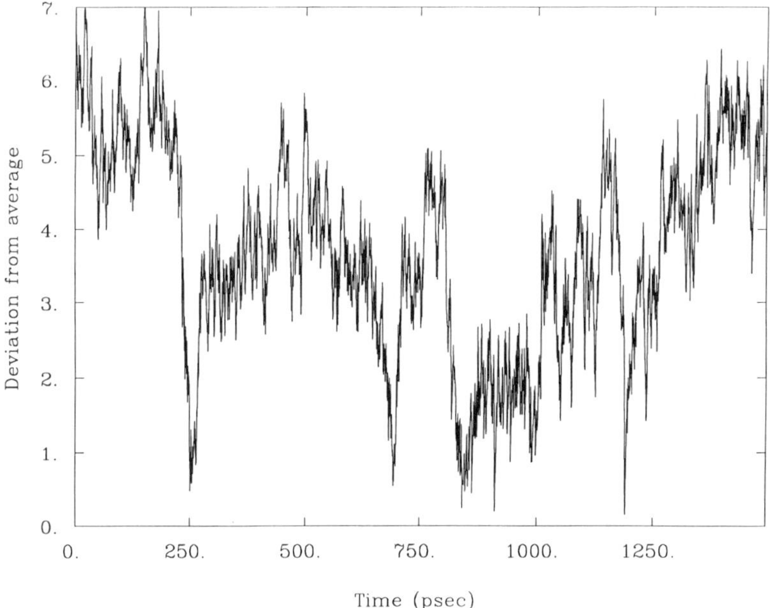

Fig. 5.13 Time-series of the deviation of Cα of loop residue 5, from its position in the average structure (Fig. 5.11).

real loop sequence there would also be an excluded volume effect from the protein, and sequence-specific loop-protein, loop-loop, and loop-solvent interactions not accounted for here. These effects would probably lower the apparent flexibility of the loop, and might make specific structures longer-lived. However, our calculations probably provide a reasonable order-of-magnitude estimate of the size of peptide backbone fluctuations, and of the rate of interconversion between structures.

5.2.2.3 Collective motions, normal modes of secondary structures   Because of the repeating pattern of hydrogen bonds found in α-helices and β-sheets, residues in these structures can undergo collective motions characteristic of breathing modes of the secondary structure of which they are a part. In this section we will discuss normal mode calculations that suggest these motions occur on time-scales on the order of tens of picoseconds.

To demonstrate the character of these normal mode motions we have constructed two model peptide systems in conformations corresponding to standard secondary structures. The first consisted of a blocked deca-alanine peptide in right-handed helical geometry (Creighton 1983), the second an antiparallel β-sheet containing three blocked strands of eight alanine

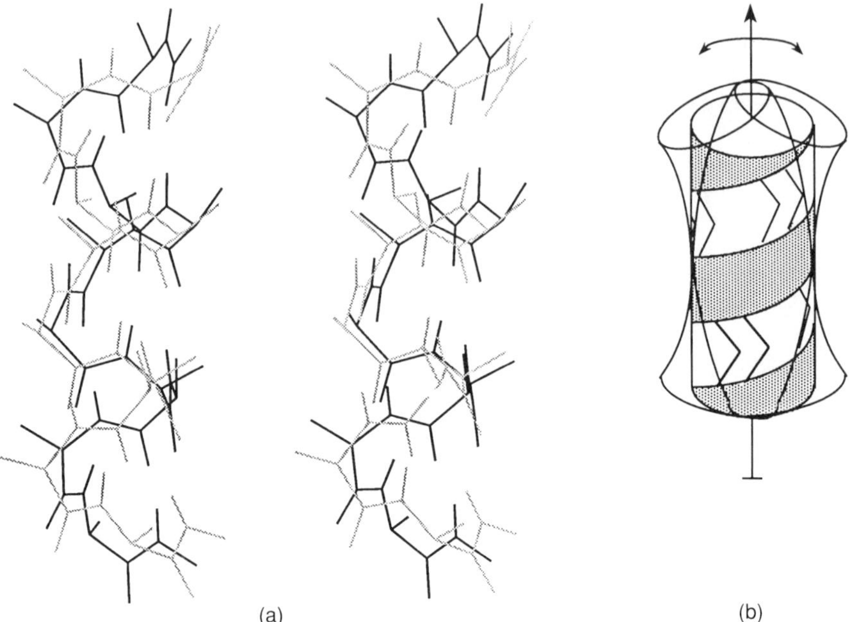

(a)   (b)

Fig. 5.14 Schematic of the α-helix bending mode, shown in stereo (a); shown schematically in (b), the top and bottom of the helix move right, and middle residues move left.

residues each (Chou and Scheraga 1982; Chou *et al.* 1982). Starting from these structural models, second derivative (Newton–Raphson) minimization was performed until the RMS gradient in the energy was less than 0.03 kcal mol^{-1}. The energy second derivative matrix was diagonalized, and the normal mode vectors and frequencies were obtained.

For the α-helical system the lowest modes were similar in character to those reported by Gō and co-workers (Gō and Gō 1976; Suezaki and Gō 1976) and by Levy and Karplus (1979) in their earlier studies of helix normal modes; Figs 5.14–5.16 show the character of a few of these low-frequency modes (the atom displacements have been exaggerated to make the character of the motions more obvious). Figure 5.14 shows a bending mode in which the helix oscillates as a flexible rod, the top and bottom of the helix move to the right in the figure while the helix centre moves to the left. Figure 5.15 shows a higher energy mode that is composed mostly of end-fraying motions. The last two backbone dihedral angles show the largest fluctuations in this mode. Among the lowest modes are combinations of bending, twisting, and end-fraying, their higher harmonics, and combination modes. Figure 5.16 shows the motion resulting from populating all of the 40 lowest frequency modes. The largest displacements along this combination of

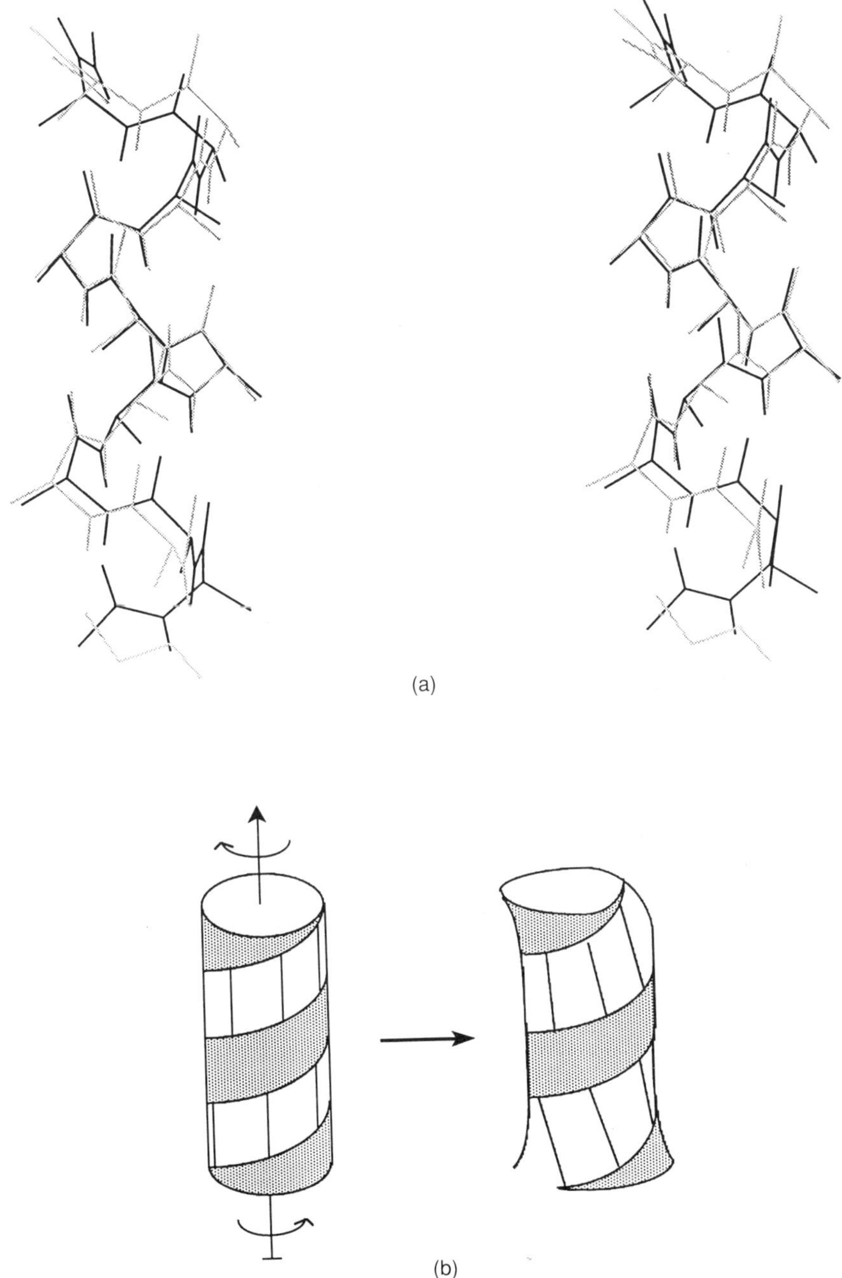

(a)

(b)

Fig. 5.15 Stereo plot of the α-helix end-fraying normal mode (a); shown schematically in (b).

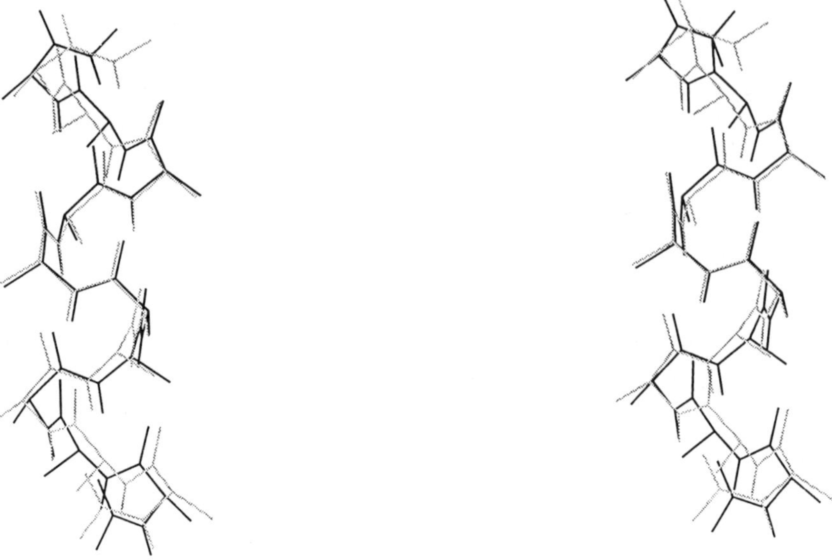

Fig. 5.16 Combination mode of the α-helix (shown in stereo), obtained by populating the first 40 modes equally.

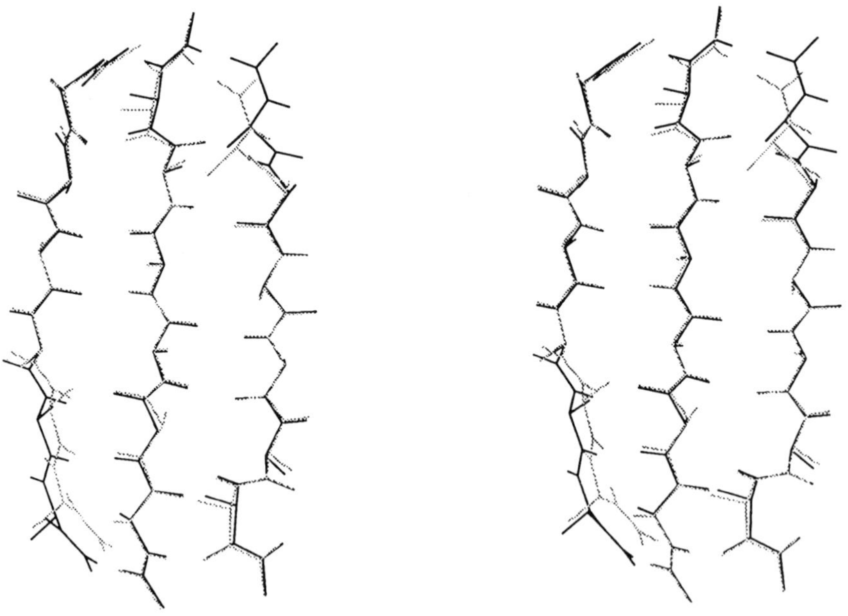

Fig. 5.17 Stereo view of the lowest frequency (6 cm^{-1}) normal mode of the β-sheet.

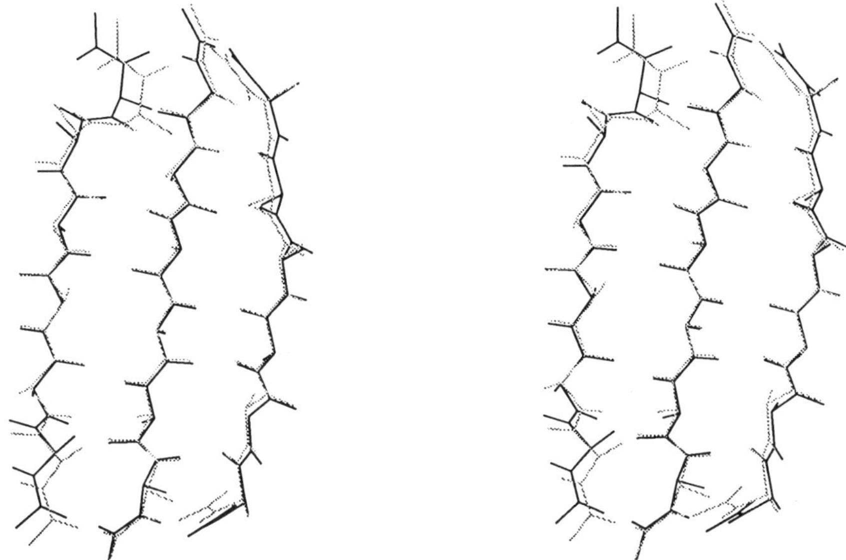

Fig. 5.18 Next higher β-sheet mode (7 cm^{-1}), the motion increases and decreases the sheet's right-handed helical twist.

modes arise from the first few low-frequency modes (as previously noticed by Brooks and Karplus 1983).

In Figs 5.17–5.19 we show stereo views of a few of the lowest frequency normal modes of a three-stranded antiparallel β-sheet. Figure 5.17 shows the lowest frequency mode. Motion along this mode is dominated by movement of the corners of the sheet towards and away from each other. In Fig. 5.18 the motions are collective along the end strands (strands 1 and 3) and this mode increases and decreases the right-handed helical twist of the sheet. Figure 5.19 shows a higher frequency mode in which the central strand expands and contracts, and the outer strands move with it (20 cm^{-1}). The frequencies of the three modes shown are 6, 7, and 20 cm^{-1}, respectively, compared to frequencies of 22, 25, and 35 cm^{-1} for the three lowest modes of the helix, this difference in the frequencies of the lowest normal modes of helices and sheets was also noticed in proteins in the work of Levitt *et al.* (1985). The frequency of a mode is related to the curvature of the potential energy surface for deformations along that mode, thus the lower frequency of the modes in the sheet calculation imply that these structures are more easily deformed (energetically) than is the helix. We attribute the difference to the fact that sheets can undergo backbone dihedral oscillations without losing as much hydrogen bond energy as in the helix, and also to the fact that the curvature of the ϕ-ψ potential energy surface is smaller in the β-region

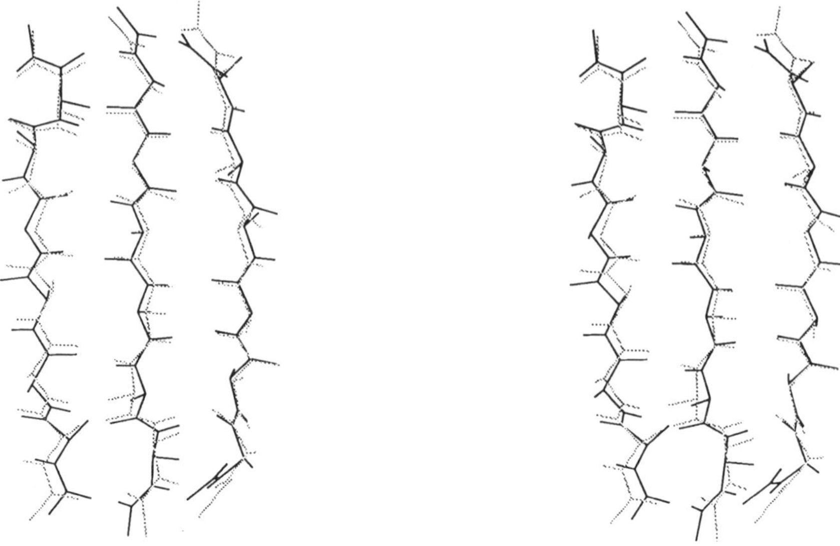

Fig. 5.19 Higher frequency normal mode (20 cm^{-1}) of the β-sheet, the central strand contracts and expands.

than in the α-region. Both of these factors make deformations of ϕ-ψ angles more costly (energetically) in α-helices than in β-sheets. It is interesting to note that these low-frequency deformations are similar to those seen in proteins. For example, in hexokinase the conformational change induced by glucose binding causes a change in β-sheet curvature similar to those shown above. The α-helices in folded proteins can also have deformations similar to the low-frequency modes calculated here. Citrate synthase provides an example; the bent non-proline helix between residues 390 and 416 resembles the bending mode discussed above.

It is important to note that the preceding calculations were all performed in gas phase and, therefore, important interactions with other protein groups and with solvent have been neglected. The α-helix and β-sheet calculations were performed with a dielectric constant of 1.0 because these structures are usually found in the protein interior where they would be in a mostly apolar environment. Previous free-energy simulations (Sneddon *et al.* 1989) have shown that the free-energy profile of hydrogen bond formation in a gas-phase dielectric constant of 1.0 and in the apolar solvent CCl_4, are very similar. Since the strength, and orientation, of hydrogen bonds are primarily what leads to the character of the normal modes, we feel that the qualitative nature of the computed modes is probably correct for secondary structures in an apolar environment. Also, sheets and helices are often amphipathic, i.e. with one face exposed to solvent and the opposite face forming hydrophobic interactions with the protein core. The competition by water for the hydrogen bonds on the polar face of these structures should greatly reduce the effective force constant for breaking the hydrogen bond (Sneddon *et al.* 1989), resulting in larger, longer time-scale fluctuations. The harmonic approximation neglects anharmonic contributions to the energy, and these have been shown to be significant in folded proteins (Levy *et al.* 1982; Noguti and Gō 1982; Brooks and Karplus 1983; Gō *et al.* 1983). These anharmonicities would change the effective force constants for motion, increasing or decreasing them depending on the environment in which the motion occurs.

None of the calculations performed in this section have included interactions with the surrounding protein matrix. Such interactions would undoubtedly lead to some damping of motions. However, these model systems were constructed to model unhindered motions, and as such are probably qualitatively correct. For motions that are damped by interactions with neighbouring groups, there would be a greater separation between the time-scale of oscillations within a given potential well, and transitions between wells. The study of such hindered motions requires the application of different theoretical techniques, which will be discussed in the next section on longer time-scale motions.

5.2.3 Long time-scale, large amplitude motions

Motions on the nanosecond to microsecond time-scale can be characterized by the requirement for structural rearrangements of many groups. In a simple transition state picture, in which the rate of such processes is given by $k \approx e^{-\beta \Delta H^{\ddagger}}$, and $\Delta H^{\ddagger}$ is the activation barrier, activation barriers of 5–10 kcal mol^{-1} would give rise to transitions between minima on a nanosecond to microsecond time-scale. Examples of such motions include: folding-unfolding transitions in short peptides, order-disorder transitions in chain termini, and ring flipping in the protein interior. We will discuss each of these types of motions briefly in the following section. The main experimental probes of these motions are again provided by NMR and intrinsic fluorescence measurements (see Section 5.2.2).

5.2.3.1 Folding of small peptides

In the previous section we considered unhindered motion of a surface peptide loop (see Sections 5.2.2.2 and 5.3.1.2). Major structural changes in that system occurred on a time-scale of between 10 and 100 ps. Transitions at this rate imply that the system was crossing barriers of roughly 1.5–3 kcal mol^{-1} in height. This system was chosen as a limiting case; the peptide has only alanine side-chains (which precludes sequence-specific interactions), and the ends of the peptide were kept from interacting by rigid constraints. Sequence-specific interactions, especially those involving charged side-chains, can be expected to introduce minima into the potential energy surface of the peptide, thus making the barriers between minima larger than in the unhindered loop system just examined. Similarly, if a peptide has charged end groups, interactions between end groups, as well as between side-chains and end groups, can be expected to slow transitions between structures. These arguments have implications for the study of short peptides which are now being shown to have substantial populations of folded conformation.

As an example, Dyson *et al.* (1988) have shown that short peptides can have significant populations of β-turn conformation as monitored by NMR. One of the peptides showing the largest proportion of turn structures in their study had the sequence Tyr-Pro-Gly-Asp-Val. To gain an understanding of the interactions stabilizing the turn conformations of this peptide a simulation study was initiated in our laboratory (Tobias *et al.* 1991). The Tyr-Pro-Gly-Asp-Val peptide sequence was simulated in a box of approximately 750 water molecules in two separate runs, one starting with the extended conformation of the backbone, the other starting with the backbone in an ideal type II β-turn. Each system was simulated at 300K for a simulation period of 2.2 ns, and the trajectories were analysed for the occurrence of turn structures, or chain reversals. The simulation that started in the turn conformation showed one transition to an extended structure. The simulation started in the extended conformation showed two transitions to structures with chain reversals, one early and one late in the simulation. During the course of both simulations there were long-lived interactions between side-chains, between side-chains and backbone, and between the charged end groups. While a small number of transitions is certainly insufficient for any statistically significant estimate of the folding-unfolding frequency, it seems clear that such transitions require a time period of at least several (perhaps tens) of nanoseconds. This time-scale estimate for peptide conformational change is supported by spectroscopic studies of trypsin and trypsinogen (Butz *et al.* 1982). In these studies a motion is found occurring on the 10 ns time-scale in trypsinogen that is absent in trypsin. The digestive enzyme trypsin is formed from its inactive apoenzyme trypsinogen through cleavage of an N-terminal hexapeptide. Crystal structures of trypsinogen lack well-defined density in several loop regions that are well resolved in the trypsin crystal; this has been ascribed to a disorder-order transition on cleaving the hexapeptide from the apoenzyme. The loop regions contain between 6 and 39 residues, the smaller of which are consistent with motions on a nanosecond time-scale.

5.2.3.2 Dihedral transitions of buried aromatic side-chains, ring flips　The protein interior, while closely packed, is not completely rigid because the interactions that hold proteins together are rather weak. Thus, thermal motions of side-chain and backbone atoms can be large enough to allow dihedral transitions at appreciable rates, even for large aromatic side-chains. NMR studies of the motion of aromatic side-chains in bovine pancreatic trypsin inhibitor (BPTI) have shown a wide range of time-scales for their rotational isomerization. Tyrosine 35 is seen in two distinct environments on the NMR time-scale, suggesting that interconversion is slow, i.e. there is a large free-energy barrier of approximately 17 kcal for rotation (Snyder *et al.* 1975; Wagner and Wüthrich 1975; Wüthrich and Wagner 1975). However, for the remaining tyrosine residues in BPTI the interconversion between rotamers is rapid, and a single proton environment is seen by symmetrically related protons on the ring.

To gain insight into the kinds of motions that are necessary to facilitate ring rotation, Gelin and Karplus (1975) performed χ^2 rotations of the tyrosine side-chains of BPTI in the field of the rigid protein, using a procedure whereby the protein was allowed to relax at each value of χ^2. Their results are summarized in Table 5.2. The energy barriers for ring rotation in the field of the rigid protein are far too large because of unfavourable van der Waals overlaps. To obtain the barrier in the field of the relaxed protein, Gelin and Karplus performed 150 steps of steepest descent minimization at each value of χ^2. The time-scale predictions based on these adiabatic energy barriers suggest that Tyr-10 would rotate on a time-scale similar to the unhindered side-chain discussed earlier (see Section 5.2.2.1), Tyr-23 would rotate on a time-scale in the nanosecond to microsecond range, Tyr-21 would rotate on the millisecond time-scale, and Tyr-35 would rotate once every few hours. The experiments suggest that Tyr-10, Tyr-21, and Tyr-23 do rotate more rapidly than 10^{-4} s, but the Tyr-35 rate is more rapid than predicted, at 0.6 s^{-1}. The experimental rate for Tyr-35 rotation implies a barrier closer to 17 kcal mol^{-1}, which has been obtained from an adiabatic energy barrier calculated by McCammon and co-workers (1983) using a different reaction coordinate. Potential of mean force (pmf) calculations have been performed in gas phase and in solvent, and these led to barriers of 10 kcal mol^{-1} (Northrup *et al.* 1982) and 13 kcal mol^{-1}, respectively (Ghosh and McCammon 1987). The reason the pmf barrier is lower than the adiabatic barrier is that the pmf is obtained

Table 5.2 Barriers and rates of ring flips in BPTI

Residue	Rigid protein barrier (kcal mol^{-1})	Adiabatic protein barrier (kcal mol^{-1})	Predicted rate (s^{-1})
Tyr-10	43	~0	1.9×10^{12}
Tyr-21	230	12	3.4×10^{3}
Tyr-23	75	7	3.3×10^{7}
Tyr-35	200	23	3.8×10^{-5}

from a dynamics calculation that allows for exploration of more ways of accommodating the strain introduced by rotating the ring over the barrier, and therefore leads to a path that is more favourable. The reason the pmf calculations underestimate the barrier suggested by the experimental rate is not clear, though it may be the result of appropximations in the model. However, it is clear that relaxations in the structure must take place for ring flips to occur at the rates seen experimentally. To understand how these motions are facilitated, Gelin and Karplus also analysed the groups exerting forces on tyrosine 35 as it crossed the barrier. In addition to interactions within Tyr-35, they found interactions with sequence adjacent residues Gly-36 and Gly-37, as well as with sequence-distant, but spatially nearby, residues Tyr-10, Arg-20, and Asn-44. McCammon and co-workers (1983) also found that local structural changes, or packing defects, introduced by thermal motion in the protein interior, are responsible for lowering the barrier to ring rotation. Thus, proteins make both local and global conformational changes that facilitate interior group motions such as ring flipping.

More recently, a potential of mean force calculation was performed for the rotational isomerization of Trp-47 in variant-3 scorpion neurotoxin (Haydock *et al.* 1990). The computed free-energy barrier to rotation (see Section 5.2.4.3) is 8.5 kcal mol^{-1} and the calculated rate of barrier crossing is 2×10^5 s^{-1}. In contrast to the cases just discussed, this ring isomerization, which occurs closer to the protein surface, requires motion of both the protein and the surrounding solvent during dihedral barrier crossing. Also, tryptophan is not symmetric for 180° rotation around χ^2, and the authors find that the isomer in the crystal structure is approximately 3 kcal mol^{-1} lower in free energy than after rotation of χ^2 by 180°.

Because of the larger volume that is swept out, dihedral transitions of aromatic groups in the protein interior require movement of several other surrounding side-chains. Transitions in smaller side-chains in the interior probably require smaller scale rearrangement and experience lower free-energy barriers (Levy 1985), as would groups closer to the surface (Olejniczak *et al.* 1981; Chen *et al.* 1988). For surface side-chains, solvent rearrangements may become important. However, structural rearrangements in water occur on a picosecond time-scale and, except for cases with strong water-side-chain interactions, water at the protein surface should not dramatically slow surface side-chain motions. The energies required for such rearrangements are small enough that they will occur rapidly at physiological temperatures.

5.2.3.3 Helix-coil transitions   The study of helix-coil kinetics provided some of the first insights into the rate of forming protein secondary structure. Helix formation is a co-operative process in which many successive residues are ordered into a specific hydrogen bonding arrangement, following formation of an initiation structure (most likely an amide-amide hydrogen bond between peptides four residues apart in the sequence). In the case of a peptide α-helix, all amide groups are hydrogen bonded only to other amides (except at the helix termini) and any strong side-chain-backbone interaction might disrupt the helical structure. In a peptide with polar or charged side-chains, a great deal of conformational sampling must occur before favourable side-chain-backbone interactions can be exchanged in favour of the backbone-backbone interactions required for the helix.

Kinetic studies of the helix-coil transition have been carried out on a variety of polypeptides, including those with strongly interacting side-chains such as Glu and Lys. In separate studies of polymers of glutamic acid and of lysine, a similar helix-coil interconversion rate of 107 s^{-1} was found (Barksdale and Stuehr 1972; Tsuji et al. 1976; Inoue et al. 1979). This corresponds to interconversion between folded and unfolded helices on the microsecond time-scale, and (along with the rate of turn formation discussed above) gives some idea of the rate at which individual fragments of structure are being formed during protein folding. This rate corresponds to crossing free-energy barriers of ≈ 10 kcal mol^{-1}, which probably involves formation and disruption of ion-hydrogen bonds between the charged side-chains and the backbone. Another study of polymers of γ-benzyl-L-glutamic acid in chloroform found a much slower rate for the helix-coil transition (Nagayama and Wada 1975). The slower rate may be due either to conformation restriction by the γ-benzyl group, or by the greater energy required to rearrange amide hydrogen bonds in apolar solvents like chloroform (Jorgensen 1989; Sneddon et al. 1989).

The motions described in this section, occurring on nanosecond to microsecond time-scales, have in common the feature that significant rearrangement of the surrounding medium occurs during the transition. For folding in short peptides (Section 5.2.3.1), significant structural change will require rearrangement of peptide-peptide, peptide-water, and water-water interactions. If these interactions are hydrogen bonds, either peptide-peptide, water-peptide, or water-water, the barrier to their rearrangement will be approximately 1–2 kcal mol^{-1} (Sneddon et al. 1989). The barrier to rearranging hydrophobic interactions is less clear, but since water-water hydrogen bonds must rearrange as the apolar cavity changes shape or size, an estimate of 1–2 kcal mol^{-1} barriers for this process seems reasonable. These barrier heights imply that peptide rearrangement occurs on the nanosecond time-scale. In the case of dihedral transitions in the interior, it is the protein 'solvent' that must rearrange. In this case, fluctuations in the core residues may introduce small vacancies into which the side-chain moves (McCammon et al. 1983). Proteins could also accommodate the strain of the transition structure by a slight lengthening of some hydrogen bonds, and a small expansion of the structure (Tilton *et al.* 1988). However, not all dihedral transitions are easily accommodated by fluctuation in the folded protein, as evidenced by the small rates of transitions for Tyr-35 in BPTI. In these cases the packing of the interior results in there being no path for ring rotation that is either low enough in energy (enthalpic) or that opens often enough (entropic) for the process to occur at appreciable rates. Because liquid water has a high mobility, and because the protein core is not completely rigid, rather large structural changes can occur at appreciable rates, the barriers to transitions being 3–5 kcal mol^{-1}, with time-scales in the nanosecond to microsecond range.

5.2.4 *Very long time-scale, very large amplitude motions; protein folding*

In this section we will discuss the largest amplitude motions proteins are capable of; those associated with protein folding. To reach its final folded conformation from the denatured state, a protein must sample an extremely large number of conformations, including significant motions in all the categories already discussed. A great many studies of protein folding have been carried out, and we will discuss only a few examples where good estimates of time-scales and structural information exist. Clearly, much remains to be understood about the stabilizing forces that govern what has been called 'the second translation of the genetic message' (Goldberg 1985), that is, the relationship between amino-acid sequence and a unique folded conformation. We will focus on three sets of investigations into the kinetics of folding, all of which attempt to characterize intermediates on the folding pathway. The first experiments involve the kinetics of proline isomerization (Section 5.2.4.1), and are aimed at determining the influence of this intrinsically slow event on folding. Next, we will discuss disulphide bond formation as a means of tracking the folding pathway (Section 5.2.4.2). Finally, we turn to some recent experiments employing hydrogen exchange, rapid-mixing, and two-dimensional NMR to gain detailed information about hydrogen bonds that form early in the folding pathway and persist in the folded conformation (Section 5.2.4.3).

5.2.4.1 *Proline isomerization* In early studies of the folding-unfolding transition it was noticed that even in small, single-domain systems, many proteins exhibited kinetics that deviate from a simple two-state, unfolded → native (U → N) process. These data imply folding events consisting of two or more kinetic phases. The reaction scheme consistent with most of the data involves two forms of the unfolded protein that differ in their rates of folding, i.e.

$$U_{slow} \leftrightarrow U_{fast} \rightarrow N.$$

If a protein is denatured and rapidly refolded, all molecules will still be in the U_{fast} form, and two-state rapid folding will be observed. If, however, the unfolded protein is allowed to reach an equilibrium of U_{fast} and U_{slow} forms, then multi-phase kinetics will be observed. Many studies have been directed at determining the nature of the U_{slow} structures and the $U_{slow} \leftrightarrow U_{fast}$ transition, and the proposal of Brandts *et al.* (1975) that the transition is controlled by proline isomerization has received much support in the literature (Lin and Brandts 1978; Schmid and Baldwin 1978; Cook *et al.* 1979; Stellwagen 1979; Kelley and Richards, 1987). The classic example of this effect is bovine ribonuclease A, whose native conformation has cis prolines at 93 and 114, and trans prolines at 47 and 117, and whose folding shows distinct fast vs. slow species.

Proline is a unique amino acid in that there is a much smaller free-energy difference between *cis* and *trans* isomers of the peptide bond of X-Pro than for X in combination with any other amino acid. This results in approximately 40 per cent *cis* isomers in Pro-containing peptides (Thomas and Williams 1972; Evans and Rabenstein 1974), versus less than 1 per cent in other amino-acid sequences. While the population of *cis* isomers is large for proline, the rate of conversion is quite slow, requiring between tens and hundreds of seconds for their interconversion (Cheng and Bovey 1977). Since the *cis-trans* isomerization causes a significant change in the direction of the peptide backbone, many proteins are unable to fully fold until the proper proline isomers have formed in the unfolded state. This is not true for all proline residues, however, as some proteins fold at rates not limited by proline isomerization (Schmid and Baldwin 1978; Levitt 1980).

The slow phase in folding introduced by the requirement of the correct proline isomer, is one of the slowest time-scale events seen in protein folding, sometimes requiring several minutes. Like dihedral transitions of buried aromatic side-chains (Section 5.2.3.2), proline isomerization is a rare event owing to the large free-energy barrier separating the *cis* and *trans* conformations. However, unlike ring flipping, the *cis-trans* barrier does not result solely from a large *effective* potential caused by packing interactions, but contains a substantial contribution from a large intrinsic barrier in proline. In addition to the large intrinsic barrier, the backbone conformational change brought about by isomerization may result in additional unfavourable interactions and an even larger effective barrier. For example, the rate of *cis-trans* isomerization in short peptides has been measured at ≈ 1 s^{-1} (Cheng and Bovey 1977), while in proteins the process can take from seconds to minutes (Stellwagen 1979), depending on the number of prolines in the sequence. This suggests that the larger scale structural rearrangements required for proline isomerization in a partially folded protein can result in a significant increase in activation free energy.

5.2.4.2 Disulphide bond formation   Another means by which the formation and interconversion of folding intermediates can be monitored involves
selective oxidation of cysteines in unfolded reduced proteins to form
disulphides. Creighton has used this technique to elucidate intermediates
along the folding pathway of RNase T_1 (1–3RNT) and BPTI (4–6PTI)
(Pace and Creighton 1986; Creighton 1988). The method involves first
reducing all disulphide bonds, then adjusting the redox potential so that
intermediates with one or two disulphides can be isolated and characterized.
Several factors influence the rate of formation and interconversion of
disulphide bonds, among which are the solvent exposure of the cysteine
groups, the distance between cysteines in the primary sequence, and their
spatial separation in the partially folded intermediate.

In studying BPTI, Creighton and Goldenberg (Creighton and Goldenberg
1984; Goldenberg and Creighton 1984) have characterized a variety of one-
and two-disulphide intermediates; native BPTI has three disulphide bonds.
Some of these disulphide bonds corresponded to the pairings in the native
structure, and some did not. In the initial stages of folding many combinations of single disulphides were found in rapid equilibrium, but the native
30–51 bond predominates. Second disulphides were formed primarily by
adding another native bond to the species with the native 30–51 disulphide.
However, the dominant folding pathway did not involve subsequent
formation of a native-like third disulphide bond, but rather involved
'incorrect' 5–14 and 5–38 disulphide species. Those cysteines are far apart in
the folded structure of BPTI, implying that these intermediates are highly
non-native. These species then undergo disulphide rearrangement in the
rate-limiting step to form the native disulphide pairings, and the folded
structure[1]. The free-energy barrier for formation of disulphides is between
11 and 16 kcal mol^{-1}, while the barrier is roughly 22 kcal mol^{-1} for
disulphide rearrangement. Thus, structural rearrangements which bring
individual cysteines together occur on a time-scale of $\approx 10^4 - 10^1$ s and
single disulphide intermediates are in rapid equilibrium. The more global
changes required for disulphide rearrangement occur on a longer time-scale
($\approx 2 \times 10^{-4}$ s^{-1}).

5.2.4.3 Amide proton exchange   Another method of characterizing folding
intermediates involves application of amide proton exchange, rapid mixing,
and two-dimensional NMR to determine which protons become protected
from exposure to solvent (presumably because they have formed intramolecular hydrogen bonds) at various stages of folding (Schmid and
Baldwin 1979; Kim and Baldwin 1980; Dobson and Evans 1988; Roder *et al.*

---

[1] While this manuscript was in press the significance of the non-native disulfide species for
folding to the native state has become a subject of debate (Weissman and Kim 1991, Darby *et
al.* 1992). Nevertheless, this work is discussed to emphasize the time-scale of large amplitude
motions during protein folding.

1988; Udgoankar and Baldwin 1988). This technique provides detailed information about which pieces of secondary structure are formed during folding, and probes time-scales as short as 10 ms. The time-scale is limited by the intrinsic rate of proton exchange in a completely exposed amide, which occurs at a rate of ≈ 1 ms^{-1} (Molday *et al.* 1972).

Recently Roder, Elove, and Englander (1988) have applied these techniques to study the folding of cytochrome c (3–5CYT). They found that amide protons on the N-and C-terminal helices become more than 60 per cent protected within the first 20 ms of folding. Two other classes of protons become protected on time-scales of 100 ms and 10 s. The terminal helices contact each other, and the hydrophobic residues of their contact region are highly conserved. This result suggests that certain essential secondary structures are formed early in folding, potentially stabilized by a local hydrophobic environment. It is important to note that these experiments determine which protons are protected from exchange, and does not provide information on the exact environment of these protons during folding. Thus, it is impossible to say whether the hydrogen bonds that protect an amide proton during folding are the same as those seen in the final folded structure. It is possible, as in the case with 'incorrect' disulphides noted above (Section 5.2.4.2), that 'incorrect' hydrogen bonds are formed during folding that later rearrange to give the secondary structure seen in the folded structure. It has been suggested that hydrogen bond rearrangement may be rate-limiting in some protein-folding processes (Sneddon *et al.* 1989).

5.3 Functional aspects of protein motions

In previous sections we discussed protein motions from a structural perspective, focusing on which kinds of structural changes take place, and on their time-scale and amplitude. We saw that protein motions can involve structural changes of: individual side-chains (see Sections 5.2.2.1 and 5.2.3.2), segments of the polypeptide chain (see Sections 5.2.2.2, 5.2.3.1 and 5.2.3.3), or entire domains (see Sections 5.2.2.3 and 5.2.4).

The amplitudes of motion range from tenths of angstroms to tens of angstroms, and the time-scale varies by many orders of magnitude (from femtoseconds to minutes). In this section we will present examples of motions that occur during protein function. Here also we will see examples of motions over many orders of magnitude in amplitude, involving structural changes similar to those discussed in the previous section. We will concentrate on three aspects of protein motion as it relates to function: motions that occur when proteins bind their substrate (Section 5.3.1), motions occurring as the substrate changes into product during catalysis (Section 5.3.2), and motions that allow ligand entry and exit in transport proteins (Section 5.3.3).

5.3.1 Ligand-induced conformational changes

The most detailed evidence about structural changes associated with ligand binding has come from X-ray structures of proteins crystallized both with and without inhibitors bound. A variety of ligand-induced conformational changes have been observed in this way, ranging from single side-chain torsional transitions (Section 5.2.2.1), through loop closing (Section 5.2.2.2), to the rigid-body motion of entire domains or subunits. Such experimental evidence does not provide direct evidence of structural change because the local environment in the crystals being compared is often different, and these differences may be the true cause of the conformational change seen. However, the structural changes seen are quite reasonable and, in those cases where the groups changing conformation are not involved in crystal contacts, they are probably representative of the types of structures available to the protein in solution.

In this section we will discuss examples of ligand-induced conformational changes involving surface side-chain motion, surface loop motion, and domain closure.

5.3.1.1 Surface side-chain motion, carboxypeptidase

Large conformational change of surface side-chains near an active site as a result of ligand binding have been detected in crystal structures of enzymes with and without substrate bound. A well-studied example of such motions is found in the structures of carboxypeptidase crystallized with (3CPA) and without (5CPA) the slowly-hydrolysed substrate Gly-Tyr (Rees and Lipscomb 1983). Carboxypeptidase is a zinc metalloenzyme that acts as an exopeptidase, removing the C-terminal amino acids of polypeptides. There is a hydrophobic pocket in the active site that makes the enzyme specific for peptide sequences with large apolar groups, such as phenylalanine (Fersht 1977). The structures with and without substrate differ mostly in the orientation of surface side-chain Tyr-248, which is extended into the solvent in the absence of substrate, and is rotated by $-140°$ about the χ^1 side-chain dihedral angle to point into the active site when substrate is bound. This change in dihedral angle results in an 8 Å change in position of the phenolic oxygen, which appears to make a hydrogen bond to the carboxy-terminus of the dipeptide substrate (Fig. 5.20). Smaller conformational changes are seen for residues Arg-145, Ile-247, and Glu-270 (Rees and Lipscomb 1981). In the structure of carboxypeptidase A at 1.54 Å resolution, Rees *et al.* (1983) note that Tyr-248 is in van der Waals contact with an adjacent molecule in the crystal. However, the phenolic hydroxyl group is hydrogen bonded to a water molecule, and the contacts found would not prevent the conformational change observed when substrate binds. The crystallographic studies cannot provide time-scale information for the motion of Tyr-248 between the two observed conformations. However, the motion appears to be unhindered,

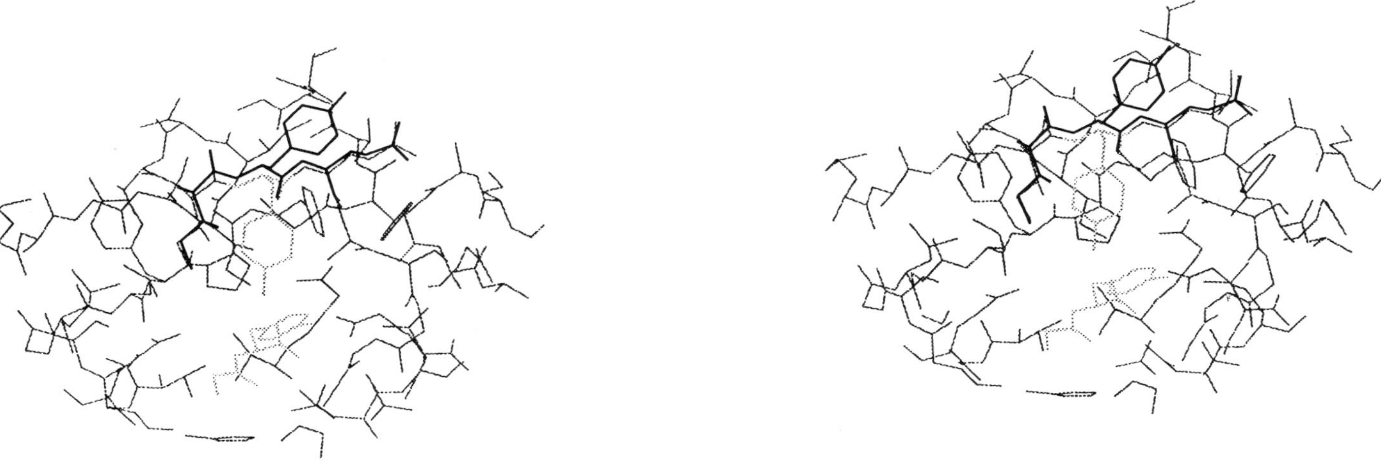

Fig. 5.20 View of residues surrounding the active site of carboxypeptidase. The dark line structure shows the conformation of Tyr-248 in the crystal without substrate. The lightest lines show the bound Gly-Tyr substrate, and the conformation of Tyr-248 in this crystal structure.

and the calculations performed in our previous section on unhindered side-chain motion suggest a time-scale of less than 0.5 ns for the transition.

The functional significance of this motion is not understood. Originally it was thought that the phenolic hydroxyl of Tyr-248 might function as a general acid catalyst in the hydrolysis of the peptide bond. However, mutation studies, which replace Tyr-248 by phenylalanine, have shown that the hydroxyl group is not necessary for catalysis. The mutated enzyme has a k_{cat} similar to that of the wild type, but a much higher K_m (Gardell *et al.* 1985). This suggests that the hydroxyl group is instead used in binding substrate. It is conceivable that the binding in the wild type is better than in the mutant because the entropy cost of donating a hydrogen bond from the tyrosine is lower than when the hydrogen bond is donated by the solvent. A crystal structure of the Phe mutant could help determine whether Phe-248 undergoes a similar dihedral transition as seen in the wild type, and how the missing hydrogen bond between the tyrosyl hydroxyl and the substrate is satisfied in the mutant enzyme.

5.3.1.2 Surface loop motion, lactate dehydrogenase The motion and adjustment of side-chains on ligand binding is quite common. Proteins can also undergo larger conformational changes involving the motion of several adjacent amino acids in a surface loop (see Section 5.2.2.2). These motions can bring several side-chains through large distances to make specific interactions with the substrate. Surface loop motion can also serve to enclose the active site to provide a non-aqueous environment for catalysis. It is difficult to determine whether substrate binding causes loop closure, or whether the substrate first forms specific interactions with loop residues, which is then followed by a conformational change in the loop-substrate complex into their final position for catalysis. In either case, such a loop-closing mechanism for catalysis requires a pre-binding phase during which the effective potential of loop residues is changed so that the loop moves from an equilibrium in the 'open' position to an equilibrium in the 'closed' position.

An illustration of loop closure upon substrate binding is provided by the lactate dehydrogenase enzyme. Lactate dehydrogenase (LDHase) catalyses the interconversion of lactate and pyruvate in the glycolytic pathway (Fersht 1977).

$$
\begin{array}{ccc}
\text{CH}_3 & & \text{CH}_3 \\
| & & | \\
\text{H}-\text{C}-\text{OH} + \text{NAD}^+ & \rightleftharpoons & \text{C}=\text{O} + \text{NADH} + \text{H}^+ \\
| & & | \\
\text{CO}_2^- & & \text{CO}_2^-
\end{array}
$$

The crystal structures of both the apoenzyme (White *et al.* 1976; Abad-Zapatero *et al.* 1987), and the complexes LDHase-NAD (Adams *et al.* 1973; Chandrasekhar *et al.* 1973), LDHase-NADH-oxamate, and LDHase-NAD-pyruvate (White *et al.* 1976) have been determined and discussed in the literature. Here we will discuss conformational differences between the apoenzyme (6LDH: crystallographic resolution 2.0 Å) and the LDHase-NADH-oxamate complex (1LDM: resolution 2.1 Å); the lower resolution LDHase-NAD-pyruvate complex (3LDH: 3.0 Å resolution) shows similar conformational changes as seen in the complex with oxamate.

Figure 5.21 shows the peptide backbone of the apoenzyme with the NADH, oxamate, and loop conformation from the complex structure superimposed onto it. The overall backbone conformation of the rest of the protein is very similar for the two structures; the only other area with significant differences is the C-terminal end where the conformational changes are less than half as large as in the loop region. The loop spans residues 95–120, and the largest main chain displacement (10 Å) is at glutamic acid 105. Other structural changes around the active site include the repositioning of the essential histidine (His-195). The closed conformation of the loop results in several specific interactions between loop residues and either the NADH coenzyme, or the oxamate moiety; these interactions include close contacts between the following groups: Arg-106 and oxamate (2.6 Å), Gln-100 and oxamate (3.1 Å), Arg-99 and NADH (3.0 Å). Loop closure also forms a hydrophobic pocket underneath the NADH coenzyme. It is interesting to note that two oxamate molecules bind to the protein in this crystal, the other binding site occurring on the opposite face of the enzyme. However, there are no significant structural changes at this site, demonstrating that large conformational changes are not *always* the result of interactions between small molecules and proteins; in fact, the crystal structure of the LDHase-NAD complex (Adams *et al.* 1973; Chandrasekhar *et al.* 1973) also shows the loop motion described above, indicating that NAD binding causes loop closure. The calculation discussed above (Section 5.2.2.2) suggests that large changes in the structure of an unhindered surface loop can occur on the sub-nanosecond time-scale. Since NADH has such an important role in inducing the conformational change, we can assume that is also influences the dynamics of the loop; whether it acts to increase or decrease the rate of transition is hard to determine.

5.3.1.3 Hinge bending, hexokinase Ligand binding can result in conformational changes of a more global nature than the side-chain and loop motions discussed above. In some instances substrate binding causes the rigid-body motion of entire protein domains. This type of structural change is common because active sites are often formed by the groove between two domains. Thus, substrate binding will alter interactions in the interdomain

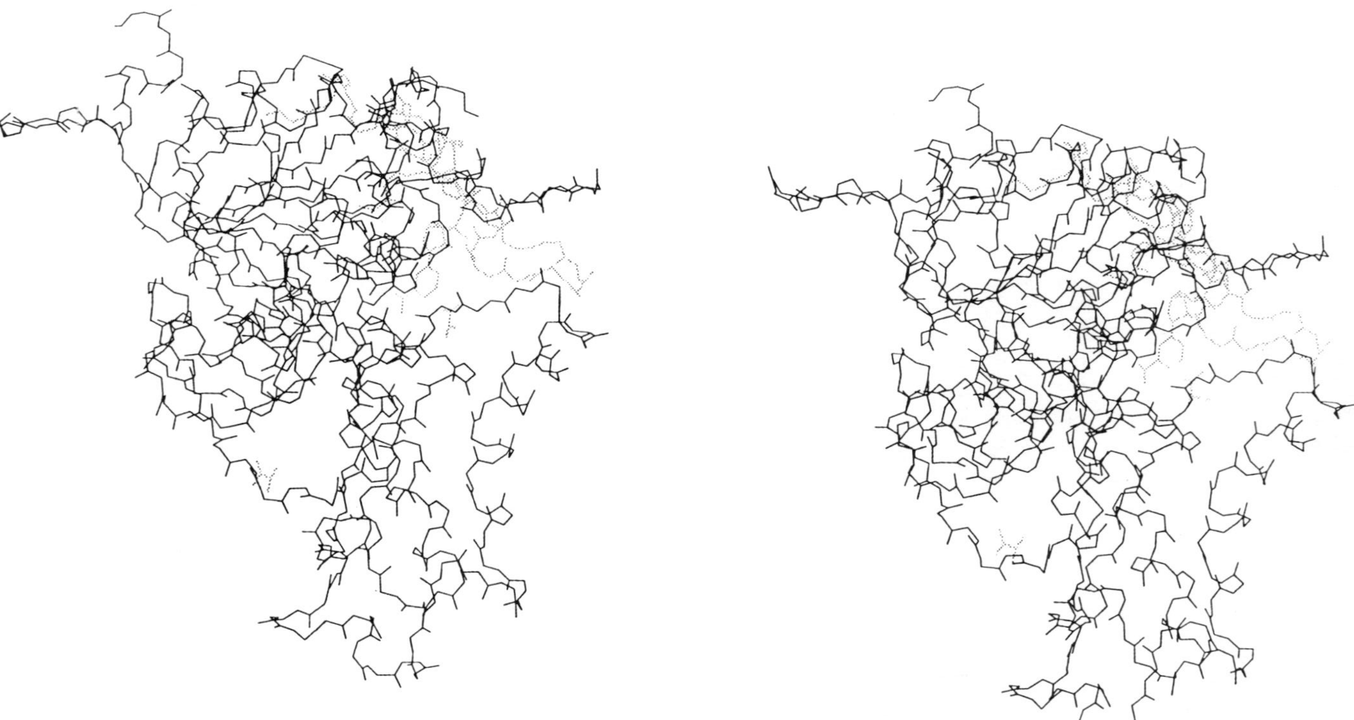

Fig. 5.21 View of the peptide backbone of lactate dehydrogenase. The solid lines are the backbone in the apoenzyme, the lightest lines show the conformation chance induced in the loop, the NADH coenzyme, and the two bound oxamate molecules (see text).

region, and can potentially alter the relative orientation of domains. In many allosteric systems, such as haemoglobin (see Chapter 6) or aspartate transcarbamylase the functional motions involve shifts of entire subunits relative to one another. As an example of domain motion, we will discuss the conformational change that takes place on binding of glucose to hexokinase.

Hexokinase is the first enzyme in glycolysis, it acts as a phosphoryl transferase, transferring the γ-phosphate of MgATP to the C-6 hydroxyl of glucose (Fersht 1977). Hexokinase is a prototype example of Koshland's induced fit hypothesis (Koshland 1959) because the conformational change is essential for activity (Anderson *et al*. 1978c; Wilkinson and Rose 1980; Shoham and Steitz 1982). Binding of glucose causes a large conformational change, as demonstrated by changes in fluorescence (Hoggett and Kellet 1976; Feldman and Kramp 1978), and by protection from proteolysis (Schultze and Colowick 1969). X-ray crystal structures of hexokinase with bound glucose and with a substrate that prevents domain closure have provided detailed information about the structural changes that facilitate domain closure (Anderson *et al*. 1978a; 1978b).

Figure 5.22 shows the peptide backbone of hexokinase with glucose and *o*-toluylglucosamine (a ligand that prevents domain closure). The conformational change can be approximated as a rigid-body rotation of one lobe by 12° with respect to the other, the domains themselves remaining largely unchanged (Anderson *et al*. 1978c; Bennet and Steitz 1980). The bulky toluyl group of *o*-toluylglucosamine prevents closure of the active site around the 6-OH groups of the sugar. This motion moves the backbone of the domains towards each other by as much as 8 Å. Small-angle X-ray studies have shown that the conformational change results in a decrease in the radius of gyration consistent with that predicted from crystal structures (McDonald *et al*. 1979). As a result of this conformational change there is a small change in the curvature of the b-sheet in the small domain that is similar to the normal mode deformations discussed above (Section 5.2.2.3). Closure of the cleft almost entirely buries the glucose molecule, only the 6-OH group remaining exposed. Model structures of hexokinase with both MgATP and glucose bound have a distance of 5.5 Å between the γ-phosphate and the 6-OH, suggesting that further conformational change is required when both ATP and glucose are bound, and that these changes are not allowed in the crystal (Shoham and Steitz 1980; Wilkinson and Rose 1980).

In discussing domain closure in a different enzyme, citrate synthetase (Remington *et al*. 1982), Lesk and Chothia (1984) describe what they termed the 'helix interface shear' mechanism for domain closure. They found that the large conformational changes seen at the active site as a result of domain closure are facilitated by the sum of many small shifts in local packing

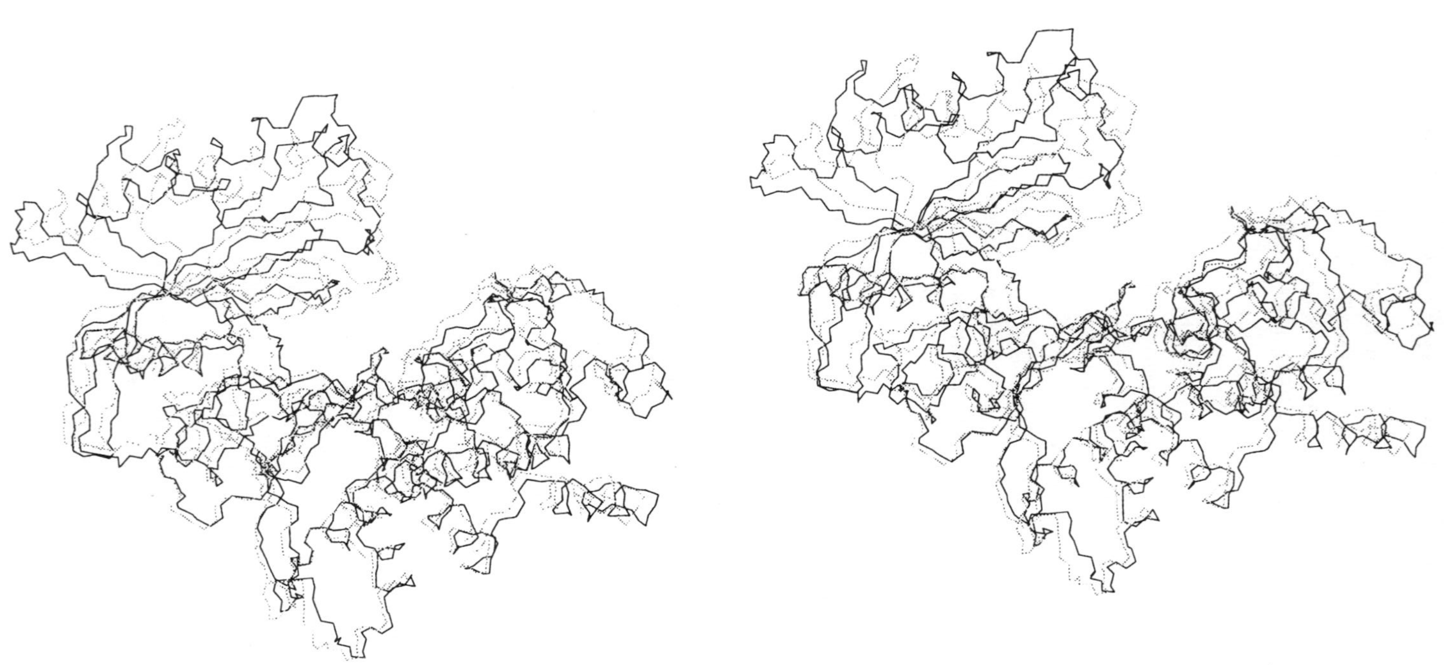

Fig. 5.22 Peptide backbone of the 'open' and 'closed' forms of hexokinase, the motion can be described as a 12° relative rotation of the domains.

between elements of secondary structure, without any dramatic changes in the overall tertiary structure within domains. They find similar features in the domain closure of both alcohol dehydrogenase and hexokinase. As discussed previously, proteins are capable of accommodating relatively large conformational changes, for example movement of entire domains relative to each other, through rather subtle changes in local structure.

5.3.2 Protein motions during catalysis, ribonuclease

Enzymes function as catalysts by lowering the free energy of the transition state relative to that of the reactants or products, i.e. proteins bind the transition state more strongly than either reactants or products (Pauling 1946). This implies that some of the free energy associated with substrate binding must go into producing strain in the enzyme-reactant complex. When the substrate reaches the transition state this strain is relaxed and the enzyme makes its most favourable interactions with the transition state structure. The enzyme again makes less favourable interactions with the product as the reaction moves to completion. The enzyme must interact in a very specific way with the reactants, transition state, and product to achieve this. One means by which the enzyme can accomplish this is by poising groups that will form favourable interactions with the transition state but not with reactants or products. In previous sections we have seen how conformational changes can result from stabilizing interactions with ligands, here we will discuss how investigating the motions of residues can provide information about interactions that stabilize the transition state. We will use as our example the binding of substrate, product, and transition state analogues to the enzyme ribonuclease.

Bovine pancreatic ribonuclease is one of the most extensively studied enzymes, it was the first enzyme to be sequenced (Potts *et al.* 1962), the first protein analysed by NMR (Saunders 1957), it is a benchmark for proton exchange experiments (Schreier and Baldwin 1976), and it has yielded both X-ray and neutron structures of high resolution with a variety of ligands (Wlodawer 1984; Campbell and Petsko 1987; Wlodawer *et al.* 1988). The enzyme hydrolyses RNA in a two-step process, during which a cyclic phosphate intermediate is produced. Pentavalent phosphorus atoms are implicated as the transition state structures for both steps in the reaction. The enzyme shows a great specificity for a pyrimidine base on the 3′ side of the phosphate backbone (Fersht 1977).

A great deal has been learned about transition state stabilization from crystal structures of RNase with the transition state analogue uridine vanidate. Of the residues forming the active site (Richards and Wyckoff 1973) only His-12, His-119, and Lys-41 are thought to be directly involved in catalysis. Lys-41 may have a significant role in stabilizing the transition state by forming strong Coulombic interactions with the highly charged transition state (pentavalent phosphorous). In the presence of substrate and product analogues, Lys-41 is in close proximity to the O-4 of the phosphate, but it has a large temperature factor, suggesting that it is not strongly bound. However, when the transition state analogue uridine vanidate is bound, the temperature factor of Lys-41 is significantly lower (Wlodawer 1984), indicating that it is interacting more strongly with, and is involved in stabilizing, the transition state structure. Here the motions of groups around the active site can indicate how the enzyme interacts differently with the reactant and the transition state (which is essential for catalysis).

5.3.3 Ligand entry and exit, myoglobin

Myoglobin is a storage protein that binds oxygen in the blood and releases it in working muscles where the oxygen demand is great. Myglobin binds its ligand at a haem site buried within the protein interior. The crystal structures of both myoglobin and oxymyoglobin are known (Takano 1977; Phillips 1978) and neither structure shows a clear pathway for ligand entry or exit. Case and Karplus (1979) carried out potential energy calculations to determine the energy barrier to ligand exit in the field of the fixed protein; they found a barrier on the order of 100 kcal mol^{-1}. The barrier to ligand exit, as determined from flash photolysis experiments, is on the order of 15 kcal mol^{-1} (Austin *et al.* 1975), indicating that motion is required to enable myoglobin to release its ligand at a biologically useful rate. The exact nature of these motions is still unclear; however, theoretical studies have suggested the kinds of motions that might take place.

Case and Karplus, in their original study of myoglobin, attempted to find likely pathways for ligand release by propagating the dynamics of the ligand in the field of the fixed protein matrix. They gave a test particle, bound at the ligand site, a large kinetic energy of 15 kcal mol^{-1} and analysed 80 trajectories of 3.75 ps duration each. The test particle was assigned a radius of 1.4 Å to allow it to overcome the high barriers presented by the fixed protein matrix. They found two pathways for ligand exit, a primary pathway between Thr-10 and Val-11, and a second, more complicated path into the space between Thr-10, Leu-4, and Phe-14 followed by escape between Leu-4 and Phe-14. To further analyse these pathways, Case and Karplus explored how distortions in side-chain dihedral angles might act to open a 'channel' for ligand exit. For the primary pathway it was found that by distorting the χ^1 dihedral angles of the distal histidine residue, Thr-10 and Val-11 could lower the barrier significantly (to ≈ 5 kcal mol^{-1}) at a cost of approximately 8.5 kcal mol^{-1} in strain energy. Dihedral distortions in the secondary pathway were too unfavourable to significantly lower the barrier along this path.

Following on this work, Case and McCammon (1987) employed similar techniques to those used in studying ring rotations (see Section 5.2.3.2) to obtain potentials of mean force for a reaction coordinate corresponding to the primary ligand entry/exit pathway of Case and Karplus. In this study, all residues within 8.0 Å of the haem pocket were allowed to move in a series of dynamics trajectories which took the ligand from the haem pocket, through the space between Thr-10 and Val-11, to the protein surface. The results of these dynamics calculations indicate that there is a barrier of ≈ 7 kcal mol^{-1} to ligand exit, again somewhat less than the experimental result and the adiabatic barrier. In contrast to the energy minimization results of Case and Karplus and the flash photolysis experiments (Austin *et al.* 1975), the free-energy change from the potential of mean force calculation appears to arise mostly from entropy; the change in potential energy is small over the reaction coordinate. Thus, it appears that when the protein matrix is allowed to dynamically relax during the potential of mean force calculation, the side-chain motions required for ligand release do not result in a large cost in strain energy. Rather, it seems that the system loses *entropy* as the ligand moves from a state of relatively large mobility in the haem pocket, through a more constrained channel between Thr-10 and Val-11.

It is important to point out that both Case and Karplus and Case and McCammon studied only a single pathway for ligand exit, and other pathways almost certainly exist. Nevertheless, these studies point out how relatively small-scale protein motions can greatly enhance the rate of processes such as ligand entry and exit.

5.4 Conclusion

Research from diverse fields in biology, chemistry, and physics has uncovered many important aspects of the relationship between protein structure and function. While a great deal remains to be discovered, much is now known about the folding patterns of proteins, how proteins bind substrates and perform catalysis, and the forces that stabilize the folded state. As we have gained information about the structure of proteins, we have also learned about their dynamics. Proteins demonstrate a variety of motions, from small-scale rattling of side-chains in the interior, to domain motion and the folding-unfolding of the protein as a whole. In this chapter we have given examples of this wide range of motions, considering both the structural aspect of these motions as well as their impact on protein function. We have surely neglected important examples of protein motions, motions of muscle fibres and conformational gating in membrane channels, for example. However, it was our aim to cover as much of the range of motions as possible, showing the range of both time-scale and amplitude. We have also tried to show the interrelationship between time-scale, amplitude, and energy for motion.

As stated in the introduction, proteins must move if they are to perform their function. Proteins bind substrates and cofactors, carry out chemical reactions, and release their products; these require motion of the active site. Storage and transport proteins, such as myoglobin and haemoglobin, bind oxygen in an area of relatively high concentration and release it where the oxygen concentration is lower. As shown, the movement of the ligand into and out of the protein requires motion. Proteins themselves are translated as linear amino-acid sequences which must undergo many complex rearrangements to fold into their final active form. Many other examples of the functional importance of protein motions exist, and a picture is now developing of protein function being determined by the motions as well as the structure of proteins.

Another aspect of protein structure that is becoming clear is that proteins are not rigid, tightly folded structures but are actually quite supple, capable of accommodating a variety of sequence alterations and still folding into the same overall structure (Lim and Sauer 1989). The folded state of most proteins is only stable by between 5 and 15 kcal mol^{-1} (Creighton 1983) and this means that substantial fluctuations in structure can occur at physiological temperature. As we have seen, since the interactions that hold the protein together are rather weak, groups are easily able to rearrange, and a small rearrangement of several groups can result in large overall motion of the protein.

The time-scale and amplitude range of motions accessible to study by theoretical and experimental methods is expanding, and our picture of protein motion is becoming more detailed. In the future the interplay between these methods, and among the scientific disciplines studying proteins, will provide an even clearer picture of how these very special molecules operate.

5.5 Acknowledgements

The authors wish to acknowledge the significant contributions of Joseph McDonald, Douglas J. Tobias, Erik Boczko, and Keith Constantine, both for commenting on the manuscript and for many important discussions regarding the concepts presented in this work.

References

Abad-Zapatero, C., Griffith, J.P., Sussman, J.L., and Rossman, M.G. (1987). Refined crystal structure of dogfish M_4 apo-lactate dehydrogenase. *Journal of Molecular Biology* **198**, 445–67.

Adams, M.J., *et al.* (1973). Structure function relationships in lactate dehydrogenase. *Proceedings of the National Academy of Sciences USA* **70**, 1968–72.

Alcala, J.R., Gratton, E., and Prendergast, F.G. (1987). Interpretation of fluorescence decays in proteins using continuous lifetime distributions. *Biophysical Journal* **51**, 925–36.

Anderson, C.M., McDonald R.C., and Steitz, T.A. (1978a). Sequencing a protein by X-ray crystallography: I. Interpretation of yeast hexokinase B at 2.4 Å resolution by model building. *Journal of Molecular Biology* **123**, 1–13.

Anderson, C.M., Stenkamp, R.E., and Steitz, T.A. (1978b). Sequencing a protein by X-ray crystallography: II. Refinement of yeast hexokinase B co-ordinates and sequence at 2.1 Å resolution. *Journal of Molecular Biology* **123**, 15–33.

Anderson, C.M., Stenkamp, R.E., McDonald, R.C., and Steitz, T.A. (1978c). A refined model of the sugar binding site of yeast hexokinase B. *Journal of Molecular Biology* **123**, 207–19.

Austin, R.H., Beeson, K.W., Eisenstein, L., Frauenfelder, H., and Gunsalus, I.C. (1975). Dynamics of ligand binding to myoglobin. *Biochemistry* **14**, 5355–73.

Axelsen, P.H. and Prendergast, F.G. (1989). Molecular dynamics of tryptophan in ribonuclease-T1: II. Correlations with fluorescence. *Biophysical Journal* **56**, 43–66.

Axelsen, P.H., Haydock, C., and Prendergast, F.G. (1988). Molecular dynamics of tryptophan in ribonuclease-T1: I. Simulation strategies and fluorescence anisotropy decay. *Biophysical Journal* **54**, 249–58.

Barksdale, A.D. and Stuehr, J.E. (1972). Kinetics of the helix-coil transition in aqueous poly(L-glutamic acid). *Journal of the American Chemical Society* **94**, 3334–8.

Beechem, J.M. and Brand, L. (1985). Time-resolved fluorescence of proteins. *Annual Review of Biochemistry* **54**, 43–71.

Bennet, W.S., Jr, and Steitz, T.A. (1980). Structure of a complex between yeast hexokinase A and glucose: II. Detailed comparisons of conformation and active site configuration with the native hexokinase B monomer and dimer. *Journal of Molecular Biology* **140**, 211–30.

Brandts, J.F., Halvorson, H.R., and Brennan, M. (1975). Consideration of the possibility that the slow step in protein denaturation reactions is due to *cis-trans* isomerism of proline residues. *Biochemistry* **14**, 4953–63.

Brooks, B.R., Bruccoleri, R.E., Olafson, B.D., States, D.J., Swaminathan, S., and Karplus, M. (1983). CHARMM: a program for macromolecular energy, minimization, and dynamics calculations. *Journal of Computational Chemistry* **4**, 187–217.

Brooks, B. R., and Karplus, M. (1983) Harmonic dynamics of proteins:

normal modes and fluctuations in bovine pancreatic trypsin inhibitor. *Proceedings of the National Academy of Sciences* **80**, 6571–5.

Brooks, C. L., III, Karplus, M., and Pettitt, B.M. (1988). Proteins: a theoretical perspective of dynamics, structure and thermodynamics. In *Advances in chemical physics*, Vol. 75, pp. 75–87 (ed. L. Prigogine and S.A. Price). Wiley and Sons, New York.

Butz, T., Lerf, A., and Huber, R. (1982). Intramolecular reorientational motion in trypsinogen studied by perturbed angular correlation of ^{199mm}Hg labels. *Physical Review Letters* **48**, 890–3.

Campbell, R.L. and Petsko, G.A. (1987). Ribonuclease structure and catalysis: crystal structure of sulfate-free native ribonuclease A at 1.5 Å resolution. *Biochemistry* **26**, 8579–84.

Case, D. and Karplus, M. (1979). Dynamics of ligand binding to heme proteins. *Journal of Molecular Biology* **132**, 343–68.

Case, D.A. and McCammon, J.A. (1987). Dynamic simulation of oxygen binding to myoglobin. *Annals of the New York Academy of Sciences* **482**, 222–33.

Chandrasekhar, K., McPherson, A., Jr, Adams, M.J., and Rossman, M.G. (1973). Conformation of coenzyme fragments when bound to lactate dehydrogenase. *Journal of Molecular Biology* **76**, 503–18.

Chen, L.X.Q., Longworth, J.W., and Fleming, G.R. (1987). Picosecond time-resolved fluorescence of ribonuclease-T1: a pH and substrate binding study. *Biophysical Journal* **51**, 865–73.

Chen, L.X.Q., Engh, R.A., Brunger, A.T., Nguyen, D.T., Karplus M., and Fleming, G.R. (1988). Dynamics simulation studies of apoazurin of *Alcaligenes denitrificans*. *Biochemistry* **27**, 6908–21.

Cheng, H.N. and Bovey, F.A. (1977). *Cis-trans* equilibrium and kinetic studies of acetyl-L-proline and glycyl-L-proline. *Biopolymers* **16**, 1465–72.

Chou, K. and Scheraga, H.A. (1982). Origin of the right-handed twist of β-sheet of poly(L-Val) chains. *Proceedings of the National Academy of Sciences USA* **79**, 7047–51.

Chou, K., Pottle, M., Nemethy, G., Ueda, Y., and Scheraga, H.A. (1982). Structure of b-sheets: origin of the right-handed twist and of the increased stability of antiparallel sheets. *Journal of Molecular Biology* **162**, 89–112.

Cook, K.H., Schmid, F.X., and Baldwin, R.L. (1979). Role of proline isomerization in folding of ribonuclease A at low temperatures. *Proceedings of the National Academy of Sciences USA* **76**, 6157–61.

Creighton, T.E. (1983). Conformational properties of polypeptide chains. In *Proteins, structures and molecular properties*, pp. 159–99. W.H. Freeman, New York.

Creighton, T.E. (1988). Toward a better understanding of protein folding pathway. *Proceedings of the National Academy of Sciences USA* **85**, 5082–6.

Creighton, T.E. and Goldenberg, D.P. (1984). Kinetic role of a meta-stable native-like two-disulphide species in the folding transition of bovine pancreatic trypsin inhibitor. *Journal of Molecular Biology* **179**, 497–526.

Darby, N.J., van Mierlo, C.P.M., Scott, G.H.E., Neuhaus, D., and Creighton, T.E. (1992). Kinetic roles and conformational properties of the non-native two-disulfide intermediates in the refolding of bovine pancreatic trypsin inhibitor. *Journal of Molecular Biology* **224**, 905–11.

Dobson, C.M. and Evans, P.A. (1988). Trapping folding intermediates. *Nature* **335**, 666–7.

Dyson, H.J., Rance, M., Houghton, R.A., Lerner, R.A., and Wright, P. E. (1988). Folding of immunogenic peptide fragments of proteins in water. I. Sequence requirements for the formation of a reverse turn. *Journal of Molecular Biology* **209**, 817–20.

Evans, C.A. and Rabenstein, D.L.(1974). Nuclear magnetic resonance studies of the acid-base chemistry of amino acids and peptide. II. Dependence of the acidity of the C-terminal carboxyl group on the conformation of the C-terminal peptide bond. *Journal of the American Chemical Society* **96**, 7312–17.

Feldman, I. and Kramp, D.C. (1978). Fluorescence-quenching study of glucose binding by yeast hexokinase isozymes. *Biochemistry* **17**, 1541–7.

Fersht, A. (1977). *Enzyme structure and mechanism.* W.H. Freeman, San Francisco.

Frauenfelder, H. and Gratton, E. (1986). Protein dynamics and hydration. *Methods in enzymology* **127**, 207–16.

Gardell, S.J., Craik,C.S., Hilvert, D., Urdea, M.S., and Rutter, W.J. (1985). Site-directed mutagenesis shows that tyrosine 248 of carboxypeptidase A does not play an essential role in catalysis. *Nature* **317**, 551–5.

Gelin, B.R. and Karplus, M. (1975). Sidechain torsional potentials and motion of amino acids in proteins: bovine pancreatic trypsin inhibitor. *Proceedings of the National Academy of Sciences USA* **72**, 2002–6.

Ghosh, I. and McCammon, J.A. (1987). Sidechain rotational isomerization in proteins: dynamic simulation with solvent surroundings. *Biophysical Journal* **51**, 637–41.

Gō, M. and Gō, N. (1976). Fluctuations of an α-helix. *Biopolymers* **15**, 1119–27.

Gō, N., Noguti, T., and Nishikawa, T. (1983). Dynamics of a small globular protein in terms of low-frequency vibrational modes. *Proceedings of the National Academy of Sciences USA* **80** 3696–700.

Goldberg, M.E. (1985). The second translation of the genetic message: protein folding and assembly. *Trends in Biochemical Sciences* **10**, 388–91.

Goldenberg, D.P. and Creighton, T.E. (1984). Folding pathway of a circular form of bovine pancreatic trypsin inhibitor. *Journal of Molecular Biology* **179**, 527–45.

Gurd, F.R.N. and Rothgeb, M.T. (1979). Motions in proteins. *Advances in Protein Chemistry* **33**, 73–165.

Haydock, C., Sharp, J.C., and Prendergast, F.G. (1990). Trytophan–47 rotational isomerization in variant–3 scorpion neurotoxin: a combination thermodynamic perturbation and umbrella sampling study. *Biophysical Journal* **57**, 1269–79.

Heatley, E. (1986). Nuclear magnetic relaxation and models for backbone motions of macromolecules in solution. *Annual Reports on NMR Spectroscopy* **17**, 179–230.

Hoggett, J.G., and Kellet, G.L. (1976). Yeast hexokinase: substrate-induced association-dissociation reaction in the binding of glucose to hexokinase P-II. *European Journal of Biochemistry* **66**, 65–77.

Huber, R. and Bennet, W.S., Jr (1983). Functional significance of flexibility in proteins. *Biopolymers* **22**, 261–79.

Ichiye, T. and Karplus, M. (1983). Fluorescence depolarization of tryptophan residues in proteins: a molecular dynamics study. *Biochemistry* **22**, 2884–93.

Inoue, S., Sano, T., Yakabe, Y., Ushio, H., and Yasunaga, T. (1979). Kinetic studies of the helix-coil transition in aqueous solutions of poly(L-lysine). *Biopolymers* **18**, 681–91.

Jorgensen, W.L. (1989). Interactions between amides in solution and the thermodynamics of weak binding. *Journal of the American Chemical Society* **111**, 3770–1.

Kay, L.E., Torchia, D.A., and Bax, A. (1989). Backbone dynamics of proteins as studied by ^{15}N inverse detected heteronuclear NMR spectroscopy: application to staphlococcal nuclease. *Biochemistry* **28**, 8972–9.

Kelley, R.F. and Richards, F.M. (1987). Replacement of proline–76 with alanine eliminates the slowest kinetic phase in thioredoxin folding. *Biochemistry* **26**, 6765–74.

Kim, P.S. and Baldwin, R.L. (1980). Structural intermediates trapped during the folding of ribonuclease A by amide proton exchange. *Biochemistry* **19**, 6124–9.

Kitamaru, R. (1986). Carbon-13 nuclear spin relaxation study as an aid to analysis of chain dynamics and conformation of macromolecules. In *NMR in stereochemical analysis*, (ed. Y. Taskeudi and A. P. Merchand) pp. 75–124. VCH Publishers, Deerfield, Florida.

Koshland, D.E., Jr (1959). In *The enzymes*, (ed. P.D. Boyer, H. Lardy, and K. MyrBäck), Vol. 1, (2nd edn), pp. 305–46. Academic Press, New York.

Lakowicz, J.R. and Maliwal, B.P. (1983). Oxygen quenching and fluorescence depolarization of tyrosine residues in proteins. *Journal of Biological Chemistry* **258**, 4794–801.

Lakowicz, J.R. and Weber, G. (1973). Quenching of protein fluorescence by oxygen. Detection of structural fluctuations in proteins on the nanosecond time scale. *Biochemistry* **12**, 4171–9.

Lakowicz, J.R., Maliwal, B.P. Cherek, H., and Balter, A. (1983). Rotational freedom of trytophan residues in proteins and peptides. *Biochemistry* **22**, 1741–52.

Lesk, A.M. and Chothia, C. (1984). Mechanism of domain closure in proteins. *Journal of Molecular Biology* **174**, 175–91.

Levitt, M. (1980). Effect of proline residues on protein folding. *Journal of Molecular Biology* **145**, 251–63.

Levitt, M., Sander, C., and Stern, P.S. (1985). Protein normal-mode dynamics: trypsin inhibitor, crambin, ribonuclease and lysozyme. *Journal of Molecular Biology* **181**, 423–47.

Levy, R.M. (1985). NMR relaxation and the dynamics of proteins. In *Molecular dynamics and protein structure*, (ed. Jan Hermans), pp. 62–4. Polycrystal Book Service, Western Springs, Illinois.

Levy, R.M. and Karplus, M. (1979). Vibrational approach to the dynamics of an α-helix. *Biopolymers* **18**, 2465–95.

Levy, R.M. and Szabo, A. (1982). Initial fluorescence depolarization in proteins. *Journal of the American Chemical Society* **104**, 2073–5.

Levy, R.M., Perahia, D., and Karplus, M. (1982). Molecular dynamics of an α-helical polypeptide: temperature dependence and deviation from harmonic behavior. *Proceedings of the National Academy of Sciences USA* **79**, 1346–50.

Levy, R.M., Sheridan, R.P., Karplus, J.W., Dubey, G.S., Swaminathan, S., and Karplus M. (1985). Molecular dynamics of myoglobin at 298K: results from a 300-psec computer simulation. *Biophysical Journal* **48**, 509–18.

Lim, W.A. and Sauer, R.T. (1989). Alternative packing arrangements in the hydrophobic core of λ. repressor. *Nature* **339**, 31–6.

Lin, L. and Brandts, J.F. (1978). Further evidence suggesting the slow phase in protein unfolding and refolding is due to proline isomerization: a kinetic study of carp parvalbumins. *Biochemistry* **17**, 4102–10.

Lipari, G. and Szabo, A. (1982a). Model-free approach to the interpretation of nuclear magnetic resonance in macromolecules. 1. Theory and range of validity. *Journal of the American Chemical Society* **104**, 4546–59.

Lipari, G. and Szabo, A. (1982b). Model-free approach to the interpretation of nuclear magnetic resonance in macromolecules. 2, Analysis of experimental results. *Journal of the American Chemical Society* **104**, 4559–70.

Lipari, G., Szabo, A., and Levy, R.M. (1982). Protein dynamics and NMR relaxation: comparisons of simulations with experiment. *Nature* **300**, 197–8.

London, R.E. (1989). Interpreting protein dynamics with nuclear magnetic resonance relaxation measurements. *Methods in Enymology* **176**, 358–75.

McCain, D.C., Ulrich, E.L., and Markley, J.L. (1988). NMR relaxation study of internal motions in staphlococcal nuclease. *Journal of Magnetic Resonance* **80**, 296–305.

McCammon, J.A. and Harvey, S.C. (1987). Short time dynamics. In *Dynamics of proteins and nucleic acids*, pp. 79–115. Cambridge University Press, Cambridge.

McCammon, J.A., Lee, C.Y., and Northrup, S.H. (1983). Side-chain rotational isomerization in proteins: a mechanism involving gating and transient packing defects. *Journal of the American Chemical Society* **105**, 2232–7.

McDonald, R.C., Steitz, T.A., and Engelman, D.M. (1979). Yeast hexokinase in solution exhibits a large conformational change upon binding glucose or glucose 6-phosphate. *Biochemistry* **18**, 338–42.

Molday, R.S., Englander, S.W., and Kallen, R.G. (1972). Primary structure effects on peptide group hydrogen exchange. *Biochemistry* **11**, 150–8.

Nagayama, K. and Wada, A. (1975). Determination of the relaxation time of the helix-coil transition of poly(γ-benzyl-L-glutamate) by the NMR technique. *Biopolymers* **14**, 2489–506.

Noguti, T. and Gō, N. (1982) Collective variable description of small-amplitude conformational fluctuations in a globular protein. *Nature* **296**, 776–8.

Northrup, S.H., Pear, M.R., Lee, C.Y., McCammon, J.A., and Karplus, M. (1982). Dynamical theory of activated processes in globular proteins. *Proceedings of the National Academy of Sciences* **79**, 4035–9.

Olejniczak, E.T., Paulsen, F.M., and Dobson, C.M. (1981). Proton nuclear Overhauser effects and protein dynamics. *Journal of the American Chemical Society* **103**, 6574–80.

Pace, C.N. and Creighton, T.E. (1986). The disulphide folding pathway of ribonuclease T1. *Journal of Molecular Biology* **188**, 477–86.

Pauling, L. (1946). Molecular architecture and biological reactions. *Chemical Engineering News* **24**, 1375–7.

Phillips, S.E.V. (1978). Structure of oxymyoglobin. *Nature* **273**, 247–8.

Post, C.B., Dobson, C.M., and Karplus, M. (1989). A molecular dynamics analysis of protein structural elements. *Proteins* **5**, 337–54.

Potts, J.T., Berger, A., Cooke, J., and Anfinsen, A.J. (1962). A reinvestigation of the sequence of residues 11 to 18 in bovine pancreatic ribonuclease. *Journal of Biological Chemistry* **237**, 1851–5.

Rees, D.C. and Lipscomb, W.N. (1981). Binding of ligands to the active site of carboxypeptidase A. *Proceedings of the National Academy of Sciences USA* **78**, 5455–9.

Rees, D.C. and Lipscomb, W.N. (1983). Crystallographic studies on carboxypeptidase A and the complex with glycyl-L-tyrosine. *Proceedings of the National Academy of Sciences USA* **80**, 7151–4.

Rees, D.C., Lewis, M., and Lipscomb, W.N. (1983). Refined crystal structure of carboxypeptidase A at 1.54 Å resolution. *Journal of Molecular Biology* **168**, 367–87.

Remington, S., Weigand, G., and Huber, R., (1982). Crystallographic refinement and atomic models of two different forms of citrate synthase at 2.7 and 1.7 Å resolution. *Journal of Molecular Biology* **158**, 111–52.

Richards, F.M. and Wyckoff, H.W. (1973). *Atlas of molecular structure 1. Ribonuclease S.* Oxford University Press, London.

Roder,H., Elove, G.A., and Englander, S.W. (1988). Structural characterization of folding intermediates in cytochrome c by H-exchange labelling and proton NMR. *Nature* **335**, 700–4.

Ruggiero, A.J., Todd, D.C., and Flemming, G.R. (1990). Subpicosecond fluorescence anisotropy studies of tryptophan in water. *Journal of the American Chemical Society* **112**, 1003–14.

Saunders, M., Wishnia, A., and Kirkwood, J. G. (1957). The nuclear magnetic resonance spectrum of ribonuclease. *Journal of the American Chemical Society* **79**, 2289–90.

Schmid, F.X. and Baldwin, R.L. (1978). Acid catalysis of the formation of the slow-folding species of RNase A: evidence that the reaction is proline isomerization. *Proceedings of the National Academy of Sciences USA* **75**, 4764–8.

Schmid, F.X. and Baldwin, R.L. (1979). Detection of an early intermediate in the folding of ribonuclease A by protection of amide protons against exchange. *Journal of Molecular Biology* **135**, 199–215.

Schreier, A.A. and Baldwin, R.L. (1976). Concentration-dependent hydrogen exchange kinetics of ^{3}H-labelled S-peptide in ribonuclease S. *Journal of Molecular Biology* **105**, 409–26.

Schultze, I.T. and Colowick, S.P. (1969). The modification of yeast hexokinases by proteases and its relationship to the dissociation of hexokinase into subunits. *Journal of Biological Chemistry* **244**, 2306–16.

Shoham, M. and Steitz, T.A. (1980). Crystallographic studies and model building of ATP at the active site of hexokinase. *Journal of Molecular Biology* **140**, 1–14.

Shoham, M. and Steitz, T.A. (1982). The 6-hydroxymethyl group of a hexose is essential for the substrate induced closure of the cleft in hexokinase. *Biochimica et Biophysica Acta* **705**, 380–4.

Sneddon, S.F., Tobias, D.J., and Brooks III, C.L. (1989). Thermodynamics of amide hydrogen bond formation in polar and apolar solvents. *Journal of Molecular Biology* **209**, 817–20.

Snyder, G.H., Rowan, R., Karplus, S., and Sykes, B.D. (1975). Complete tyrosine assignments in the high field ^{1}H nuclear magnetic resonance spectrum of the bovine pancreatic trypsin inhibitor. *Biochemistry* **14**, 3775–7.

Stellwagen, E. (1979). Proline peptide isomerization and the reactivity of denatured enzymes. *Journal of Molecular Biology* **135**, 217–29.

Suezaki, Y. and Go, N. (1976). Fluctuations and mechanical strength of α-helices of polyglycine and poly(L-alanine). *Biopolymers* **15**, 2137–53.

Takano, T. (1977). Structure of myoglobin refined at 2.0 Å resolution II. Structure of deoxymyoglobin from sperm whale. *Journal of Molecular Biology* **110**, 569–84.

Thomas, W.A. and Williams, M.K. (1972). ^{13}C Nuclear magnetic resonance spectroscopy and *cis/trans* isomerism in dipeptides containing proline. *Journal of the Chemical Society — Chemical Communications* p. 994.

Tilton, R.F., Jr, Singh, U.C., Kuntz, I.D., Jr, and Kollman, P.A. (1988). Protein-ligand dynamics: a 96 picosecond simulation of a myoglobin-xenon complex. *Journal of Molecular Biology* **199**, 195–211.

Tobias, D.J., Mertz, J., and Brooks III, C.L. (1990). Nanosecond folding dynamics of a pentapeptide in water. *Biochemistry* **30**, 6054.

Torda, A.E. and Norton, R.S. (1989). Protein NMR relaxation study of the dynamics of anthopleurin-A in solution. *Biopolymers* **28**, 703–16.

Tsuji, Y., Yasunaga, T., Sano, T., and Ushio, H. (1976). Kinetic studies of the helix-coil transition in aqueous solutions of poly(α-L-glutamic acid) using the electric pulse method. *Journal of the American Chemical Society* **98**, 813–18.

Udgaonkar, J.B. and Baldwin, R.L. (1988). NMR evidence for an early framework intermediate on the folding pathway of ribonuclease. *Nature* **335**, 694–9.

Wagner, G. and Wüthrich, K. (1975). Proton NMR studies of the aromatic residues in the basic pancreatic trypsin inhibitor (BPTI). *Journal of Magnetic Resonance* **20**, 435–45.

Weaver, A.J., Kemple, M.D., and Prendergast, F.G. (1988). Tryptophan sidechain dynamics in hydrophobic oligopeptides determined by use of ^{13}C nuclear magnetic resonance spectroscopy. *Biophysical Journal* **54**, 1–15.

Weissman, J.S. and Kim, P.S. (1991). Reexamination of the folding of BPTI: predominance of native intermediates. *Science* **253**, 1386–93.

White, J.L., *et al.* (1976). A comparison of the structures of apo dogfish M_4 lactate dehydrogenase and its ternary complexes. *Journal of Molecular Biology* **102**, 759–79.

Wilkinson, K.D. and Rose, I.A. (1980). Glucose exchange and catalysis by two crystalline hexokinase glucose complexes: evidence for an obligatory

ATP-dependent conformational change in catalysis. *Journal of Biological Chemistry* **255**, 7569–74.

Wlodawer, A. (1984). Structure of bovine pancreatic ribonuclease by X-ray and neutron diffraction. *Biological Macromolecular Assemblies* **2**, 393–439.

Wlodawer, A., Svensson, L.A., Sjolin, L., and Gilliland, G.L. (1988). Structure of phosphate-free ribonuclease A refined at 1.26 Å resolution. *Biochemistry* **27**, 2705–17.

Wüthrich, K. and Wagner, G. (1975). NMR investigation of the dynamics of the aromatic amino acid residues in the basic pancreatic trypsin inhibitor. *FEBS Letters* **50**, 265–8.

6

Myoglobin and Haemoglobin

Giulio Fermi

6.1 Introduction

This chapter is intended for the novice; the topics introduced here have been discussed in greater depth in the reviews of Baldwin (1975), Imai (1982), Perutz *et al.* (1987), and Perutz (1989). In general, additional references are cited in the text only when they would not be found readily in these reviews. The structures of a wide variety of globins have been determined but are not considered here. These include human fetal haemoglobin, haemoglobins with ligands other than oxygen, natural and artificial mutants, chemically modified variants, and globins from other species. The structure and function of a large number of variant human haemoglobins has been compiled by Perutz (Fermi and Perutz 1981); species differences in the globins have been subjected to structural analysis by Lesk and Chothia (1980).

6.2 Overview of globin structure and function

6.2.1 General nature of globin structure

Myoglobin is the simpler of the two globins. Its structure, the first three-dimensional protein structure to be determined, was solved, initially at low resolution, by Kendrew and co-workers in 1958 (see Table 6.1). Myoglobin has a molecular weight of about 18 000 and consists of a single polypeptide chain of 153 amino acids folded about a prosthetic group, the haem (Figs 6.1 and 6.2). The haem is a protoporphyrin ring with an iron atom at its centre. The amino-acid sequence (primary structure) of sperm whale myoglobin, which was not completed until 1965, is shown in Table 6.2. The table also indicates the helix notation for residues, which is useful for structural comparisons and is based on the positions of residues in elements of secondary structure (helical and non-helical regions) in the protein.

Haemoglobin has a molecular weight of about 65 000 and consists of four subunits. The structure was solved, initially at low resolution, by Perutz and co-workers in 1960. Each subunit is similar to myoglobin, with a polypeptide

Table 6.1 Selected myoglobin and haemoglobin structure determinations

A. Sperm whale myoglobin

Chemical form	Resolution	Protein Data Bank designation	Reference
met	6.0		Kendrew *et al.* (1958)
met	2.0	1MBN	Kendrew *et al.* (1960)
met	2.0 (refined structure)	4MBN	Takano (1977a)
oxy	1.6 (pH 8.4)	1MBO	Phillips (1980)
deoxy	2.8	1MBN	Nobbs *et al.* (1966)
deoxy	2.0 (pH 5.75)	5MBN	Takano (1977b)
deoxy	1.8 (pH 8.4)	1MBD	Phillips (unpublished)

B. Haemoglobin (normal)

Species and chemical form	Resolution Å	Space group	Protein Data Bank designation	Reference
horse met	5.5	C2		Perutz *et al.* (1960)
horse met	2.8	C2		Perutz *et al.* (1968a, 1968b)
horse met	2.0	C2	2MHB	Ladner *et al.* (1977)
horse CO	2.8	C2		Heidner *et al.* (1976)
human CO	2.8	$P4_12_12$	1HCO 2HCO	Baldwin (1980)
human oxy	2.1	$P4_12_12$	1HHO	Shaanan (1983)
horse deoxy	5.5	$C222_1$		Bolton *et al.* (1968)
horse deoxy	2.8	$C222_1$	2DHB	Bolton and Perutz (1970)
human deoxy	5.5	$P2_1$		Muirhead *et al.* (1967)
human deoxy	3.5	$P2_1$		Muirhead and Greer (1970)
human deoxy	2.5	$P2_1$		Fermi (1975) Ten Eyck and Arnone (1976)
human deoxy	1.74	$P2_1$	2HHB 3HHB 4HHB	Fermi *et al.* (1984)

C. Analogues of intermediate ligation states of haemoglobin

Species, chemical form, and quaternary structure	Resolution Å	Space group	Protein Data Bank designation	Reference
human met T	3.5	$P2_1$		Anderson (1973)
human F-met T	3.5	$P2_1$		Fermi and Perutz (1977)
human α-CO β:Mn[†] T	3.0	$P2_1$		Arnone *et al.* (1986)
human α-CO β:Co[†] T	2.8	$P2_1$	4COH	Luisi (1986)
human α-oxy β-deoxy[§] T	2.1	$P2_12_12$[‡]		Brzozowski *et al.* (1984)
human met T	2.1	$P2_12_12$[‡]		Liddington *et al.* (1988)
human α:Ni[†] β-CO T	2.6	$P2_1$[‡]		Luisi *et al.* (1990)
horse deoxy[¶] R	1.8	C2		Luisi (1986)

* The table lists only the best-refined structures of the oxy and deoxy forms of sperm whale myoglobin and human adult haemoglobin and their historical antecedants, and the structures of analogues of the intermediate ligation states of haemoglobin. A large number of structure determinations of globins are omitted (variants, derivatives, other species).

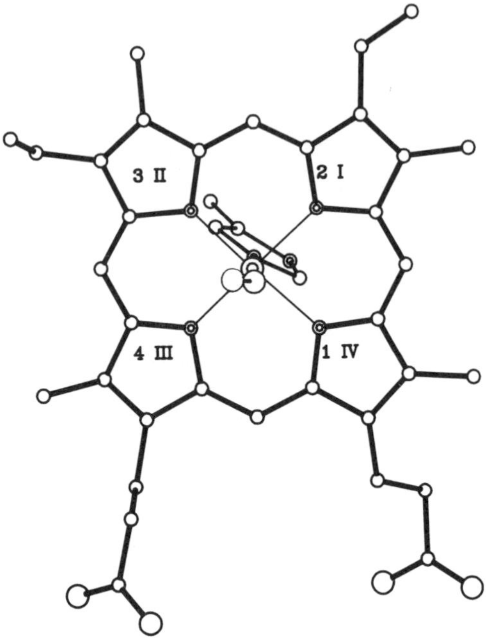

Fig. 6.1 The α-haem of oxyhaemoglobin with the haem-linked histidine F8, typical of ligated haems. Concentric double circles represent iron atoms (larger radius) or nitrogen atoms (smaller radius); single circles represent oxygen atoms (larger radius) or carbon atoms (smaller radius); hydrogen atoms are omitted. The histidine is below the plane of the screen and the bound O_2 molecule above the plane. The arabic and roman numerals indicate different numberings of the ring that have been used by various authors.

chain folded about a haem group. The subunits are identical in pairs: the polypeptide chains of the two subunits designated α_1 and α_2 each have 141 amino acids; those of the β_1 and β_2 subunits each have 146 amino acids. The amino-acid sequences of the α- and β-chains of human and horse haemoglobin are shown in Table 1, which indicates the helix notation for residues based on structural homology with myoglobin. The amino-acid sequences of the α- and β-chains of haemoglobin differ from each other and

† Fe is replaced by the designated metal so that CO does not bind.

‡ Crystallized in medium of low ionic strength with polyethylene glycol (PEG); the other structures were determined from crystals grown from high salt medium. The molecular packing of the $P2_1$ space groups in PEG and high salt mediums are quite different.

§ β-haems partially oxidized.

¶ Reacted with BME [*bis*(*N*-maleimidomethyl) ether] prior to crystallization.

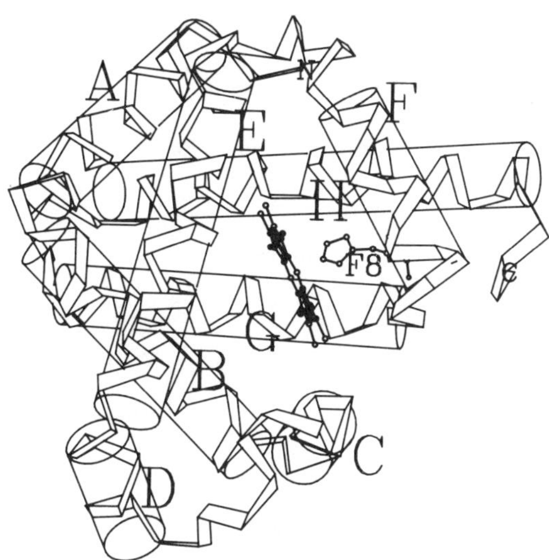

Fig. 6.2 Schematic of the main chain of myoglobin, illustrating the 'globin fold', which is typical of all globins. Each rectangle along the ribbon representing the structure is fitted to the peptide plane of an amino acid residue. Cylinders, labelled with large letters, show the helical regions (Table 6.2). The small 'N' and 'C' designate the amino- and carboxyl-termini, respectively. The haem, seen edge on, is represented by heavy bonds, omitting the side-chains; the distal histidine F8 is represented by light bonds. In the α-chains of human haemoglobin, the D helix is missing.

from that of myoglobin, but the three-dimensional folding of the peptide chain about the haem, illustrated in Fig. 6.2, is similar in all three monomers.

The tetrahedral arrangement of the four subunits of haemoglobin is shown schematically in Fig. 6.3, which shows the axes of the molecular coordinate system. The molecule has a twofold symmetry axis (dyad; the y-axis) relating the two α-chains to each other and the two β-chains to each other. There are also two pseudo-dyads, which are perpendicular to each other and to the true dyad, and give the molecule an approximate 222 point-group symmetry: 180° rotation about one of these (the x-axis) sends the subunit α_1 into the similar, but not identical, subunit β_1 (and α_2 into β_2); a similar rotation about the other pseudo-dyad (the z-axis) sends α_1 into β_2 (and α_2 into β_1). The overall shape of the haemoglobin tetramer is torroidal, as there is a gap between the subunits along the dyad (Fig. 6.3a), called the central cavity.

The site of oxygen binding in myoglobin and haemoglobin is the iron atom in the haem, which is also linked to a histidine residue, F8, known as the proximal histidine. In physiologically active globin the iron is in the ferrous state. In free haems the ferrous iron is oxidized to the ferric state by reaction

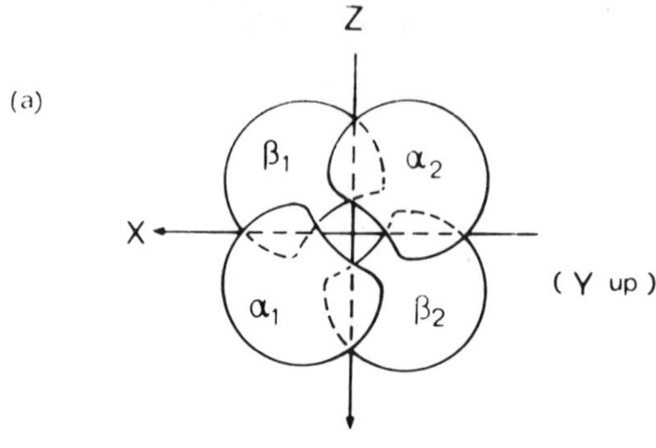

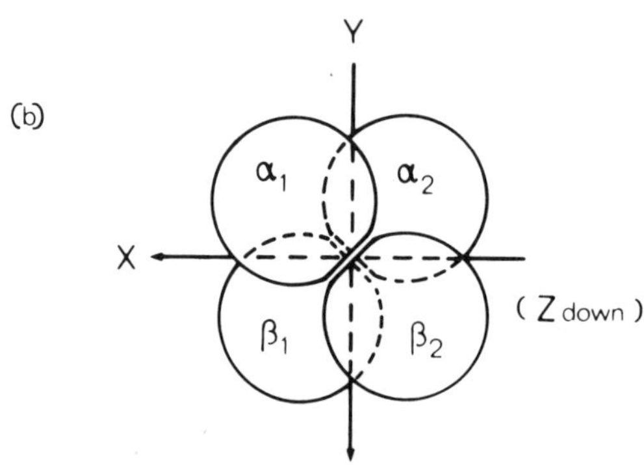

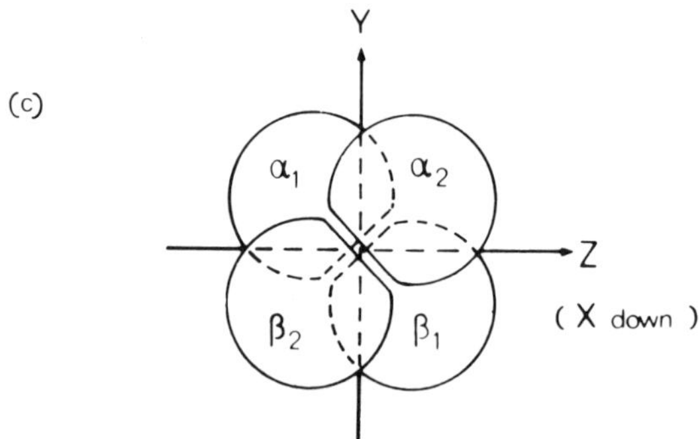

Fig. 6.3 Schematic diagram showing the quaternary arrangements of the subunits of haemoglobin and the molecular coordinate system. y is the molecular dyad relating the two identical α-subunits to each other and the two identical β-subunits to each other. x and z are the pseudo-dyads (approximate dyads) relating unlike subunits: x relates α_1 to β_1 and α_2 to β_2; z relates α_1 to β_2 and α_2 to β_1. (From Fermi and Perutz 1981.)

with oxygen and water, but in the globins the iron atom combines reversibly with the one molecule of oxygen (O_2) and remains ferrous in both the oxygenated form (oxymyoglobin and oxyhaemoglobin) and the unliganded form (deoxymyoglobin and deoxyhaemoglobin). The reversibility of the reaction, which is basic to the physiological function of the globins, is aided by enclosure of the haem in a pocket within the protein. *In vitro*, however, the iron atom gradually oxidizes to the ferric state and a water molecule replaces oxygen as the iron ligand (autoxidation); the oxidized forms are known as metmyoglobin and methaemoglobin. *In vivo*, accumulation of the

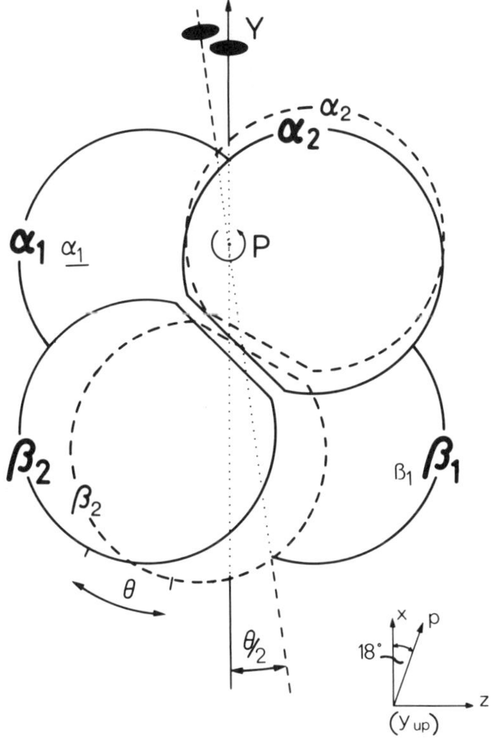

Fig. 6.4 Schematic diagram showing the change in quaternary structure that accompanies ligation of haemoglobin (the T to R transition). Bold symbols refer to deoxyhaemoglobin and light symbols to oxyhaemoglobin. The diagram shows the relative positions of the deoxy and oxy $\alpha_2\beta_2$ dimers when the $\alpha_1\beta_1$ dimers are superimposed. In going from the deoxy to the oxy form, the $\alpha_2\beta_2$ dimer rotates by an angle $\phi = 12$–$15°$ about the axis p (perpendicular to the plane of the drawing) and shifts about 1 Å into the page. The axis p intersects and is perpendicular to the dyads of both the oxy and deoxy tetramers; the point of intersection is between the α-subunits, about 13–15 Å above the centre of the molecule; the relationship of the axis to the coordinate system of Fig. 6.3 is shown in the inset. (Adapted from Baldwin and Chothia 1979.)

met form is prevented by an enzyme system that reduces ferric haems to the ferrous state.

The oxygen molecule in the oxy forms, and the water molecule in the met forms, can be replaced by other ligands. One ferrous ligand is CO, for which the globins have a higher affinity than for O_2; ferric ligands include OH^-, F^-, N_3^-, CN^-, and NO_2^- (Lemberg and Legge 1949). The structures of many of these globin compounds have been determined to varying degrees of precision.

The structure of myoglobin varies very little with ligation state. In haemoglobin there is relatively little change in the structure of the individual subunits (tertiary structure) with ligation state, but the spatial arrangement of the subunits (quaternary structure) of deoxyhaemoglobin differs from that of the liganded forms; the latter all have a common quaternary structure. The two quaternary structures are illustrated schematically in Fig. 6.4. Both structures have a dyad relating the $\alpha_1\beta_1$ dimer to the $\alpha_2\beta_2$ dimer, and the relationship between α_1 and β_1 (and that between α_2 and β_2) changes relatively little. The main difference is that the $\alpha_2\beta_2$ dimer is rotated (by about 12–15°) and translated (by about 1 Å) relative to the $\alpha_1\beta_1$ dimer, so that in deoxyhaemoglobin the two β-subunits are about 5 Å further apart, and the two α-subunits are about 1 Å further apart, than they are in the liganded form. Thus the entire central cavity is narrower in ligated haemoglobin than it is in the unligated form.

The difference in tertiary structure between the oxy and deoxy forms of haemoglobin is shown in Plate 1, which shows the relative positions of the helices when the two $\alpha_1\beta_1$ dimers are superimposed. The largest differences are in the regions furthest from the $\alpha_1\beta_1$ contact, namely, at the haems, the F helices, and the FG regions. The distance between the α-carbons of residues FG1 α and FG1 β is about 4 Å less in oxyhaemoglobin than it is in deoxyhaemoglobin. The positions of helices E and F relative to each other, in both the α- and β-subunits, is also markedly different in the oxy and deoxy forms.

6.2.2 Physiological properties

Interest in the structures of the globins results largely from the vast amount of physiological and biochemical information that has been collected on haemoglobin and, to a lesser extent, on myoglobin. We outline here a few of the more fundamental properties, knowledge of which is required for an appreciation of the structures.

6.2.2.1 Oxygen uptake and co-operativity
The most fundamental physiological property of myoglobin and haemoglobin is their ability to combine reversibly with oxygen. The basic measurement *in vitro* of this property is the oxygen equilibrium curve, wherein the fraction of haem groups bearing

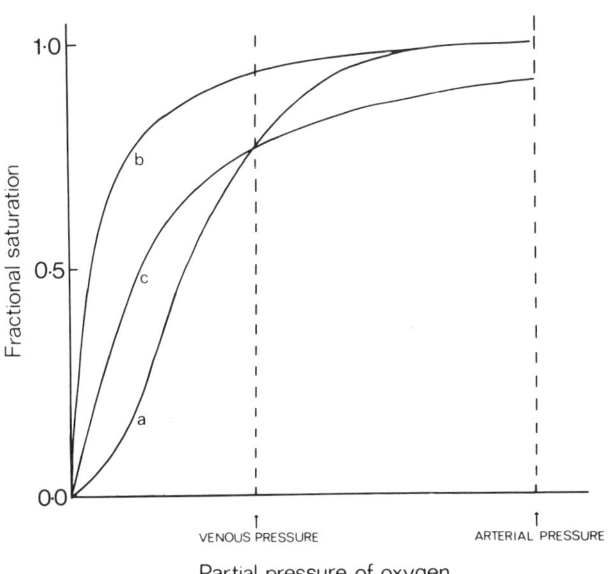

Fig. 6.5 Typical oxygen equilibrium curves for haemoglobin (curve a) and myoglobin (curve b); curve c indicates a myoglobin-type (i.e. a non-cooperative oxygenation curve) with the same percentage saturation as haemoglobin at venous oxygen partial pressure. (Adapted from Baldwin 1975.)

oxygen (the fractional saturation) is measured as a function of the partial pressure of oxygen. Typical oxygen equilibrium curves for haemoglobin and myoglobin are shown in Fig. 6.5. The curve for myoglobin is hyperbolic, as expected for the combination of one molecule of myoglobin with one molecule of oxygen, but the curve for haemoglobin has a more complex sigmoid shape. The physiological advantage of the sigmoid curve is that it allows haemoglobin to deliver more oxygen to the tissues. As oxygen partial pressure is reduced from arterial to venous levels, haemoglobin saturation is reduced from nearly full saturation to about three-quarters saturation, so that haemoglobin can deliver about a quarter of a mole of oxygen per mole of haem to the tissues, where it is taken up by myoglobin, which has a higher oxygen affinity. A hypothetical haem compound with the same fractional saturation as haemoglobin at venous pressure, but with a hyperbolic oxygen equilibrium curve (curve c in Fig. 6.5), would be so much less than fully saturated at arterial pressure that it could deliver only about one-eighth of a mole of O_2 per mole of haem, so that it would be only about one-half as efficient as haemoglobin in oxygen transport. A further advantage of the sigmoid curve is that its steepness at venous pressure enables the amount of oxygen delivered to the tissues to adjust sensitively to changes in oxygen partial pressure there.

The sigmoid shape of the oxygen equilibrium curve for haemoglobin indicates that the oxygen affinity of haemoglobin rises with uptake of oxygen. This phenomenon is known as co-operativity or haem-haem interaction. Since haemoglobin has four binding sites for oxygen, its equilibrium with oxygen is defined by four equations:

$$\text{Hb}(O_2)_{n-1} + O_2 \rightleftharpoons \text{Hb}(O_2)_n, \ (n = 1,2,3,4).$$

The four corresponding equilibrium constants, K_n, are known as the Adair constants. The sigmoid oxygen equilibrium curve indicates that K_1, the equilibrium constant for uptake of the first oxygen molecule, is less than K_4, the constant for uptake of the last oxygen molecule. This means that the ligation state of one haem in a molecule affects the affinity of the other haems in the molecule. The energy of haem-haem interaction is defined as $\Delta G_I = \text{RTln}(K_4/K_1)$ and is a measure of the free energy associated with the structural changes that increase the affinity of the haems when any one haem takes up oxygen. The value of ΔG_I is only about 3–4 kilocalories per mole of haem, which could correspond, say, to the rupture of only one or two hydrogen bonds per subunit.

In broadest outline, the structural basis of haem-haem interaction was already guessed by Perutz and associates in 1960 when they first obtained a low-resolution (5.5 Å) structure for methaemoglobin, even before the structure of the unliganded form was solved and at a time when the amino-acid sequence of haemoglobin was largely unknown. The low-resolution structure indicated the similarity of the haemoglobin subunits to myoglobin and showed that each iron atom is at least 25 Å from its nearest neighbour. The latter fact argued against a direct effect of ligation at one haem upon the affinity at another haem, since changes in structure at any haem would be expected to dissipate over the long distance to a neighbouring haem. It was known that in myoglobin there is no appreciable difference in structure between the liganded and unliganded forms, while for haemoglobin there was evidence that the still unknown structure of the deoxy form must differ rather grossly from that of the liganded form, since crystals of the two forms exhibit different lattice structures. From these facts it was hypothesized that the structure of deoxyhaemoglobin differs from that of the liganded form primarily in the spatial arrangement of its subunits, with relatively little difference in tertiary structure, and that the dependence of oxygen affinity upon degree of ligation is related to that change in quaternary structure.

In 1965, Monod, Wyman, and Changeux put these ideas on firm thermodynamic ground with their allosteric theory of co-operativity in proteins, also called the MWC, or two-state, model. As applied to haemoglobin, the basic features of their theory are as follows. The arrangement of subunits in fully liganded haemoglobin is known as the R, or relaxed, quaternary structure; that of unliganded (deoxy) haemoglobin is known as the T, or tense, quaternary structure. The reason for this terminology is that the subunits are joined more loosely in R than in T; for example, liganded haemoglobin dissociates into dimers more readily than does deoxyhaemoglobin. There is also evidence that the structure of individual subunits in the T state is distorted relative to that of R-state subunits or of monomers; for example, the oxygen affinity of subunits in the R structure (measured by the Adair constant K_4 is similar to that of α- or β-subunits, while the oxygen affinity of subunits in the T state (measured by the Adair constant K_1) is much less. The T structure is stabilized by closer packing of, and an increased number of bonds between, subunits. The R and T quaternary structures appear to be the main stable configurations of the subunits of haemoglobin, so that the arrangement of subunits in intermediate ligation states resembles one or the other of these two stable configurations. The two quaternary structures are in equilibrium in all ligation states, although the conformational equilibrium between R and T varies enormously with the number of haems bearing ligand. The ligand affinity of any particular haem depends primarily upon the quaternary structure of the molecule as a whole, and not (or only to a minor extent) on the ligation state of other haems in the molecule so long as the quaternary structure does not change. A change of quaternary structure from R to T reduces the ligand affinity of all four haems in the molecule by inducing tertiary structure changes in the subunits, including changes in the vicinity of the haems. It is a thermodynamic necessity that if a change from R to T results in lower oxygen affinity of the haems, then ligation of the haems must reduce the stability of the T state relative to that of R (Fig. 6.6). Thus, as oxygen partial pressure is increased in a solution of haemoglobin molecules, as more and more haems take up ligand, so more and more molecules take up the high-affinity R conformation, so that overall oxygen affinity increases. Thus the allosteric model accounts for haem-haem interaction without invoking the propagation of structural changes over long distances within the haemoglobin molecule.

The allosteric theory also accounts for the regulation of the oxygen affinity of haemoglobin by molecules that do not interact directly with the haem; some examples of these are given in the next two sections. Any molecule that binds preferentially to either the R or the T structure will alter the equilibrium between the high- and low-affinity states, thereby increasing or reducing the overall oxygen affinity of haemoglobin. Such molecules are called allosteric effectors.

The overall oxygen affinity of haemoglobin is often given in terms of p_{50}, the partial pressure of oxygen required to produce oxygenation of 50 per cent of the haems. (This is of course a reciprocal measure of affinity: oxygen affinity decreases as p_{50} increases.) The degree of co-operativity

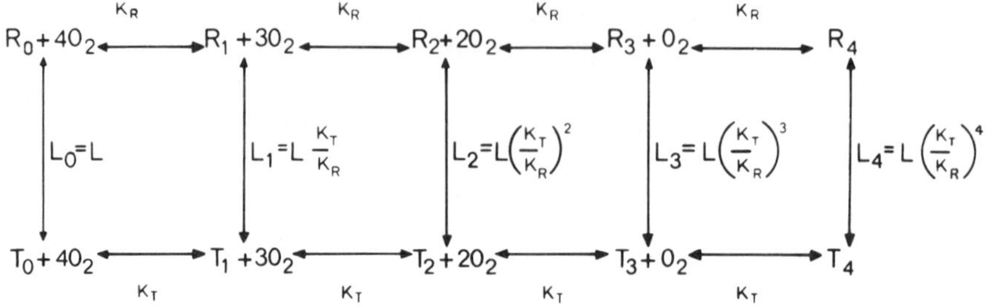

Fig. 6.6 Thermodynamics of the allosteric model of Monod *et al.* (1965). R_n and T_n represent haemoglobin in the R and T states with *n* molecules of O_2 bound. K_R and K_T are the oxygen association constants for the R and T states, respectively, and are assumed independent of ligation state. $L = [T_0]/[R_0]$ is the constant for the equilibrium between R and T in the absence of oxygen; the subscripted L_ns are the corresponding conformational equilibrium constants at the various degrees of ligation. The entropic factors, corresponding the number of different ways of attaining partial ligation states, are omitted. Within a wide range of typical conditions, L is so large and the ratio $c = K_T/K_R$ is so small that L_0, $L_1 \gg 1$ and L_3, $L_4 \ll 1$. It then follows that at low oxygen concentration most molecules (even among those with some oxygen bound) are in the T state, while at high oxygen concentration most molecules are in the R state; the Adair constant $K_1 \approx K_T$ and the Adair constant $K_4 \approx K_R$, so that the free energy of haem-haem interaction is approximately $\Delta G_I \approx RT\ln K_R - RT\ln K_T$ and hence represents the difference in free energy of binding oxygen to R-state and T-state molecules. (From Fermi and Perutz 1981.)

is often expressed in terms of Hill's coefficient n_{50}, which can be derived from the slope of the oxygen equilibrium curve at 50 per cent saturation. Theoretically, Hill's coefficient can vary between 1 (no co-operativity) and 4 (maximum co-operativity); for haemoglobin in physiological conditions n_{50} is about 3. Determination of the four Adair constants, or of the three constants of the MWC model, requires precise measurement of the oxygen equilibrium curve from zero to nearly total saturation, a procedure that is experimentally difficult (Imai 1982; Marden *et al.* 1989). Hence for some haemoglobins (e.g. some mutants) only p_{50} and n_{50} have been measured.

6.2.2.2 Carbon dioxide transport and Bohr effect   Haemoglobin has a second physiological function, which is to facilitate carbon dioxide transport from the tissues to the lungs. Haemoglobin carries out this function by two mechanisms: directly, by binding CO_2 at specific sites, and indirectly, by altering CO_2 solubility through an effect on the hydrogen ion concentration (pH) in the blood. Both mechanisms interact favourably with the role of haemoglobin as an oxygen carrier. They are examples of allosteric control as envisaged by the MWC theory.

Direct binding of CO_2 to haemoglobin accounts for only about 10 per cent of the physiological transport of CO_2. Carbon dioxide binds to both oxy- and deoxyhaemoglobin by interaction with amino groups in the protein to form carbamate groups:

$$Hb\text{-}NH_2 + CO_2 \rightleftharpoons Hb\text{-}NHCOOH$$

Deoxyhaemoglobin has a higher affinity for CO_2 than does the oxy form. Two sites are responsible for the differential CO_2 affinity. These are the N-terminal amino groups of both the α- and β-chains (so four sites per molecule), whose environments differ in the oxy- and deoxyhaemoglobin structures. A thermodynamic consequence of the higher affinity of deoxy-haemoglobin for CO_2 is that an increase in CO_2 concentration must cause a lowering of the oxygen affinity of haemoglobin. These two related effects increase the efficiency of both oxygen and CO_2 transport: the high concentration of CO_2 in the tissues as compared with the arteries favours the unloading of oxygen from oxyhaemoglobin to form deoxyhaemoglobin, and the latter then carries CO_2 away more efficiently toward the lungs.

The more important, indirect mechanism of CO_2 transport is a consequence of the Bohr effect, or pH dependence of haemoglobin oxygen affinity. In the physiological pH range the oxygen affinity is increased by increasing pH (lowering hydrogen-ion concentration) and, equivalently, the release of oxygen by oxyhaemoglobin tends to raise pH, i.e. the transition from oxy- to deoxyhaemoglobin is accompanied by the uptake of hydrogen ions by the protein. Most of the CO_2 in the tissues and the blood is in the form of bicarbonate ion,

$$CO_2 + H_2O = HCO_3^- + H^+$$

so that pH tends to be low where CO_2 concentration is high. As in the case of direct CO_2 binding, these effects couple to increase the efficiency of both oxygen and CO_2 transport by haemoglobin. The high CO_2 concentration in the tissues causes the pH there to be lower than in arterial blood; the lower pH favours the release of oxygen by oxyhaemoglobin as it passes from the arteries into the capillaries; and the transition from the oxy to the deoxy form raises the pH of venous blood so that it can carry more CO_2 in the form of bicarbonate ions.

The Bohr effect in the physiological pH range results from changes that occur in the ionizability (pK) of specific groups during oxygenation of haemoglobin. Two groups, accounting for most of the effect, have been identified: the imidazole group of the C-terminal histidine of the β-chain and (in the presence of chloride ion) the N-terminal amino group of the α-chain. In the T structure these groups form salt bridges (His HC3(146)β to Asp(94)FG1β, Val(1)NA1 to Arg(141)HC3α) that are absent in the R structure. The effect of oxygenation on the pK of these Bohr groups thus may be understood as resulting from the rupture of the salt bridges as the quaternary structure switches from T to R upon oxygenation. However, there is evidence that the Bohr effect must occur within the T state. The measured parameters of the MWC model indicate that at sufficiently low oxygen concentrations virtually all oxygen-bearing haemoglobin molecules are in the T state, i.e. the R state is favoured only among molecules bearing two or more oxygen molecules. Yet the release of hydrogen ions is linear with oxygen uptake even within the range of low oxygen concentrations where virtually all molecules remain in the T state, indicating that the pK of Bohr groups in T-state haemoglobin molecules must decrease upon oxygenation even when no quaternary structure change occurs. This in turn implies that the Bohr-group salt bridges should rupture upon oxygenation of T-state molecules. The crystal structures of analogues of liganded haemoglobin in the T state show no large perturbations of these salt bridges, but it has been argued that this might be due to constraints imposed by the crystal lattice.

In acid medium, below about pH 6.0, the Bohr effect is reversed (acid Bohr effect), i.e. oxygen affinity is lowered by increasing pH. The structural basis of the acid Bohr effect has been examined by Perutz *et al.* (1980).

6.2.2.3 Organic phosphate and other allosteric effectors The organic phosphate, 2,3-diphosphoglycerate (DPG), is a normal constituent of red cells that serves as an allosteric effector of haemoglobin oxygen affinity; oxygen affinity is lowered as DPG concentration is increased. When the level of DPG in the blood rises, more oxygen is delivered from the lungs to the tissues. This is because the reduction in oxygen affinity has a greater effect at venous pressure, where the oxygen equilibrium curve is steep (Fig. 6.6), than it does at arterial pressure, where the curve is flatter.

Both oxy- and deoxyhaemoglobin can bind DPG, but the deoxy form has a much higher affinity for the phosphate (as must be the case from the effect of DPG in reducing haemoglobin oxygen affinity). Under physiological conditions deoxyhaemoglobin can bind a single molecule of DPG; the binding site, shown in Plate 2, is between the two β-chains (Arnone 1972). The principal mechanism of action of DPG in lowering haemoglobin oxygen affinity is by inhibiting the transition from the low-affinity T quaternary structure to the high-affinity R structure: in the latter the gap between the β-chains is too small to accommodate DPG.

DPG plays a crucial role in the transfer of oxygen from mother to fetus during mammalian development. Human fetal haemoglobin (HbF) differs from that of adults (HbA) in that the β-chains are replaced by γ-chains. The amino-acid sequence of the γ-chains differs from that of the β-chains at 39 positions, but the three-dimensional structures of HbF and HbA are nearly identical (Frier and Perutz 1977). Under physiological conditions fetal haemoglobin has a higher affinity for oxygen than does the adult form, as would be expected from the need for the fetus to obtain its oxygen from the mother. In the absence of DPG, however, fetal haemoglobin actually has a lower oxygen affinity than adult. The high oxygen affinity *in vivo* of HbF results entirely from its having a low affinity for DPG, compared with HbA. This in turn results from amino-acid sequence differences between β- and γ-chains at sites involved in DPG binding.

Chloride is a fourth naturally occurring allosteric effector. It binds between the α-chains, whereby it alters the Bohr effect, and and also interacts with residues involved in DPG binding. Hence, under physiological conditions, there is a complex interaction between the allosteric effectors and oxygen in binding to haemoglobin. A brief discussion of such heterotropic effects is contained in the reviews by Kilmartin (1976, 1977).

Certain substances have been found that are more powerful allosteric effectors than those that occur naturally; some of these may eventually become useful medically. Inositol hexaphosphate (IHP) is a compound similar to DPG; it binds at the same site as DPG in the T quaternary structure (Arnone and Perutz 1974), but it has a greater effect on oxygen affinity because it binds more tightly. Recently, certain chlorobenzene compounds have been found to be potent allosteric effectors (Perutz *et al.* 1986; Lalezari *et al.* 1988, 1990). These reduce oxygen affinity by binding preferentially to the T structure. Their binding sites are in the central cavity near the centre of the molecule.

6.2.3 Overview of allostery in haemoglobin

The allosteric theory of Monod *et al.* (1965) is primarily a thermodynamic model and does not address the mechanism by which quaternary structure influences the oxygen affinity of the haems. The stereochemical basis of the allosteric model was first proposed by Perutz (1970) and extended by Baldwin and Chothia (1979). Their ideas, based mainly on the structures of oxy- and deoxyhaemoglobin, have largely been confirmed by X-ray crystallography of analogues of intermediate ligation states and by chemical, spectroscopic, and magnetic studies (see Perutz *et al.* 1987; Perutz 1989). Only a brief outline of the mechanism is given here, followed by mention of some empirical extensions and open questions related to it.

6.2.3.1 Outline of the stereochemical model

6.2.3.1 Outline of the stereochemical model The regulation of oxygen affinity at the α-haem is illustrated schematically in Fig. 6.7. The figure indicates that the geometry of the haem and the iron-linked proximal histidine F8 depends both on the ligation state of that haem and on the quaternary structure of the whole haemoglobin molecule. The tertiary structure of other residues surrounding the haem is determined mainly by the constraints imposed by the quaternary structure; this results from the proximity of some residues, particularly those of the FG corner, to both the haem

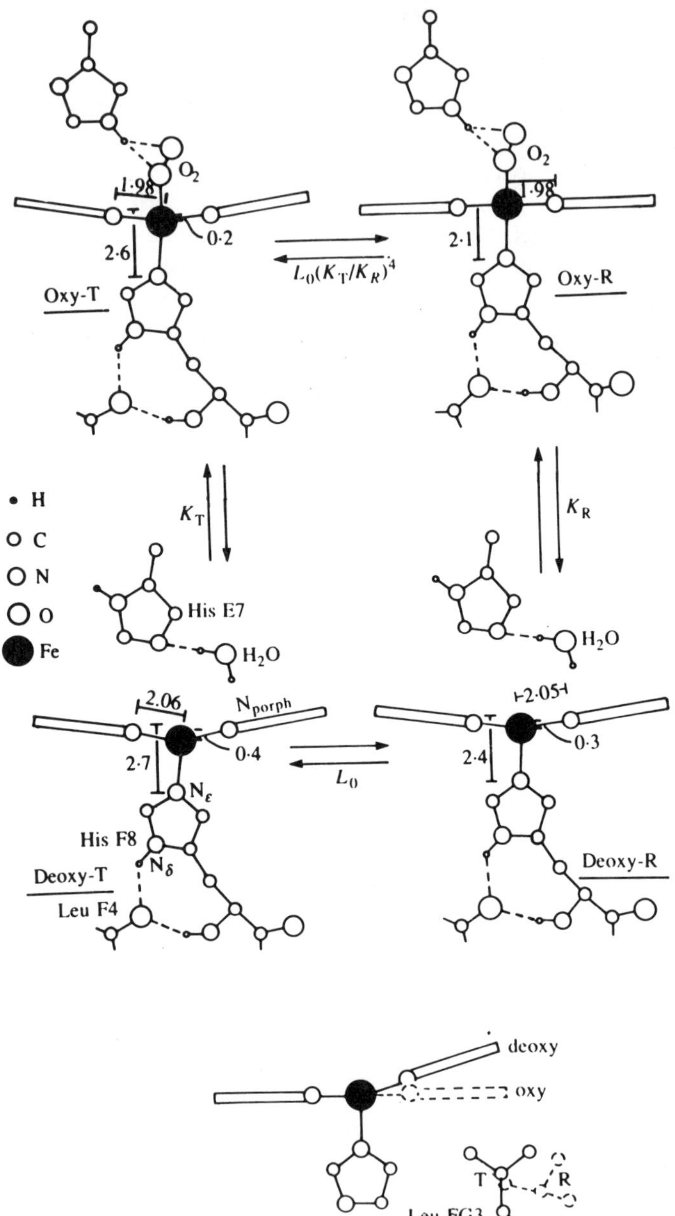

Fig. 6.7 Regulation of the oxygen affinity of the α-haem in intermediate and end-state haemoglobin. The numbers give distances in Å: the distance between Nϵ of the proximal histidine F8 and the mean plane of the prophyrin (vertical bars), the mean bond length between iron and haem nitrogens (horizontal bars), and the displacement

and to the interface between the $\alpha_1\beta_1$ and $\alpha_2\beta_2$ dimers. As is suggested by the lower part of the diagram, residues surrounding the haem tightly constrain relative movements of the haem and His F8 in the T structure but leave the complex relatively unhindered in the R structure. Thus in the R structure the haem-F8 complex can take up the geometry preferred by either the oxy or deoxy state, and the R-state oxygen affinity is similar to that of myoglobin or of haemoglobin monomers. In the T quaternary structure, the constraints of surrounding residues force the haem-F8 complex to maintain a geometry far from that preferred by the oxy form of the complex, so that oxygen affinity is low.

Similar principles apply also to the β-haem, but there the situation is more complicated. The oxygen affinity of the β-haem is regulated also by steric hindrance of the O_2-binding site by a residue on the distal side of the haem (the side opposite to His F8), namely valine E11. The position of this residue relative to the haem depends about equally on quaternary structure and ligation state (Luisi *et al.* 1990).

Nagai and associates (Nagai *et al.* 1987; Olson *et al.* 1988; Mathews *et al.* 1989; Tame 1990) have examined the roles of Val E11 and of the distal histidine, E7, in the α- and β-subunits of haemoglobin and in myoglobin by engineering various amino-acid substitutions at these sites. They confirmed that Val E11β, but not Val E11α, makes a significant contribution to the mechanism of co-operativity through steric hindrance of the binding site. The distal histidine appears to play an important role in discriminating between O_2 and CO_2 in myoglobin and in haemoglobin α-subunits, but not in β-subunits.

The validity of the hypothesis that there are only two stable quaternary structures of haemoglobin rests experimentally on the fact that all mammalian haemoglobins whose structure has been determined have either the R or T quaternary structure. This includes the analogues of intermediate ligation states, and is not an artefact of crystal packing forces, as the structures include at least five different crystallographic space groups and are similar in media of high and low ionic strength. The reason that quaternary structure cannot vary continuously between R and T is that one part of the interface between $\alpha_1\beta_1$ and $\alpha_2\beta_2$, illustrated in Fig. 6.8, acts as a two-way switch. The shape of the subunit surfaces and the placement of polar residues in this region preclude close packing and hydrogen-bond formation in quaternary arrangements intermediate between R and T.

---

of the iron from the mean plane of haem nitrogens (figures to right of iron). The equilibrium constants are defined in Fig. 6.6. The lower part of the diagram schematically indicates the coupling between haem conformation and the positions of residues surrounding the haem; the latter are determined largely by the quaternary structure. (From Perutz *et al.* 1987.)

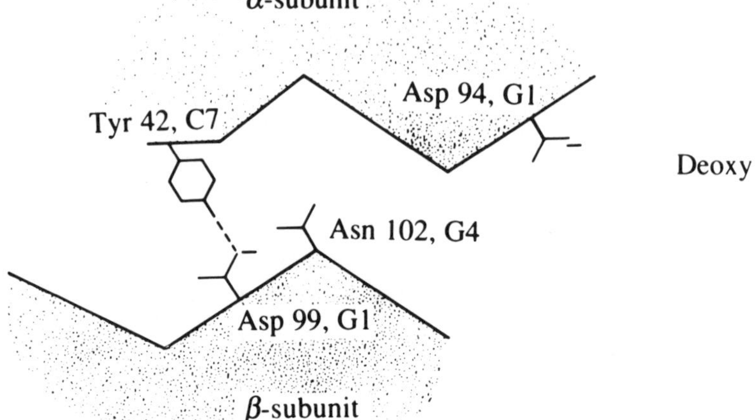

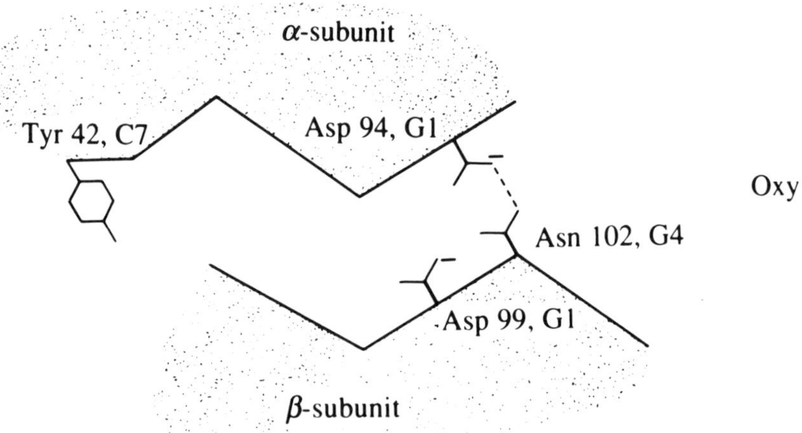

Fig. 6.8 The $\alpha_1\beta_2$ contact as a two-way switch, showing the alternative hydrogen bonds and packing arrangements in the T and R quaternary structures (from Perutz 1989).

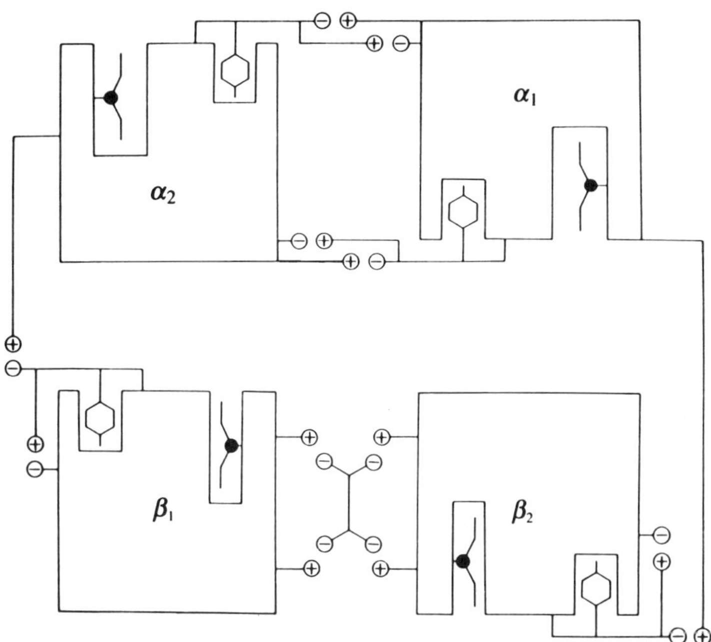

Fig. 6.9 Diagrammatic representation of the salt bridges in the T structures. Those at the top link the C-Terminal Arg HC3(141)α_2 to Asp H9(126)α_1 and Lys H10(127)α_1; the others link the C-terminal His HC3(146)β_1 to Asp FG1(94)β_1 and Lys C5(40)α_2. The bridge between the β-subunits represents DPG. (From Perutz 1970.)

Finally, what stabilizes the T quaternary structure relative to R in the absence of oxygen? Perutz (1970) proposed that the stabilization is due mainly to salt bridges that are present in T but absent in R. These include salt bridges both between and within the subunits (Fig. 6.9). Strong evidence for the importance of these salt bridges comes from mutant haemoglobins. Several of these have substitutions that preclude formation of one or another of the salt bridges, and all of these exhibit reduced stability of T relative to R (Fermi and Perutz 1981). Chothia and others (Chothia *et al.* 1976; Lesk *et al.* 1985) have proposed that hydrophobic forces also play a role, since the amount of subunit surface made inaccessible to water in close-packed subunit interfaces is greater in T than it is in R. It is not possible at present to evaluate quantitatively the relative contributions of salt bridges and hydrophobic or other forces to the free energy of T-state stabilization.

Table 6.2 Amino-acid sequence of human haemoglobin and sperm whole myoglobin*

Helix notation	Haemoglobin		Myoglobin
	alpha	beta	
NA1	1 Val* (a)	1 Val	1 Val
NA2	2 Leu* (a)	2 His	2 Leu
NA3	–	3 Leu	–
A1	3 Ser	4 Thr	3 Ser
A2	4 Pro	5 Pro	4 Glu
A3	5 Ala	6 Glu	5 Gly
A4	6 Asp	7 Glu	6 Glu
A5	7 Lys	8 Lys	7 Trp
A6	8 Thr	9 Ser	8 Gln
A7	9 Asn	10 Ala	9 Leu
A8	10 Val	11 Val	10 Val
A9	11 Lys	12 Thr	11 Leu
A10	12 Ala	13 Ala	12 His
A11	13 Ala	14 Leu	13 Val
A12	14 Trp	15 Trp	14 Trp
A13	15 Gly	16 Gly	15 Ala
A14	16 Lys	17 Lys	16 Lys
A15	17 Val	18 Val	17 Val
A16	18 Gly	–	18 Glu
AB1	19 Ala	–	19 Ala
B1	20 His	19 Asn	20 Asp
B2	21 Val	20 Val	21 Val
B3	22 Gly	21 Asp	22 Ala
B4	23 Glu	22 Glu	23 Gly
B5	24 Tyr	23 Val	24 His
B6	25 Gly	24 Gly	25 Gly
B7	26 Ala	25 Gly	26 Gln
B8	27 Glu	26 Glu (1)	27 Asp
B9	28 Ala	27 Ala	28 Ile
B10	29 Leu	28 Leu	29 Leu
B11	30 Glu (1)	29 Gly	30 Ile
B12	31 Arg 1	30 Arg 1	31 Arg
B13	32 Met (h)	31 Leu (h)	32 Leu
B14	33 Phe	32 Leu	33 Phe
B15	34 Leu 1	33 Val (1)	34 Lys
B16	35 Ser 1	34 Val (1)(2)	35 Ser
C1	36 Phe (1)	35 Tyr 1(2)	36 His
C2	37 Pro (2)	36 Pro 2	37 Pro
C3	38 Thr (2)	37 Trp 2	38 Glu
C4	39 Thr (h)	38 Thr (h)	39 Thr h
C5	40 Lys (2)	39 Gln (2)	40 Leu
C6	41 Thr 2	40 Arg 2	41 Glu
C7	42 Tyr h2	41 Phe h	42 Lys h
CD1	43 Phe h	42 Phe h	43 Phe h
CD2	44 Pro (2)	43 Glu (2)	44 Asp
CD3	45 His h	44 Ser	45 Arg h

Table 6.2 *Continued*

Helix notation	Haemoglobin		Myoglobin
	alpha	beta	
CD4	46 Phe (h)	45 Phe	46 Phe
CD5	47 Asp	46 Gly	47 Lys
CD6	–	47 Asp	48 His
CD7	48 Leu	48 Leu	49 Leu
CD8	49 Ser	49 Ser	50 Lys
D1	–	50 Thr	51 Thr
D2	–	51 Pro	52 Glu
D3	–	52 Asp	53 Ala
D4	–	53 Ala	54 Glu
D5	–	54 Val	55 Met
D6	50 His	55 Met (1)	56 Lys
D7	51 Gly	56 Gly	57 Ala
E1	52 Ser	57 Asn	58 Ser
E2	53 Ala	58 Pro	59 Glu
E3	54 Gln	59 Lys	60 Asp
E4	55 Val	60 Val	61 Leu
E5	56 Lys	61 Lys	62 Lys
E6	57 Gly	62 Ala	63 Lys
E7	58 His h	63 His h	64 His h
E8	59 Gly	64 Gly	65 Gly
E9	60 Lys	65 Lys	66 Val
E10	61 Lys h	66 Lys h	67 Thr (h)
E11	62 Val (h)	67 Val h	68 Val h
E12	63 Ala	68 Leu	69 Leu
E13	64 Asp	69 Gly	70 Thr
E14	65 Ala (h)	70 Ala (h)	71 Ala (h)
E15	66 Leu (h)	71 Phe (h)	72 Leu (h)
E16	67 Thr	72 Ser	73 Gly
E17	68 Asn	73 Asp	74 Ala
E18	69 Ala	74 Gly	75 Ile
E19	70 Val	75 Leu	76 Leu
E20	71 Ala	76 Ala	77 Lys
EF1	72 His	77 His	78 Lys
EF2	73 Val	78 Leu	79 Lys
EF3	74 Asp	79 Asp	80 Gly
EF4	75 Asp	80 Asn	81 His
EF5	76 Met	81 Leu	82 His
EF6	77 Pro	82 Lys	83 Glu
EF7	78 Asn	83 Gly	84 Ala
EF8	79 Ala	84 Thr	85 Glu
F1	80 Leu	85 Phe (h)	86 Leu
F2	81 Ser	86 Ala	87 Lys
F3	82 Ala	87 Thr	88 Pro
F4	83 Leu h	88 Leu (h)	89 Leu (h)
F5	84 Ser	89 Ser	90 Ala
F6	85 Asp	90 Glu	91 Gln
F7	86 Leu h	91 Leu (h)	92 Ser h

Table 6.2 *Continued*

Helix notation	Haemoglobin		Myoglobin
	alpha	beta	
F8	87 His h	92 His h	93 His h
F9	88 Ala	93 Cys	94 Ala
FG1	89 His	94 Asp	95 Thr*
FG2	90 Lys	95 Lys	96 Lys*
FG3	91 Leu h(2)	96 Leu h	97 His* h
FG4	92 Arg 2	97 His 2	98 Lys*
FG5	93 Val h(2)	98 Val (h)(2)	99 Ile* h
G1	94 Asp 2	99 Asp 2	100 Pro
G2	95 Pro (2)	100 Pro (2)	101 Ile
G3	96 Val 2	101 Glu (1)(2)	102 Lys
G4	97 Asn h(2)	102 Asn h(2)	103 Tyr h
G5	98 Phe h	103 Phe h	104 Leu h
G6	99 Lys (1)	104 Arg	105 Glu
G7	100 Leu	105 Leu 2	106 Phe
G8	101 Leu h	106 Leu h	107 Ile (h)
G9	102 Ser	107 Gly	108 Ser
G10	103 His 1	108 Asn 1	109 Glu
G11	104 Cys	109 Val	110 Ala
G12	105 Leu	110 Leu	111 Ile
G13	106 Leu	111 Val (1)	112 Ile
G14	107 Val 1	112 Cys (1)	113 His
G15	108 Thr	113 Val	114 Val
G16	109 Leu	114 Leu	115 Leu
G17	110 Ala 1	115 Ala 1	116 His
G18	111 Ala 1	116 His 1	117 Ser
G19	112 His (1)	117 His	118 Arg
GH1	113 Leu	118 Phe	119 His
GH2	114 Pro 1	119 Gly 1	120 Pro
GH3	115 Ala	120 Lys (1)	121 Gly
GH4	116 Glu	121 Glu	122 Asp
GH5	117 Phe 1	122 Phe 1	123 Phe
H1	118 Thr (1)	123 Thr 1	124 Gly
H2	119 Pro 1	124 Pro 1	125 Ala
H3	120 Ala	125 Pro (1)	126 Asp
H4	121 Val	126 Val	127 Ala
H5	122 His 1	127 Gln 1	128 Gln
H6	123 Ala (1)	128 Ala 1	129 Gly
H7	124 Ser	129 Ala	130 Ala
H8	125 Leu	130 Tyr	131 Met
H9	126 Asp (a) 1	131 Gln 1	132 Asn
H10	127 Lys a	132 Lys	133 Lys
H11	128 Phe	133 Val	134 Ala
H12	129 Leu	134 Val	135 Leu
H13	130 Ala	135 Ala	136 Glu
H14	131 Ser	136 Gly	137 Leu
H15	132 Val (h)	137 Val	138 Phe (h)
H16	133 Ser	138 Ala	139 Arg

Table 6.2 *Continued*

Helix notation	Haemoglobin		Myoglobin
	alpha	beta	
H17	134 Thr (a)	139 Asn (b)	140 Lys
H18	135 Val	140 Ala	141 Asp
H19	136 Leu h	141 Leu h	141 Ile
H20	137 Thr	142 Ala	143 Ala
H21	138 Ser (a)	143 His	144 Ala
H22	–	–	145 Lys
H23	–	–	146 Tyr
H24	–	–	147 Lys
H25	–	–	148 Glu
H26	–	–	149 Leu
HC1	139 Lys (a)	144 Lys	150 Gly
HC2	140 Tyr 2(a)	145 Tyr (2)	151 Tyr
HC3	141 Arg a(2)	146 His (2)(b)	152 Gln
HC4	–	–	153 Gly

The symbols h,1,2,a,b indicate, respectively, residues within the haem-globin, $\alpha_1\beta_1$, $\alpha_1\beta_2$, $\alpha_1\alpha_2$, $\beta_1\beta_2$ interfaces. Given r is the minimum interatomic distance across the contact for a given residue in a given form (i.e. oxy or deoxy), the residue is counted within the interface if $r < 4.2$ Å for either form; the corresponding symbol is in paraentheses if $r > 3.8$ Å for either form. Interatomic distances are calculated from the corrdinates with Protein Data Bank designations 1MBO,1MBD,2HHB,1HHO (see Table 6.1).

* Changes made in the helix notation from that given by Perutz (1969): haemoglobin residues 1 and 2; previously designated NA2 and NA3; myoglobin residue 95 previously designated F10; myoglobin residues 96 through 99 previously designated FG1 through FG4.

6.2.3.2 Extensions and limitations   The MWC model contains three parameters that, in principle, can vary independently under changing conditions of the medium or through structural modifications of the globin (see Fig. 6.9). Imai (1983) has shown empirically that over a wide range of concentrations of the naturally occurring allosteric effectors (at neutral and alkaline pH), only one parameter varies independently: in this range K_R and the quantity $L_4 = L(K_T/K_R)^4$ are constant, while K_T varies by an order of magnitude. No satisfying explanation of this tantalizingly simple relationship has yet come forth. At acid pH, K_R is no longer constant; it is a function of pH and of the concentration of such effectors as DPG and IHP and the chlorobenzene compounds (Imai 1983; Lalezari *et al.* 1990). Attempts have been made to explain these results with thermodynamic models with more than two affinity states (Imai 1982; Kister *et al.* 1987), and to demonstrate the existence of a third, distinct structural state of haemoglobin (Smith and Ackers 1985). These matters are still controversial (Ferrone 1986; Marden *et al.* 1989). Allosteric models with more than two affinity states have been of limited utility, because they lack predictive power and because there is no known structural basis for the additional states.

Among the naturally occurring abnormal haemoglobins that have been studied, the majority of those with altered oxygen affinities are mutants with single-site replacements of residues distant from the haem (Fermi and Perutz 1981). The effect of one class of these mutations can be understood as an effect on the equilibrium between the R and T states; they disrupt bonds that stabilize specifically either the R or T quaternary structures. Another large class of these mutations raise oxygen affinity even though they occur in regions of the molecule, such as the $\alpha_1\beta_1$ contact, that would be expected to affect the R and T structures equally, yet are too far from the haem for direct effects on the haem to be likely. Mutations in either of these two classes usually affect K_T and L in opposite directions and have little effect on K_R. Thus the reciprocal relationship between L and K_T and relative constancy of K_R that is observed for the allosteric effector compounds appears to hold qualitatively also for structural alterations caused by mutation. From these observations, Perutz (Fermi and Perutz 1981; Perutz 1989) formulated the semi-empirical rule that any structural change that relaxes T structure constraints will raise K_T and lower L, even if R structure constraints are also relaxed. This rule can be understood in a general way from the structural mechanism outlined in the preceding section: since the low oxygen affinity of the T structure results from constraints on the movement of the haem-F8 complex, relaxation of the T structure may be expected to reduce these constraints so that K_T is raised; since the haem-F8 complex is relatively unconstrained in the R structure, further relaxation will have little effect on K_R. A reciprocal linkage between L and K_T may be expected, since structural changes (including the binding of effectors) that loosen bonds between subunits, thereby reducing L, may be expected to lessen the quaternary constraints on T-state tertiary structure, and thereby reduce constraints on the haem-F8 complex, so increasing K_T. However, there is no detailed understanding of this mechanism. In particular, crystal structures of deoxy-haemoglobins with altered oxygen affinity, due either to binding of effectors or to mutations that alter subunit contacts, generally show no structural changes in the immediate vicinity of the haems (Fermi and Perutz 1981; Perutz *et al.* 1986; Lalezari *et al.* 1988, 1990). Such changes may occur in the ligated T-state forms of these haemoglobins, but the appropriate structures are not available.

If Perutz's rule is quite general, then co-operativity should occur within the T quaternary structure, since oxygenation is itself a structural change that destabilizes the T structure. Perutz (1989) has suggested that there may be a degree of co-operativity within the T structure resulting from a sequential mechanism (Koshland *et al.* 1966) mediated through rupture of the Bohr-group salt bridges (see Section 6.2.2.2). Co-operativity within the

T state has not been observed in haemoglobins that remain in the T state upon oxygenation. Two well-studied examples are fish haemoglobins (Tan and Noble 1973; Perutz and Brunori 1982) and haemoglobin Kansas (Anderson 1975; Kilmartin *et al.* 1975, 1978). In the presence of IHP at low pH many fish haemoglobins bind oxygen non-cooperatively and remain in the T state when oxygenated. This is presumed to result from a strengthening of the salt bridges that stabilize the T structure relative to R under these conditions. Since the strengthening of the salt bridges would also reduce any co-operativity within the T structure that they may mediate, these observations are not inconsistent with Perutz's ideas. Haemoglobin Kansas (Asn $(102)G4\alpha \rightarrow$ Thr), at high haemoglobin concentration in the presence of IHP, binds oxygen non-cooperatively with low affinity corresponding to K_T. This behaviour is presumed to result from its remaining in the T state upon oxygenation because the mutation precludes formation of a hydrogen bond that is formed only in the R state (Fig. 6.8), thereby stabilizing R relative to T. Lack of co-operativity in haemoglobin Kansas under these conditions is not readily understood within the framework of Perutz's ideas, although it could be argued, as in the case of fish haemoglobin above, that IHP constrains the T structure of Kansas in such a way as to prevent sequential co-operativity.

These semi-empirical extensions of the allostery of haemoglobin illustrate the limits of our understanding, since their stereochemical basis is not known in detail, but they greatly extend the predictive power of the original theories as set out by Monod *et al.* (1965) and Perutz (1970).

References

Anderson, L. (1973). Intermediate structure of normal human haemoglobin: methaemoglobin in the deoxy quaternary conformation. *Journal of Molecular Biology* **79**, 495–506.

Anderson, L. (1975). Structures of deoxy and carbonmonoxy haemoglobin Kansas in the deoxy quaternary conformation. *Journal of Molecular Biology* **94**, 33–49.

Arnone, A. (1972). X-ray diffraction study of binding of 2,3-diphosphoglycerate to human deoxyhaemoglobin. *Nature* **237**, 146–9.

Arnone, A. and Perutz, M.F. (1974). Structure of inositol hexaphosphate–human deoxyhaemoglobin complex. *Nature* **249**, 34–6.

Arnone, A., Rogers, P., Blough, N.V., McGourty, J.L., and Hoffman, B.M. (1986). X-ray diffraction studies of a partially liganded hemoglobin, α(FeII-CO)β(MnII)]$_2$. *Journal of Molecular Biology* **188**, 693–706.

Baldwin, J.M. (1975). Structure and function of haemoglobin. *Progress in Biophysics and Molecular Biology* **29**, 225–320.

Baldwin, J.M. (1980). The structure of human carbonmonoxy haemoglobin at 2.7 Å resolution. *Journal of Molecular Biology* **136**, 103–28.

Baldwin, J. and Chothia, C. (1979). Haemoglobin: the structural changes related to ligand binding and its allosteric mechanism. *Journal of Molecular Biology* **129**, 175–220.

Bolton, W. and Perutz, M.F. (1970). Three dimensional fourier synthesis of horse deoxyhaemoglobin at 2.8 Å resolution. *Nature* **228**, 551–2.

Bolton, W., Cox, J.M., and Perutz, M.F. (1968). Structure and function of haemoglobin. IV. A three-dimensional fourier synthesis of horse deoxyhaemoglobin at 5.5 Å resolution. *Journal of Molecular Biology* **33**, 283–97.

Brzosowski, A., *et al.* (1984). Bonding of molecular oxygen to T state human haemoglobin. *Nature* **307**, 74–6.

Chothia, C., Wodak, S., and Janin, J. (1976). Role of subunit interfaces in the allosteric mechanism of hemoglobin. *Proceedings of the National Academy of Sciences, USA* **73**, 3793–7.

Fermi, G. (1975). Three-dimensional fourier synthesis of human deoxyhaemoglobin at 2.5 Å resolution: refinement of the atomic model. *Journal of Molecular Biology* **97**, 237–56.

Fermi, G. and Perutz, M.F. (1977). Structure of human fluoromethaemoglobin with inositol hexaphosphate. *Journal of Molecular Biology* **114**, 421–31.

Fermi, G. and Perutz, M.F. (1981). Haemogobin and Myoglobin. In *Atlas of biological structures*, (ed. D.C. Phillips and F.M. Richards). Clarendon Press, Oxford.

Fermi, G., Perutz, M.F., Shaanan, B., and Fourme, R. (1984). The crystal structure of human deoxyhaemoglobin at 1.74 Å resolution. *Journal of Molecular Biology* **175**, 159–74.

Ferrone, F.A. (1986). Allosteric interpretation of cooperative free energy in cyanomethemoglobin. *Proceedings of the National Academy of Sciences, USA* **83**, 6412–14.

Frier, J.A. and Perutz, M.F. (1977). Structure of human foetal deoxyhaemoglobin. *Journal of Molecular Biology* **112**, 97–112.

Heidner, E.J., Ladner, R.C., and Perutz, M.F. (1976). Structure of horse carbonmonoxyhaemoglobin. *Journal of Molecular Biology* **104**, 707–22.

Imai, K. (1982). *Allosteric effects in haemoglobin*. Cambridge University Press.

Imai, K. (1983). The Monod-Wyman-Changeux allosteric model describes haemoglobin oxygenation with only one adjustable parameter. *Journal of Molecular Biology* **167**, 741–9.

Kendrew, J.C., Bodo, G., Dintzis, H.M., Parrish, R.G., Wycoff, H., and Phillips, D.C. (1958). A three-dimensional model of the myoglobin molecule obtained by X-ray analysis. *Nature* **181**, 662–6.

Kendrew, J.C., *et al.* (1960). Structure of myoglobin: a three-dimensional fourier synthesis at 2 Å resolution. *Nature* **185**, 422-7.

Kilmartin, J.V. (1976). Interaction of hemoglobin with protons, CO_2, and 2,3-diphosphoglycerate. *British Medical Bulletin* **32**, 209-12.

Kilmartin, J.V. (1977). The Bohr effect of human hemoglobin. *Trends in Biochemical Sciences*, 237-9 (Nov.).

Kilmartin, J.V., Imai, K. and Jones, R.T. (1975). Functional role of the salt bridges in hemoglobin. In *Erythrocyte structure and function*, pp 21-35. Alan R. Liss, New York.

Kilmartin, J.V., Anderson, N.L., and Ogawa, N. (1978). Response of the Bohr groups salt bridges to ligation of the T state of haemoglobin Kansas. *Journal of Molecular Biology* **123**, 71-87.

Kister, J., Poyart, C., and Edelstein, S.J. (1987). An expanded two-state allosteric model for interactions of human hemoglobin A with nonsaturating concentrations of 2,3-diphosphoglycerate. *Journal of Biological Chemistry* **262**, 12085-91.

Koshland, D.E., Nemethy, G., and Filmer, D. (1966). Comparison of experimental binding data and theoretical models in proteins containing subunits. *Biochemistry* **5**, 365-85.

Ladner, R.C., Heidner, E.J., and Perutz, M.F. (1977). The structure of horse methaemoglobin at 2.0 Å resolution. *Journal of Molecular Biology* **114**, 385-414.

Lalezari, I., Rahbar, S., Lalezari, P., Fermi, G., and Perutz, M.F. (1988). LR16, a compound with potent effects on the oxygen affinity of hemoglobin, on blood cholesterol, and on low density lipoprotein. *Proceedings of the National Academy of Sciences, USA* **85**, 6117-21.

Lalezari, I., Lalezari, P., Poyart, C., Marden, M., Bohn, B., Fermi, G., and Perutz, M.F. (1990). New effectors of human hemoglobin: structure and function. *Biochemistry* **29**, 1515-23.

Lemberg, R. and Legge, J.W. (1949). *Hematin compounds and bile pigments*. Interscience, New York.

Lesk, A.M. and Chothia, C. (1980). How different amino acid sequences determine similar protein structures: the structure and evolutionary dynamics of the globins. *Journal of Molecular Biology* **136**, 225-70.

Lesk, A.M., Janin, J., Wodak, S., and Chothia, C. (1985). Haemoglobin: the surface buried between the $\alpha_1\beta_1$ and $\alpha_2\beta_2$ dimers in the deoxy and oxy structures. *Journal of Molecular Biology* **183**, 267-70.

Liddington, R., Derewenda, Z., Dodson, G., and Harris, D. (1988). Structure of the liganded T state of haemoglobin identifies the origin of cooperative oxygen binding. *Nature* **331**, 725-8.

Luisi, B.F. (1986). Crystallographic studies of intermediate states of haemoglobin. Unpublished Doctoral thesis, Cambridge University.

Luisi, B., Liddington, R., Fermi, G., and Shibayama, N. (1990). Structure of deoxy-quaternary haemoglobin with liganded beta subunits. *Journal of Molecular Biology* **214**, 7–14.

Marden, M.C., Kister, J., Poyart, C., and Edelstein, S.J. (1989). Analysis of hemoglobin oxygen equilibrium curves: are unique solutions possible? *Journal of Molecular Biology* **208**, 341–5.

Mathews, A.J., Rohlfs, R.J., Olson, J.S., Tame, J., Renaud, J.-P., and Nagai, K. (1989). The effects of E7 and E11 mutations on the kinetics of ligand binding to R state human hemoglobin. *Journal of Biological Chemistry* **264**, 16573–83.

Monod, J., Wyman, J., and Changeux, J.P. (1965). On the nature of allosteric transitions: a plausible model. *Journal of Molecular Biology* **12**, 88–118.

Muirhead, H. and Greer, J. (1970). Three-dimensional fourier synthesis of human deoxyhaemoglobin at 3.5 Å resolution. *Nature* **228**, 516–19.

Muirhead, H., Cox, J.M., Mazzarella, L. and Perutz, M.F. (1967). Structure and function of haemoglobin: III. A three-dimensional fourier synthesis of human deoxyhaemoglobin at 5.5 Å resolution. *Journal of Molecular Biology* **28**, 117–50.

Nagai, K., *et al.* (1987). Distal residues in the oxygen binding site of haemoglobin studied by protein engineering. *Nature* **329**, 858–60.

Nobbs, C.L., Watson, H.C., and Kendrew, J.C. (1966). Structure of deoxymyoglobin: a crystallographic study. *Nature* **209**, 339–41.

Olson, J.S., *et al.* (1988). The role of the distal histidine in myoglobin and haemoglobin. *Nature* **336**, 265–6.

Perutz, M.F. (1969). The Croonian lecture 1968: The haemoglobin molecule. *Proceedings of the Royal Society of Biochemistry* **173**, 113–40.

Perutz, M.F. (1970). Stereochemistry of cooperative effects in haemoglobin. *Nature* **228**, 726–39.

Perutz, M.F. (1989). Mechanisms of cooperativity and allosteric regulation in proteins. *Quarterly Reviews of Biophysics* **22**, (2), 139–236.

Perutz, M.F. and Brunori, M. (1982). Stereochemistry of cooperative effects in fish and amphibian haemoglobins. *Nature* **299**, 421–6.

Perutz, M.F., Rossman, M.G., Cullis, A.F., Muirhead, H., Will, G., and North, A.C.T. (1960). Structure of haemoglobin: a three-dimensional fourier synthesis at 5.5 Å resolution obtained by X-ray analysis. *Nature* **185**, 416–22.

Perutz, M.F., *et al.* (1968a). Three-dimensional fourier synthesis of horse oxyhaemoglobin at 2.8 Å resolution: (I) X-ray analysis. *Nature* **219**, 29–32.

Perutz, M.F., Muirhead, H., Cox, J.M., and Goaman, L.C.G. (1968b). Three-dimensional fourier synthesis of horse oxyhaemoglobin at 2.8 Å resolution: the atomic model. *Nature* **219**, 131–9.

Perutz, M.F., Kilmartin, J.V., Nishikura, K., Fogg, J.H., Butler, P.J.G., and Rollema, H.S. (1980). Identification of residues contributing to the Bohr effect of human haemoglobin. *Journal of Molecular Biology* **138**, 649–70.

Perutz, M.F., Fermi, G., Abraham, D.J., Poyart, C., and Bursaux, E. (1986). Hemoglobin as a receptor of drugs and peptides: X-ray studies of the stereochemistry of binding. *Journal of the American Chemical Society* **108**, 1064–78.

Perutz, M.F., Fermi, G., Luisi, B., Shaanan, B., and Liddington, R.C. (1987). Stereochemistry of cooperative mechanisms in hemoglobin. *Accounts of Chemical Research* **20**, 309–21.

Phillips, S.E.V. (1980). Structure and refinement of oxymyoglobin at 1.6 Å resolution. *Journal of Molecular Biology* **142**, 531–54.

Shaanan, B. (1983). Structure of human oxyhaemoglobin at 2.1 Å resolution. *Journal of Molecular Biology* **171**, 31–59.

Smith, F.R. and Ackers, G.K. (1985). Experimental resolution of cooperative free energies for the ten ligation states of human hemoglobin. *Proceedings of the National Academy of Sciences, USA* **82**, 5347–51.

Takano, T. (1977a). Structure of myoglobin refined at 2.0 Å resolution. I. Crystallographic refinement of metmyoglobin from sperm whale. *Journal of Molecular Biology* **110**, 537–68.

Takano, T. (1977b). Structure of myoglobin refined at 2.0 Å resolution. II. Structure of deoxymyoglobin from sperm whale. *Journal of Molecular Biology* **110**, 569–84.

Tame, J. (1990). The role of the distal residues in haemoglobin, in Protein structural analysis, folding and design. Japan Scientific Societies Press, Tokyo.

Tan, A.L. and Noble, R.W. (1973). Conditions restricting allosteric transitions in Carp hemoglobin. *Journal of Biological Chemistry* **248**, 2880–8.

Ten Eyck, L.F. and Arnone, A. (1976). Three-dimensional fourier synthesis of human deoxyhaemoglobin at 2.5 Å resolution: I. X-ray analysis. *Journal of Molecular Biology* **100**, 3–11.

7

Electron transporters

F. Scott Mathews

7.1 Introduction

Electron transporters can be defined for our present purposes as low molecular weight (8–15 kDa) protein molecules that contain a redox active cofactor and have no catalytic function other than as an electron sink. There are four types of cofactors utilized by this protein class: haem, iron-sulphur centres, flavin mononucleotide (FMN), and copper. The cofactor, when bound to the protein, can normally exist in only two stable oxidation states. The protein environment can mediate the midpoint oxidation/reduction (redox) potential and establish the equilibrium level of the two states. In some cases (e.g. FMN or [4Fe–4S]) the cofactor is capable of existing in more than two redox states and the protein surroundings will regulate which pair of redox potentials are available under physiological conditions for the electron transfer process.

The primary function of the electron transporters is to transfer a single electron from one protein or protein complex to another. The electron donor protein is often a soluble protein while the acceptor is usually membrane bound. The acceptor or donor is also usually an enzyme catalysing an oxidation or reduction reaction, but another electron transporter can often serve as acceptor or donor.

7.1.1 Overview

In this chapter four major classes of electron transporters will be discussed. These are the cytochromes, iron-sulphur proteins, copper proteins (cupredoxins), and flavodoxins. (For a general review of electron transporters, see Adman 1979.) As the oxidation/reduction cofactors, these proteins contain, respectively, haems, iron-sulphur clusters, a single copper ion, and FMN. They span all major types of tertiary structure organization. In most cases, these classes are subdivided into subclasses with distinct structural and functional properties. One or more examples of these individual subclasses will be described and discussed, depending on the availability of accurate structural data and on the distinctive properties of individual members of the subclasses. The relevant set of Protein Data Bank files can be selected by the keyword 'electron transport'.

A number of enzymes also contain the oxidation/reduction cofactors described above. For example, catalase (7CAT), P_{450} (2CPP) and cytochrome c peroxidase (2CYP) contain haem, aconitase contains iron-sulphur, succinate dehydrogenase contains both iron-sulphur and FMN, and sulphite reductase contains an iron-sulphur cluster and haem very close together, plus flavin in another subunit. Many of these enzymes also carry out electron transfer reactions. However, these enzymes are larger and more complex both in structure and function than the simple electron transporters and will not be described in this chapter.

The cytochromes are subdivided into four main structural subclasses, mitochrondrial c-type cytochromes, b_5-type cytochromes, four-α-helical cytochromes, and multihaem cytochromes. The main distinguishing features and important properties of these different subclasses will be discussed in the next section.

The iron-sulphur-containing electron transporters are likewise divided into three major categories, depending on the type of cluster contained. These are the one-, two-, and four-iron clusters, respectively. In some cases more than one cluster of a given type is contained in an electron transporter, and sometimes one of the iron atoms of a cluster is labile. However, no example of a protein containing two distinct types of clusters is known or structurally characterized.

The copper proteins discussed here are of much more uniform type, and only contain a single copper ion. However, there is considerable structural variability among these copper proteins, leading to definition of four or five subclasses. Only two subclasses, azurin and plastocyanin, are discussed in detail. Likewise, there is only one basic folding pattern for the flavodoxins, although some distinct variations in chain length and detailed folding patterns exist. Only the most well-characterized flavodoxin will be described.

7.2 Cytochromes

7.2.1 Overview

Classically, cytochromes have been divided into three types, a, b, and c, based on their spectroscopic properties. More recently, the distinction between cytochromes has come to depend, in part, on the nature and mode of binding of the haem cofactor (for a general review of cytochromes, see Mathews 1985). The cytochromes discussed in this chapter all contain the same haem group, the iron-bound form of protoporphyrin IX shown in Fig. 7.1. The haem group is asymmetrical. One edge, containing pyrrole

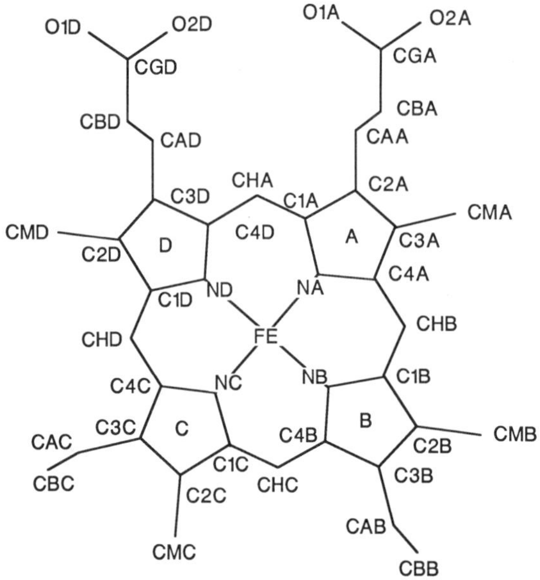

Fig. 7.1 Haem group showing the Protein Data Bank convention for the numbering of the individual atoms.

rings A and D, is quite polar, with two methyl groups and two propionic acid groups. The other edge, rings B and C, is much less polar, containing two methyl groups and two vinyl groups. Double bonds connect atom C3B to CAB and atom C3C to CAC. At the present time, the only distinction between a *b*-type and a *c*-type cytochrome is in the mode of attachment of the haem group to the protein. In a *b*-type cytochrome the haem is bound non-covalently, while in a *c*-type cytochrome the haem is bound covalently, usually at both the CAB and CAC positions (see Fig. 7.1) to cysteine side-chains of the protein.

In this chapter, four classes of cytochromes are described. The classification is based on common structural features, not just on the mode of haem binding. Some structural classes of cytochromes contain both *b*- and *c*-type haem binding, and *b*- and *c*-type haem binding is found in more than one structural class.

The mitochondrial-type cytochromes are all *c*-type cytochromes with covalently bound haem groups. They are found in mitochondria of eukaryotic organisms as well as in many bacteria (for a review of mitochondrial-type cytochromes, see Dickerson and Timkovich 1975; see also Meyer and Kamen 1982). The iron atom of the haem group is coordinated through a histidine side-chain on one side and a methionine side-chain on the other. They all have a common structural motif, consisting of a polypeptide chain wrapping around the haem group in a characteristic fashion with the haem attached to the chain in a similar manner. The only secondary structure is helical, with the number of helices ranging from three to five.

The eukaryotic mitochondrial cytochromes have a very highly conserved structure with little variation in chain length, amino-acid sequence, or placement of conserved side-chains. Considerably more variation occurs in the bacterial cytochromes c. Only two classes of bacterial cytochromes have been well characterized structurally, and these will be discussed bellow. The so called large bacterial c_2-type cytochromes are structurally similar to the eukaryotic c-type cytochromes. The small bacterial c-type cytochromes, on the other hand, have considerably less structural and biophysical similarities to the eukaryotic cytochromes.

The b_5-type cytochromes all have non-covalently bound haem groups. The coordination of the haem iron atom at the fifth and sixth positions is quite symmetrical, with both ligands being histidines. The folding of the polypeptide chain, which defines this cytochrome class, consists of several α-helices and a mixed β-sheet.

The four-α-helix class of cytochromes contains either non-covalently bound haem or haem bound covalently in a c-type manner. One of the haem ligands is a histidine. In the non-covalently bound case (cytochrome b_{562}) the other iron ligand is methionine, while in the covalently bound case (cytochrome c') the sixth coordination site of the iron is vacant. The secondary structure is almost entirely helical, consisting mostly of four long, antiparallel helices.

The multihaem cytochromes are all c-type. This class does not include certain mitochondrial-type cytochromes which are composed of two single-haem cytochromes fused together, e.g. cytochrome c_4 (Meyer and Kamen 1982). The iron coordination of the multihaem cytochromes consists of histidine side-chains at both positions. The folding of the polypeptide chains is largely helical with some β-structure.

7.2.2 Mitochondrial c-type cytochromes

7.2.2.1 Eukaryotic cytochromes c   Mitochondrial cytochrome c is located in the inner membrane space of mitochondria. It is synthesized in the cytoplasm of the cell and transported across the eukaryotic outer membrane. The physiological function is to transfer electrons from the cytochrome reductase b-c_1 complex to the cytochrome oxidase a-a_3 complex. Both donor and acceptor complexes are integral membrane proteins. However, the cytochrome c molecule is quite soluble.

About 60 different eukaryotic cytochrome c molecules have been studied in detail. The amino-acid sequences are highly homologous, with about 60 per cent sequence identity on the average. They are very similar in their physical and biochemical properties. Their redox potentials are about

260 mV $\pm$ 10 mV and are constant over the pH range 5–8. This means the reduced form is energetically stable under physiological conditions. They are all quite basic, with a pI of about 10.

Plant and animal cytochrome c molecules differ to some extent. Those from animals are about 103 residues in length. Those from plants, including yeast, contain several additional residues at the N-terminal end of the molecule. In addition, two or three lysine residues have undergone post-translational modification, being trimethylated at the NE positions.

The structure of oxidized tuna cytochrome c is shown in Plate 3. The polypeptide chain is wrapped around the haem group in a simple zigzag manner, starting at the top and covering the right, bottom, left, and top rear portions of the molecule. There are five short helices, comprising about 45 per cent of the structure. The two longest helices lie at the N-and C-termini of the chain, containing 10 and 14 residues, respectively. The remaining three helices are in the left half of the molecule. There is no β-structure.

Most of the haem group is in a hydrophobic environment. The covalent attachment of the haem to the protein occurs through thioether bonds from Cys-14 and Cys-17 to the vinyl methylene atoms CAB and CAC, respectively (Fig. 7.1). His-18, immediately following the second covalent linkage, is one of the axial ligands to the haem iron atom, coordinated through the NE atom of the His ring. These three residues are located on the right side of the molecule, as shown in Plate 3, just at the C-terminal end of helix 1. The other axial iron ligand is Met-80, on the left side of the molecule, which is coordinated through the SD atom of the side-chain. The positions of the axial ligands are further stabilized by hydrogen bonding to other parts of the protein. ND of His-18 is hydrogen bonded to the backbone carbonyl of Pro-30 and the SD of Met-80 forms an additional hydrogen bond to the OH of Tyr-67. The two propionic acid groups are located at the bottom of the molecule and are both shielded from solvent. One of the propionate carboxyl groups, toward the rear of the molecule, is quite buried and forms a salt bridge to the side-chain of Arg 38, as well as hydrogen bonds to two other side-chains and a buried water molecule. The other carboxyl group, toward the front of the molecule, forms three internal hydrogen bonds. These hydrogen bonds and salt bridges serve to stabilize the buried charges on these groups and to modify their dissociation constants.

The overall charge on cytochrome c is quite imbalanced, with 18 positively charged and seven negatively charged residues on the surface of the tuna protein. This accounts for the very high isoelectric point of the protein. Furthermore, the distribution of acidic and basic residues over the surface of the molecule is not uniform. Many excess basic residues are localized around the mouth of the haem crevice at the front of the molecule. Elsewhere on the protein surface the two types of charges are more evenly distributed. This clustering of positive charges can be seen in Plate 4. It may be important for recognition by cytochrome *c* oxidase and cytochrome *c* reductase, which tend to be more acidic in nature.

Reduced and oxidized cytochrome c
The relative stabilities of the reduced and oxidized forms of cytochrome *c* differ to some extent (Dickerson and Timkovich 1975). In comparison with the oxidized forms, the reduced forms are generally more stable toward thermal and surface denaturation and to protease digestion. A number of studies suggest that the reduced form is more rigid in solution. It is also known that the methionine to iron bond in the oxidized cytochrome *c* is about a thousandfold weaker than in the reduced form.

The structure of tuna cytochrome *c* is known in both the oxidized and reduced states. The structures in these two states are very similar. After superimposing the structures, the root mean squared (rms) deviation between main-chain atoms is 0.43 Å when main-chain plus CB atoms are compared. The positions of three internal water molecules and nine surface waters are also conserved in the two states. The value of the rms deviation indicates a barely significant difference between the two oxidation states, since two crystallographically independent copies of the reduced protein deviate by only 0.28 Å in rms main-chain plus CB positions. The largest changes occur in the orientations of three side-chains which are hydrogen bonded to an internal water molecule located within 5A of Met-80, the haem ligand. The water molecule is about 0.7 Å closer to SD of Met-80 in the oxidized than in the reduced protein, and causes movements of up to 1.4 Å in some of the side-chain atom positions and a slight inward movement of the second helix.

The structures of other eukaryotic cytochrome *c* molecules from animal sources are very similar to the tuna structure. Cytochrome *c* from horse heart has 18 amino-acid differences, but is folded in a very similar manner (Bushnell *et al.* 1990). Detailed comparison of the refined horse and tuna structures indicates an rms difference in main-chain position of only 0.41 Å.

Plant cytochromes c
The structures of two plant cytochromes *c* have been determined, one from rice and the other from yeast (Louie and Brayer 1990). Rice cytochrome *c* contains 41 amino-acid substitutions with respect to the tuna protein, in addition to seven extra C-terminal residues and the trimethylation of two lysine residues. Yet the common structural features are very similar. The rms deviation of equivalent main-chain atoms after optimal alignment of the oxidized rice and tuna structures is 0.59 Å, compared to 0.28 Å for the two independent tuna molecules. All main-chain hydrogen bonds and nearly all internal hydrogen bonds are identical in the two proteins. The two trimethyllysine residues, at positions 72 and 86, are near the front of the

molecule on the left side and are involved in the cluster of positively charged lysine residues thought to be important for interaction with the redox partners. The seven N-terminal residues are in extended conformation and are wrapped around the back of the molecule, far from the exposed haem edge on the front of the molecule.

Yeast contains two isoforms of cytochrome c. Isoform 1, the more abundant form, contains five extra C-terminal residues as well as trimethyl-lysine residues at positions 72 and 86. It differs in sequence from rice cytochrome c at 39 places. The rms deviation in main-chain positions between rice and tuna is 0.52 Å, excluding the extra C-terminal residues which differ somewhat more in conformation.

7.2.2.2 Bacterial cytochrome c_2 (large bacterial cytochromes)

The mono-haem cytochromes c_2 are found mostly in phototrophic bacteria, where they serve as electron carriers during the cyclic process of photosynthesis. In some bacteria they serve as electron donors to the photosynthetic reaction centre, to complete the electron and proton membrane translocation process triggered by light absorption. [There is another class of cytochrome c_2, the tetrahaem cytochrome c_2 found in *Rhodopseudomonas viridis*, which also donates electrons to a photosynthetic reaction centre.] In *Pseudomonas denitrificans*, the cytochrome c_2 (cytochrome c_{550}) donates electrons to nitrite reductase (Meyer and Kamen 1982).

The molecular weights of the cytochromes c_2 are generally the same as or somewhat greater than the eukaryotic cytochrome c molecules. The sequence identity between selected pair of cytochromes c_2 is about 50 per cent, and is about 40 per cent when compared to the eukaryotic proteins. The redox potential of the cytochrome c_2 molecules are generally higher and more varied than among the eukaryotic cytochrome c molecules, ranging from 250 to 400 mV. In some cases the redox potential varies with pH in ranges near neutrality. The isoelectric points are generally lower, with pI averaging about 6.

The backbone structure of *Rhodospirillum rubrum* cytochrome c_2 is shown in Plate 5. Its structure is compared with that of tuna cytochrome c in Fig. 7.2. The structures are very similar, with three or four short insertions in the sequence of the *R. rubrum* cytochrome compared with that of the tuna protein. These insertions are localized primarily to the left side of the molecule as viewed in Plate 3. One short insertion is located just after helix 2, at the bottom rear of the molecule, while another, longer insertion occurs after helix 4, at the lower front left of the molecule.

The haem environment of *R. rubrum* cytochrome c_2 is also very similar to that of tuna cytochrome c. The pattern of haem ligation is identical, being common in almost all eukaryotic and bacterial cytochrome c molecules. The axial iron ligands also form equivalent secondary hydrogen bonds to the protein; His-18 ND to Pro-30 carbonyl oxygen, and Met-80 SD to Tyr-67

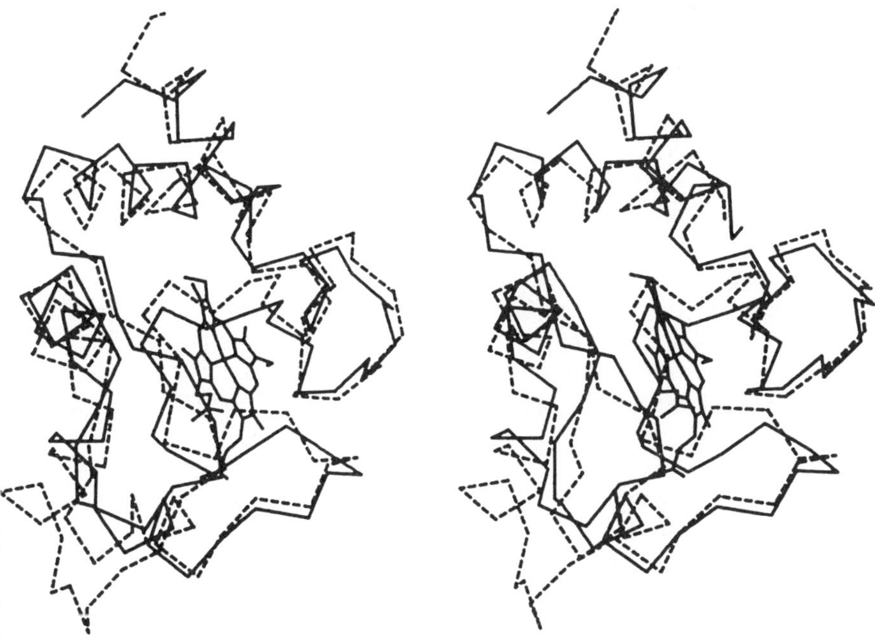

Fig. 7.2 Comparison of the α-carbon backbones of *R. rubrum* cytochrome c_2 (dashed lines) and tuna cytochrome c (solid lines).

hydroxyl. Conservation of Pro-30 is universal among cytochrome c molecules, while Tyr-67 is replaced by phenylalanine in some cytochrome c_2 molecules.

The environment of the haem propionate carboxyl groups is also similar in the tuna and *R. rubrum* structures, the groups being quite buried in both cases. Each forms several hydrogen bonds to neighbouring side- and main-chain atoms, although the salt bridge made by the propionate group of ring D to Arg 38 in tuna is replaced by a hydrogen bond to Asn 38 in *R. rubrum*.

The comparison of tuna cytochrome c to *P. denitrificans* cytochrome c_{550} (not shown) reveals a similar pattern of structural conservation with one or two additional insertions in sequence. The environment of the haem group is also very similar, with some variation in the hydrogen-bonding pattern of the haem propionates.

7.2.2.3 Small bacterial cytochromes The small bacterial cytochromes are a heterogeneous collection of cytochromes with certain common, highly conserved structural features, but with considerable variation of bio-chemical and physical properties (Meyer and Kamen 1982). They fall into about six distinct classes with similar properties and amino-acid sequences, with 40–50 per cent sequence identity within groups, and 10–20 per cent identity between groups. Their redox potentials range from highly positive values, close to 400 mV for phototrophic bacteria, to -220 mV for sulphate-reducing bacteria. The physiological electron donors and acceptors are quite varied, ranging from membrane-bound complexes to soluble redox enzymes or other small electron transporters.

One or more structures are known for each class of small bacterial cytochrome. Only two are highly refined, however, cytochrome c_{551} from *Pseudomonas aeruginosa* and cytochrome c_5 from *Azotobacter vinlandii*. Only cytochrome c_{551} will be discussed.

Cytochrome c_{551} contains 82 amino acids. It is involved in nitrite reduction or in aerobic metabolism, interacting with membrane-bound proteins as electron donors or acceptors. Its redox potential is about 280 mV at neutral pH and is pH dependent, falling by about 30 mV per pH unit over the pH range 4–9.

The structure of cytochrome c_{551} is shown schematically in Plate 6. The haem group is bound to the molecule by two thioether linkages through cysteine side-chains, and by coordination to the iron by a histidine and a methionine side-chain, just as in the mitochondrial cytochromes c. There are four helices in cytochrome c_{551}. The two longest helices occur at the N-and C-terminal ends of the molecule, as in eukaryotic cytochromes c. Helices 2 and 3 lie in the middle of the sequence, with helix 2 at the bottom rear of the molecule and helix 3 on the left side, just before the occurrence of the methionine axial ligand.

The backbone structures of cytochromes c_{551} and tuna cytochrome c are compared in Fig. 7.3. The right half of the cytochrome c_{551} molecule, including the two Cys and one His attachment sites to the haem, is quite similar to tuna cytochrome c. The major difference is a 15-residue deletion in the middle of the cytochrome c_{551} sequence relative to tuna cytochrome c, between the two axial ligands to the haem group. This deletion manifests itself in a much shorter connecting region at the bottom of the molecule, causing a loss of the peptide loop and helix covering the haem propionate groups. The propionates remain covered, however, by another portion of the protein chain, also located between the axial haem ligands.

The orientation and environment of the haem group in c_{551} is similar to that in tuna cytochrome c, except that some alternative hydrogen bonds occur and its surroundings are somewhat more polar. His-18 ND is hydrogen bonded to the carbonyl of Pro-30, but Met-80 SD forms a hydrogen bond to an asparagine side-chain located differently from the tyrosine side-chain in tuna cytochrome c. There is also a buried water molecule adjacent to pyrrole ring A, absent in the tuna structure. The propionate on ring A (Fig. 7.1), in the rear of the molecule, is excluded from solvent and forms hydrogen bonds and a salt bridge to protein side-chains. The salt bridge partner is not equivalent to the partner in tuna cytochrome c. The other

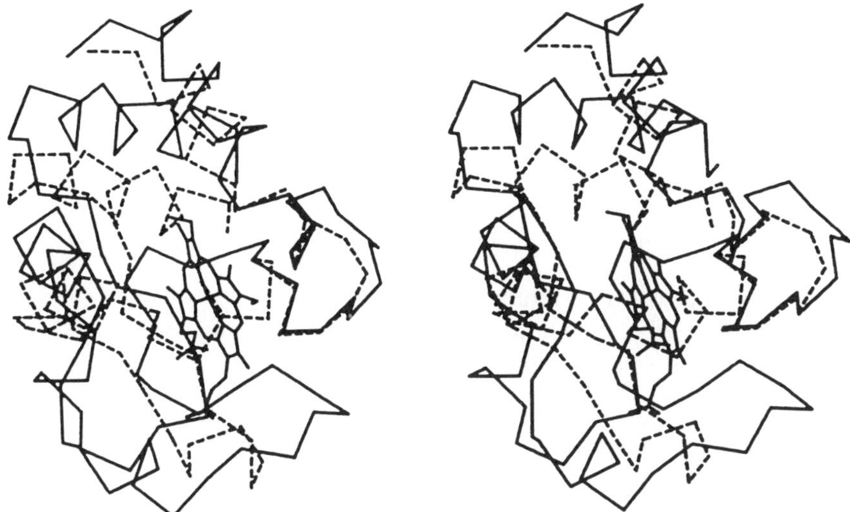

Fig. 7.3 Comparison of the α-carbon backbones of cytochrome c_{551} from *Pseudomonas aeruginosa* (dashed lines) and tuna cytochrome c (solid lines).

propionate, on ring D, is more exposed to solvent and interacts with a buried water molecule as well as with side-chain oxygen and nitrogen atoms.

The structure of reduced cytochrome c_{551} is almost identical to the oxidized form. The rms deviation of α-carbon coordinates in the two oxidation states is only 0.08 Å. The only significant perturbation in structure is a slight movement of the buried water molecule away from pyrrole ring A.

7.2.3 Cytochrome b_5

7.2.3.1 Background Cytochrome b_5 is found in liver cells in birds and mammals. In these cells it is localized in the endoplasmic reticulum and is an integral membrane protein. Its best-studied function is in fatty acid desaturation, where it acts as an electron carrier between the NADH-dependent cytochrome b_5 reductase and a fatty acid desaturase enzyme. Both donor and acceptor proteins are also integral membrane proteins. Cytochrome b_5 can also serve as an electron donor to P_{450}, a haem-containing detoxification enzyme which is also an integral membrane protein.

Native cytochrome b_5 consists of two domains, a hydrophilic N-terminal domain of about 90 residues and a hydrophobic C-terminal domain of about 40 residues. The haem group is located in the hydrophilic domain. This domain can be released from the membrane readily by treatment with proteases, and has been extensively studied in this highly water-soluble form.

Cytochrome b_5 is also found in mammalian reticulocytes, where it acts primarily as a methaemoglobin reductase. In this case as well, the electron donor is the NADH cytochrome b_5 reductase. However, in contrast to the microsomal system, both proteins function in a water-soluble form. They are synthesized as membrane-bound precursors and processed to the soluble form by proteolysis during erythroid maturation.

7.2.3.2 Structure of cytochrome b_5 The structure of the soluble fragment of cytochrome b_5 is shown in Plate 7. It contains 93 amino-acid residues, of which only 85 (from 3 to 87) could be located in the crystal structure. The protein contains six short helices and a mixed five-stranded β-sheet. The β-sheet divides the molecule into two parts, both of which contain a hydrophobic core of aliphatic and aromatic side-chains. One part, slightly larger, binds the haem group and the other seems to play a largely structural role.

The haem group is oriented with the propionic acid portions directed toward the exterior of the molecule and the remainder buried in the protein interior. It is bound non-covalently, by coordination to two histidine side-chains located at positions 39 and 63 along the sequence. Coordination to the iron occurs through the NE atoms of the histidines, while the protonated ND atoms form hydrogen bonds to main-chain carbonyl oxygen atoms. Twelve other side-chains make contact with the buried part of the haem group. Eleven of these are hydrophobic and the twelfth is neutral polar, a serine residue whose hydroxyl forms a hydrogen bond to a peptide carbonyl oxygen. One of the haem propionate groups extends into the solvent region. The other bends back onto the molecule, forming hydrogen bonds to a peptide nitrogen and a second serine hydroxyl side-chain. The latter propionate group is partially shielded from solvent by the protein.

The charge distribution of cytochrome b_5 is very uneven. The top part of the molecule, as oriented in Plate 7 is very acidic and carries a net negative charge of about -9. The bottom part of the molecule is nearly neutral. The pronounced localization of negative charge may be important for interaction with its electron transfer partners.

The structure of reduced cytochrome b_5 (Argos and Mathews 1975) indicates that there is little structural change on reduction. However, it appears that a cation becomes bound near the buried propionate in the reduced form of the protein. This cation would substitute for the net positive charge on the haem iron lost upon reduction, and suggests that the propionate group serves to stabilize positive charge in the oxidized protein.

7.2.3.3 The cytochrome b_5 fold There are a number of multifunctional enzymes that contain a cytochrome domain which is closely related in structure, sequence, and physical properties to cytochrome b_5. Individual members of this 'cytochrome b_5 family' are homologous at the level of 30–40 per cent identity in sequence and are clearly related by divergent evolution (Guiard and Lederer 1979). Yet each is part of a larger polypeptide chain containing unrelated domains.

Flavocytochrome b_2 contains two domains, a b_5-like domain, called 'b_2 core', and an FMN-containing β/α-barrel domain. The enzyme has a subunit molecular weight of 58 000 Da, is localized in yeast mitochondria, and functions as a lactate dehydrogenase. Substrate oxidation occurs in the flavin domain, after which electrons are transferred from the flavin to the cytochrome domain and are subsequently passed on to a separate molecule of cytochrome c.

Sulphite oxidase also contains a b_5-like domain. The other domain is a molybdo-enzyme. The enzyme is found in liver cells and has a subunit molecular weight of about 58 000 Da. The electrons extracted during oxidation of sulphite in the molybdenum-containing domain are passed to the cytochrome domain and then to cytochrome c.

Nitrate reductase, from *Neurospora crassa*, consists of three domains in a polypeptide chain of molecular weight about 60 000 Da. In addition to a cytochrome b_5-like domain, the enzyme contains flavin adenine dinucleotide (FAD) and a molybdenum cofactor in the remaining domains.

7.2.4 Four-α-helix cytochromes

7.2.4.1 Cytochrome b_{562}   Cytochrome b_{562} is a protein of molecular weight about 12 000 Da, containing non-covalently bound haem, found in the periplasm of *Escherichia coli*. Its function is unknown, but it may serve as a soluble electron carrier. Its redox potential is about 185 mV at neutral pH, and is pH dependent. The haem iron atom is low spin and cannot bind any exogenous ligands. The protein contains very few aromatic residues. There are two histidines, two phenylalanines, and two tyrosines.

The structure of cytochrome b_{562} (156B) is shown in Plate 8. It consists of four nearly antiparallel α-helices, which pack together into a bundle with a left-handed twist of 15–20°. The four helices form a simple up-down-up-down pattern, with two short connections of two residues each between helices A and B and helices C and D. On the other hand, the connection between helices B and C is longer, containing 15 residues, four of which are proline. In Plate 8, the B-C loop shows little regular secondary structure. Recent high-resolution refinement has shown, however, that it contains one turn of left-handed 3–10 helix.

The haem group of cytochrome b_{562} is inserted into the helix bundle near the centre and is in an asymmetrical environment. There are two ligands to the haem, Met-7 near the amino terminus, and His-102 near the carboxyl terminus. One face of the haem, adjacent to Met-7, is in a very hydrophobic environment. The other face is much more exposed to solvent. Pyrrole rings B and C (Fig. 7.1) are quite buried, but parts of the back face of rings A and D are exposed to solvent. Part of His-102 is also exposed to solvent, including atom ND, which does not form any hydrogen bond to the protein nor to any fixed water molecule. This situation differs from all other cytochromes where ND forms a hydrogen bond to a protein main- or side-chain atom. The propionate groups of the haem are also exposed to solvent and extend into solution.

The charged residues appear to be distributed uniformly over the protein surface. However, there are four acidic groups contained within the first three turns of helix A and four basic groups within the last three turns of the helix D. All six aromatic groups lie in the vicinity of the haem group.

7.2.4.2 Cytochrome c' Cytochrome c' is a dimeric protein of identical subunits, each with a molecular weight of 14 000 Da. It is found in photosynthetic and denitrifying bacteria, and is produced in higher yield in organisms grown anaerobically. It is located in the periplasmic space of photosynthetic bacteria, but its function is unknown. The iron atom of the haem is in a mixed spin state, of spin about 3/2. The midpoint redox potential ranges from 0 mV to about 150 mV. The reduced protein will bind CO or NO, but not O_2.

The structure of the monomer of cytochrome c' (2CCY) is shown in Plate 9. The enzyme consists of four α-helices oriented nearly antiparallel to one another, forming a helix bundle twisted by about 19° in a left-handed manner. The loops connecting helices A and B and helices C and D are quite short, only one or two residues in length. The BC loop, however, is about 15 residues in length and serves to cover the haem propionates.

The haem group is bound covalently by thioether linkage to the Cys-118 and Cys-121, and the iron atom is coordinated to the ND of His-122. The other coordination site is vacant, lacking even a bound water molecule. There is a methionine residue, Met-16, situated near the sixth coordination site. However, the iron to sulphur distance of 3.75 Å is too long to allow bond formation. One face of the haem group is buried in a hydrophobic environment. The other face, where the histidine coordination occurs, is partially exposed to solvent. The extent of solvent exposure of the haem group and its His ligand is shown in Plate 10. The propionate groups are partially buried, forming hydrogen-bonding and salt-bridge networks to the protein atoms. The propionate on ring D is covered by the BC interhelix loop and forms a salt bridge to Arg 12 on the same molecule. The other propionate, on ring A, extends away from the cytochrome monomer and forms a salt bridge with Arg10 of the other subunit.

The dimer of cytochrome c' is formed by the antiparallel interaction of helices A and B with helices A' and B' to form another four-stranded antiparallel bundle. The bundle is twisted in a left-handed manner similar to the four-stranded helix bundle of the monomer.

7.2.4.3 Similarities of cytochrome b_{562} and cytochrome c' The three-dimensional structures of cytochromes b_{562} and c' are remarkably similar. When the haem groups of the two proteins are aligned, the four-helix bundles are also aligned, as shown in Fig. 7.4. The first half of helix A and most of helices B, C, and D are in nearly equivalent positions. Of the 77 amino-acid residues in equivalent position in the two proteins, 14 are identical in sequence. Four of the six aromatic residues contained in cytochrome b_{562} are unchanged or replaced by other aromatic residues. These are Phe-61, Phe-65, His-102, and Tyr-105, which correspond to Phe-82, Trp-86, His-122, and Phe-125 in cytochrome c'.

The most curious feature of the structure comparison is the near super-position of Met-16 of cytochrome c' with Met-7 of b_{562}. The latter is a ligand to the haem iron, but the former is too far from the iron atom to allow coordination by the SD atom. This difference in function of the two methionine groups arises from the fact that helix A of cytochrome c' is more divergent from helices B, C, and D than it is in b_{562}, and that this helix in c' is rotated to some extent compared to b_{562}. A compensatory movement of the haem of cytochrome c' to a position where it could form a bond to Met-16 is prevented by the covalent attachment of the haem to helix D through Cys-118 and Cys-121, as well as by ligation with His-122.

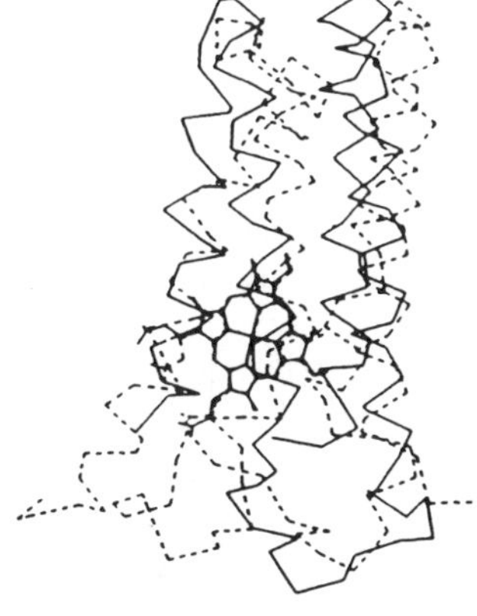

Fig. 7.4 Superposition of the α-carbon tracings of cytochrome b_{562} (solid lines) with cytochrome c' (dashed lines).

7.2.5 Multihaem c_3-type cytochromes

Multihaem cytochromes of only two types have been studied by X-ray crystallography. One class, cytochrome c_3, is found only in sulphate-reducing bacteria. In these organisms, they serve to transfer electrons from hydrogenase to ferredoxin. The other class, the tetrahaem cytochrome c_2 from the photosynthetic reaction centre of *Rhodopseudomonas viridis*, serves to replenish a deficient electron to the reaction centre at the end of the photosynthetic event. The structural features of the latter class of multihaem cytochrome are not described further in this chapter, except to note that these cytochromes contain four haems bound to a fourfold internal repeat of helices.

The cytochromes c_3 from *Desulfovibrio* bacteria are about 15 000 Da and contain four haem groups, all bound covalently by thioether linkages to cysteine side-chains. However, both haem ligands are histidine for all four haem groups, in contrast to most c-type cytochromes, which have a histidine and a methionine as iron ligands. Most of the covalent linkages involve the sequence Cys-X-Y-Cys-His characteristic of c-type cytochromes. However, in some cases the linkage involves the sequence variant Cys-X-Y-X$'$-Y$'$-Cys-His, i.e. with four residues inserted between the two cysteine groups instead of two.

The redox potentials of cytochromes c_3 are quite negative, averaging approximately $-300\,\mathrm{mV}$. There is a spread of about 200 mV among the haem groups and reduction usually occurs in four one-electron steps.

The structure of cytochrome c_3 from a strain of *Desulfovibrio desulfuricans* (DD) is shown schematically in Plate 11. It contains 118 amino-acid residues folded into two domains of approximately equal size. Each domain is wrapped about a single haem group, numbers 1 and 4, and the two central haem groups are located in a large groove between the domains. There is little secondary structure, except for a two-stranded hairpin loop near the N-terminus and a single helix near the C-terminus.

The first three haem groups in the sequence are nearly perpendicular to each other. The fourth haem group is approximately parallel to the first. Each of the histidine ligands to the haem is held in place by hydrogen bonds to the protein, either to peptide carbonyl oxygens or to the side-chains of Ser, Thr, or Gln. The average separation of the iron atoms is about 14 Å, ranging from 10 to 17 Å. The charge distribution over the surface of the protein is rather even, except for a basic cluster of amino acids near the C-terminus of the helix.

The haem groups are all relatively buried, with the first haem the most buried and the second the most exposed. Haem 2 also contains the two inserted residues in the thioether linkage peptide. In addition to the histidine ligands, each haem is in contact with at least two aromatic side-chains, as well as with several aliphatic side-chains. The propionate groups of haem 1 are also buried, interacting with protein side-chains, while the propionates

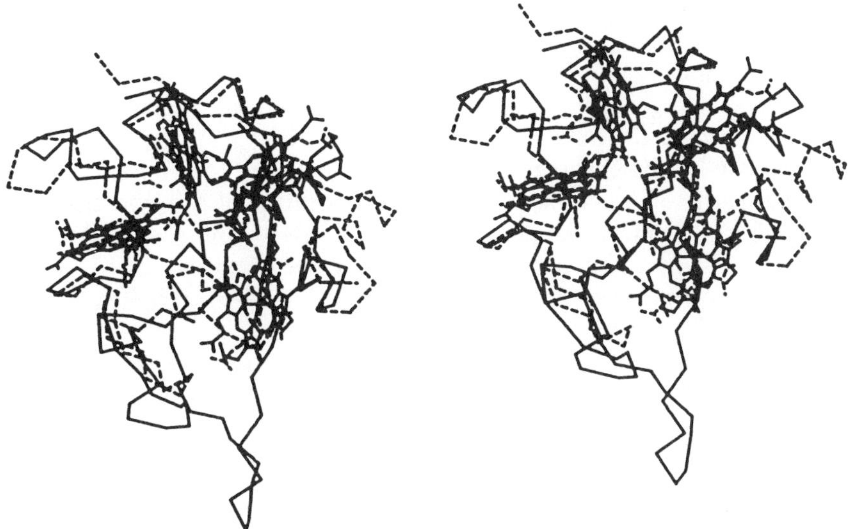

Fig. 7.5 Superposition of the α-carbon backbones of cytochrome c_3 from *Desulfovibrio desulfuricans* (dashed lines) and cytochrome c_3 from *Desulfovibrio vulgaris* (solid lines).

of the other haems are more exposed to solvent.

The crystal structure of cytochrome c_3 from a strain of *Desulfovibrio vulgaris* (DV) has also been determined. Its structure is shown in Plate 12. The arrangement of the haem groups is very similar in the two proteins. However, the DV protein contains only 107 residues, so the polypeptide chains differ to some extent in conformation. The structures of the two cytochromes c_3 are compared in Fig. 7.5. There are three short helices in the DV protein, two of which are located at approximately the same position as the single helix of the DD cytochrome, near the C-terminus. There is also a shorter antiparallel β-hairpin near the N-terminus.

The haem environments are also quite similar in the two proteins. However, in the DD structure, the haem propionates are less exposed to solvent; only two haem groups are exposed, with exposure limited to a single propionate each.

7.3 Iron-sulphur proteins

7.3.1 Overview

Iron-sulphur proteins are a class of proteins, found in a wide variety of organisms, which are mostly involved in electron transfer processes. They contain one or more iron-sulphur centres consisting of iron, inorganic sulphur, and/or cysteinyl sulphur atoms in a cluster. (For a review of iron-sulphur proteins, see Carter 1977.) The simplest of the iron-sulphur proteins are often quite small, consisting of about 50 amino acids, but can be considerably larger. The redox potentials for the iron-sulphur centres is quite varied, covering the range of $-400\,\mathrm{mV}$ to $+350\,\mathrm{mV}$.

The iron-sulphur proteins are classified according to the type of iron-sulphur cluster. The simplest type of cluster consists of one iron atom and four cysteine ligands. This cluster is found in rubredoxin. A second type of cluster is the two-iron cluster, consisting of two atoms each of iron and inorganic sulphur and two cysteine ligands [2Fe2S*2SG]. It is found in the two-iron plant ferredoxins.

The third type of cluster is the so-called four-iron cluster, containing usually four iron atoms, four inorganic sulphur atoms, and four cysteine sulphur atoms, [4Fe4S*4SG]. The shape of this cluster (the [4Fe4S*] part) is a distorted cube with iron and inorganic sulphur at alternate corners, and is called a cubane structure. In some proteins the cubane cluster contains only three iron atoms, i.e. [3Fe4S*]. Of the simple iron-sulphur electron transfer proteins (having no enzymatic activity and containing no other prosthetic groups), the one-and two-iron proteins contain only a single cluster per molecule. However, some of the cubane-containing proteins have two iron-sulphur clusters per molecule.

7.3.2 Rubredoxin

Rubredoxins are found mostly in anaerobic bacteria, although some are found in aerobic bacteria as well. They are thought to be involved in electron transfer, although their exact role is not known. Their molecular weight is about 6 kDa and they consist of 50–55 amino acids. They contain a single iron atom bound by four cysteine side-chain ligands. The redox potential is about $-50\,\mathrm{mV}$.

The structure of rubredoxin from *Clostridium pasteurianum* is shown in Plate 13. The protein contains 54 amino acids and has a relatively small amount of secondary structure. There is a short three-stranded β-sheet, containing residues 3–7, 11–13, and 48–52, and an antiparallel β-loop, consisting of residues 19–24.

The iron atom is bound to two antiparallel strands of the polypeptide chain, residues 3–13 and 36–46, which form two hydrophobic arms in very similar conformation. The binding site for the iron atom consists of two pairs of loops with the sequence Cys-X-Y-Cys-Gly. These loops are related by an approximate twofold axis. The four cysteine residues are arranged with approximately tetrahedral symmetry. The iron to sulphur distances range from 2.24 Å to 2.33 Å, with an average value of 2.29 Å.

There is a total of six hydrogen bonds between peptide NH groups and the four SG ligands of the iron atom. The atoms involved in these bonds are arranged in two groups, which are related in a twofold symmetrical manner. For each Cys-X-Y-Cys-Gly-Z ligand pair, the two NH groups from residues 'X' and 'Y' form hydrogen bonds to the first Cys, and the third NH from residue 'Z' forms a hydrogen bond to the second Cys. Also, for each of the peptide groups of 'X', 'Y', and 'Z' the carbonyl oxygen atom lies on the surface of the protein molecule. Thus, the dipoles of all six peptide groups are arranged with their positive ends pointing inwards toward the [Fe4SG] group. This arrangement of dipoles appears to enable the negative charge of the buried iron-sulphur cluster (-2 or -1 in the reduced or oxidized states, respectively) to be delocalized to the surface of the protein. Thus the buried charge can become partially neutralized, despite the fact that there are no positively charged side-chains near the iron group.

The interior of the rubredoxin molecule is packed with six aromatic side-chains, forming a hydrophobic core, some of which are also packed against the [Fe4SG] centre. There are also a number of other hydrophobic residues on the protein surface surrounding the iron centre, forming a 'knob' at one end of the molecule. This hydrophobic 'knob' is delineated by a ring of negatively charged side-chains distributed circumferentially about the centre of the molecule. The remainder of the charged amino-acid side-chains, including most of the positively charged ones, are located on the surface opposite the iron centre, near the antiparallel β-loop formed by residues 19–24. This asymmetrical distribution of negative charges surrounding the hydrophobic 'knob' may be important for recognition of and electron transfer to the redox partners of rubredoxin.

7.3.3 Two-iron ferredoxin

The two-iron ferredoxins are generally found in plants and algae. These proteins have molecular weights of about 12 kDa and contain approximately 100 amino acids. They operate at a rather low redox potential, about -400 mV. Their principle function is in the photoreduction of cytochrome c in chloroplasts, which is catalysed by ferredoxin-NADP$^+$ reductase.

The structure of the ferredoxin from *Spirulina platensis*, a blue-green alga, is shown in Plate 14. It contains 98 amino-acid residues and has a midpoint redox potential of about -380 mV. It is an extremely acidic protein, with 25 Asp and Glu residues and only three Lys and Arg residues.

The protein contains 10 extended strands and one two-turn α-helix. Seven of the β-strands are divided into a four-stranded mixed-sheet and a three-stranded antiparallel β-sheet, which together form a barrel structure. The single helix extends from one of the β-strands and helps complete the barrel.

The active site containing the [2Fe-2S*] centre is held in a loop structure which is folded external to the barrel. The structure of the active-site region of the protein is shown in Plate 15. Each of the two iron atoms is covalently bound to two inorganic sulphur atoms, which bridge the two iron atoms. Each iron atom is also coordinated by two cysteine SG atoms. One of the irons is coordinated to SG 41 and SG 46 and the other to SG 49 and SG 79. Both iron atoms are located within 5 Å of the surface of the protein. The SG of residue 79 is the most exposed iron ligand, while the ligands at positions 41 and 46 lie slightly beneath the protein surface. Cys-49 is located in the protein interior. The inorganic sulphur atoms are also buried, thereby being protected from attack by solvent.

There are five hydrogen bonds to the sulphur ligands of the iron atoms. SG 41 receives two hydrogen bonds from main-chain NH groups (Gly-43 and Ala-45) while SG 46 receives hydrogen bonds from one peptide NH (Thr-48) and one side-chain hydroxyl (Ser-40). SG 79 and one of the inorganic sulphur atoms each receive a single NH hydrogen bond, one from Gly-44 and the other from Ser-40.

The interior of the β-barrel is packed with non-polar aliphatic amino acids almost exclusively, and contains only one aromatic residue (Phe). The remaining aromatic residues are located on the protein surface or protrude into the solvent region. The charged amino acids are distributed more or less evenly over the surface of the protein.

7.3.4 Four-iron-four-sulphur 'cubane' clusters

7.3.4.1 High-potential iron protein The family of high-potential iron proteins (HIPIP) is found in purple phototrophic bacteria. The proteins in this family appear to function in electron transport during photochemical processes, although their exact metabolic role is unknown. They have molecular weights in the range 6.5–9 kDa and contain a single [4Fe-4S*] cluster. The redox potentials of the iron centres are high but vary considerably, ranging from +50 to +450 mV. The proteins in this family also have a wide range of isoelectric point, suggesting little conservation of charge distribution.

The HIPIP from *Chromatium vinosum* has 85 residues, being one of the larger members of this family. It has a redox potential of +350 mV and is quite acidic, with pI = 3.8.

The structure of HIPIP is shown in Plate 16. It consists of two closely associated segments of nearly equal size, made up of residues 1–42 and 43–85, respectively. The first half contains two short helical segments, 12–16 and 28–31. The second half contains a three-stranded antiparallel β-sheet within the segment 48–72. The attachment sites for the iron-sulphur cluster are contained within this second half. There is also a two-stranded anti-parallel β-sheet formed by segments 17–20 and 75–80 holding the two halves of the molecule together. The interface between the two halves of the molecule is made up of non-polar amino acids.

The [4Fe-4S*] cluster is covalently bound to Cys-43, Cys-46, Cys-63, and Cys-77. A view of HIPIP showing the Cys ligands (yellow) to the iron-sulphur cluster (red) is shown in Plate 17. The first two cysteines are contained within a pair of adjacent hairpin turns which form one side of the cluster-binding cavity. The second pair of cysteines are part of the three-stranded β-sheet which forms the other side of the cavity.

The iron-sulphur cluster is quite buried. It is in contact with an elaborate arrangement of aromatic and other non-polar amino-acid side-chains as well as with part of the polypeptide backbone. Tyr-19 is located quite close to one of the S* atoms of the cluster, with part of the aromatic ring in van der Waals contact with the sulphur atom. One of the cysteine SG ligands is located at the interface between the two halves of the molecule.

The [4Fe-4S*] cluster forms a distorted cube. It consists of two inter-penetrating tetrahedra, one consisting of 4Fe and the other of 4S*. The iron atoms are separated by about 2.72 Å while the sulphur atoms are separated by about 3.55 Å. Consequently the Fe-S*-Fe angles are about 74° while the S*-Fe-S* angles are about 104°. The mean Fe-S* distance is about 2.26 Å.

There are about six hydrogen bonds from peptide NH groups to sulphur atoms of the iron-sulphur cluster. Four of these are directed to three of the cysteine SG ligands and two of these are to two inorganic sulphur atoms. Three of the cysteine-directed hydrogen bonds follow the pattern in which the amide group two positions beyond a cysteine side-chain forms a hydrogen bond to the SG of that cysteine. Thus Cys-46, 63, and 77 receive hydrogen bonds from NH of Phe-48, Leu-65, and Ser-79, respectively. In addition, the SG of Cys-46 is hydrogen bonded to the NH of Thr 81. One of the two inorganic sulphur atoms is hydrogen bonded to the NH of Met-49 and the other to the NH of Cys-77. However, the hydrogen-bond distance for the former is longer than usual, suggesting a weak interaction.

When HIPIP is reduced, the protein structure is very nearly the same as in the oxidized state. The biggest change in structure is a slight expansion of the iron-sulphur cluster. The Fe-S* and Fe-SG distances increase about 0.1 Å. At the same time, hydrogen-bond distances from the NH to the SG and S* atoms decrease by about 0.2 Å. In particular, the long hydrogen-bond distance from Met-49 NH to the S* becomes shorter, to make a stronger interaction. These structural changes suggest that the increased negative charge on the reduced iron-sulphur cluster is stabilized by the positive dipoles of the peptide groups close to the cluster, leading to stronger hydrogen bonding.

7.3.4.2 Ferredoxin Bacterial ferredoxins are low molecular weight electron transfer proteins containing one or more iron-sulphur centres of the cubane type. They have low redox potentials, usually about $-400\,\mathrm{mV}$, and are similar in physiological function to the two-iron-two-sulphur chloroplast ferredoxins. They are involved in oxidation-reduction reactions in the

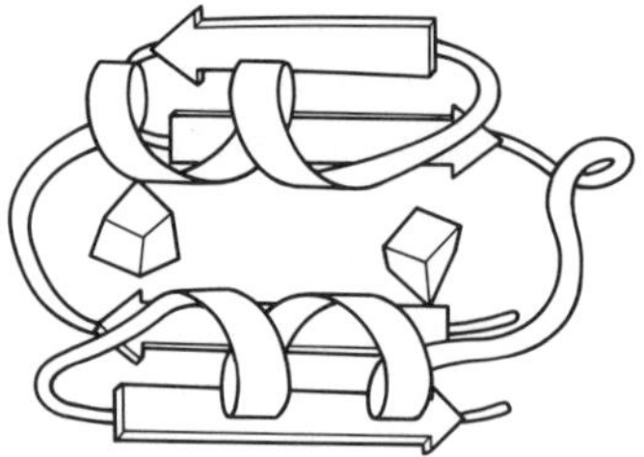

Fig. 7.6

metabolism of various substances, such as organic material and elemental sulphur, nitrogen, and hydrogen.

Ferredoxin from *Peptococcus aerogenes* is made up of 54 amino acids and contains two [4Fe4S*] clusters. The redox potentials for both clusters is about -400 mV.

The structure of this eight-iron ferredoxin is shown in Plate 18. The protein contains four β-strands and two short helices. The β-strands are made up of residues 2–5, 22–25, 28–30, and 49–52. The helices run from residues 13 to 18 and from 39 to 45. The secondary structural elements form two $\beta\alpha\beta$-units which interdigitate and are related by approximate twofold symmetry. The amino-acid sequence also reflects the approximate twofold symmetry. The folding pattern results in two pairs of antiparallel strands, 1,4 and 2,3, with an α-helix on top of each pair. This structural motif is illustrated in Fig. 7.6. The two [4Fe-4S*] clusters located between the two structural elements also possess approximate twofold symmetry.

The cysteine ligands to the two [4Fe-4S*] clusters are also, to a great extent, symmetrically related to each other. The ligand-binding pattern can be seen in Plate 19. The ligand binding for each pair of clusters follows the sequential pattern Cys-X-X-Cys-X-X-Cys-....-Cys. The first three cysteines of each pair bind to one cluster and the fourth binds to the other. Thus the ligand pattern for each cluster is Cys-8, Cys-11, Cys-14, Cys-43 for the first, and Cys-35, Cys-38, Cys-41, Cys-18 for the second. The conformations of the polypeptide loops containing these sequences of residues are also quite similar.

The geometry of the two four-iron clusters in *Peptococcus* ferredoxin are quite similar to each other and to the cluster in HIPIP. Each cluster is a distorted cube. The average Fe-S* distances are 2.22 Å and the Fe-SG distances are 2.23 Å. The distortion of the cube arises because it is made up of two interdigitating tetrahedra of different size, the 4Fe and 4S* groups. The average Fe-Fe distance is 2.70 Å, while the average S*-S* distance is 3.5 Å. This leads to average values of the Fe-S*-Fe and S*-Fe-S* bond angles of 74.1° and 104.2°, respectively. The two iron-sulphur clusters are separated by about 12 Å.

The two iron-sulphur clusters are involved in extensive hydrogen bonding by peptide NH groups. For each cluster there are a total of nine such hydrogen bonds, five to SG atoms and four to S* atoms. Two SG atoms and one S* atom receive two hydrogen bonds each, while one SG and two S* atoms receive one hydrogen bond each. One SG and one S* atom of each cluster are not involved in hydrogen bonding.

The seven-residue peptide fragments which sequentially bind each iron-sulphur cluster, Cys-X-X-Cys-X-X-Cys, have all of their CO groups pointing away from the cluster, while their peptide NH groups point toward the clusters. This arrangement may contribute to dipolar stabilization of negative charges on the clusters.

The two iron-sulphur clusters are surrounded by mostly aliphatic amino-acid side-chains. Only one aromatic group, a tyrosine, is in contact with each group. This leads to a rather hydrophobic environment for the clusters. However, one SG ligand for each cluster, from Cys-11 or Cys-38, is exposed to solvent.

There are three regions that have nearly invariant acidic side-chains. These are at position Glu-17, positions Asp-31 and Asp-33, and at position Asp-37. These negatively charged groups may be involved in interactions with the redox partners of the ferredoxins.

7.3.4.3 Comparison of ferredoxin and HIPIP   The major functional difference between the four-iron-type ferredoxins and the high-potential iron protein is the difference of approximately 750 mV in their redox potentials. Examination of their protein structures indicates that the geometries of their clusters are extremely similar, and their protein surroundings are somewhat similar.

The most obvious structural difference is that the iron-sulphur cluster in HIPIP is more deeply buried than it is in ferredoxin. However, the amino-acid side-chains surrounding the clusters in both proteins are generally hydrophobic. In ferredoxin, one SG ligand in each cluster is exposed to solvent, while in HIPIP, all atoms and ligands are buried. One of the SG ligands in HIPIP is located in the interface between domains, but in a hydrophobic environment.

Ferredoxin also has more hydrogen bonds from peptide NH groups to sulphur atoms of the clusters. In ferredoxin, there are nine such hydrogen bonds per cluster, while there are only six in HIPIP. Also the NH-S hydrogen-bonding pattern is more regular in ferredoxin. All 12 intrapeptide NH groups in the two sequences Cys-X-X-Cys-X-X-Cys are bonded to an SG or S* atom.

7.3.4.4 The three-state hypothesis   Carter (1977) has proposed that the [4Fe-4S*] cluster is capable of existing and operating in three oxidation

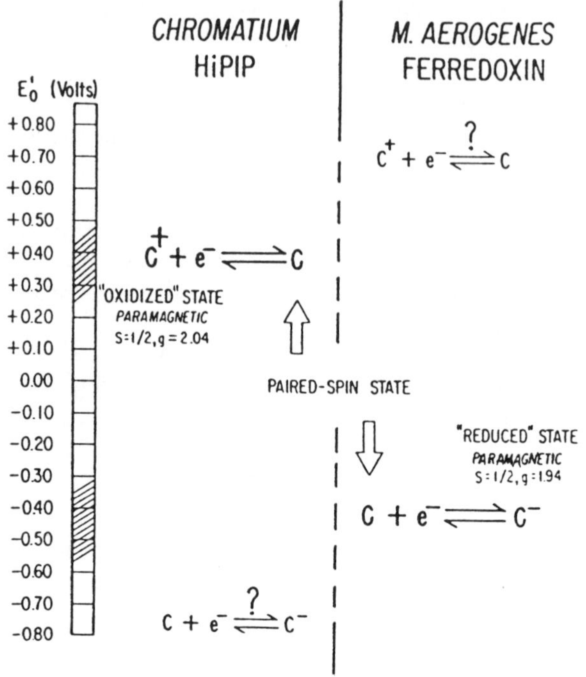

Fig. 7.7 The three-state hypothesis for relative oxidation levels of the [4Fe-4S*] cluster in iron-sulphur proteins. The paired-spin state, *C*, occurs in both reduced HIPIP and oxidized ferredoxin. (Reproduced by permission from Carter 1977.)

states, $[4Fe\text{-}4S^*]^{-1}$, $[4Fe\text{-}4S^*]^{-2}$, and $[4Fe\text{-}4S^*]^{-3}$. This situation is illustrated in the energy-level diagram shown in Fig. 7.7. HIPIP operates by cycling between the -1 and -2 states, while ferredoxin cycles between the -2 and -3 states. Thus, the intermediate -2 state is common to both proteins, i.e. reduced HIPIP and oxidized ferredoxin are in the same electronic state. The protein environment of the cluster is able to mediate its redox potentials so that one or the other cycle is in the physiological range. This hypothesis helps explain the wide difference in redox potential between HIPIP and ferredoxin despite the similarity of structure. The hypothesis has been supported by redox studies of partially denatured proteins and model compounds.

7.4 Blue copper proteins

Blue copper proteins are a class of electron transfer proteins found in a variety of plants and bacteria. They are often called 'Type I' copper proteins

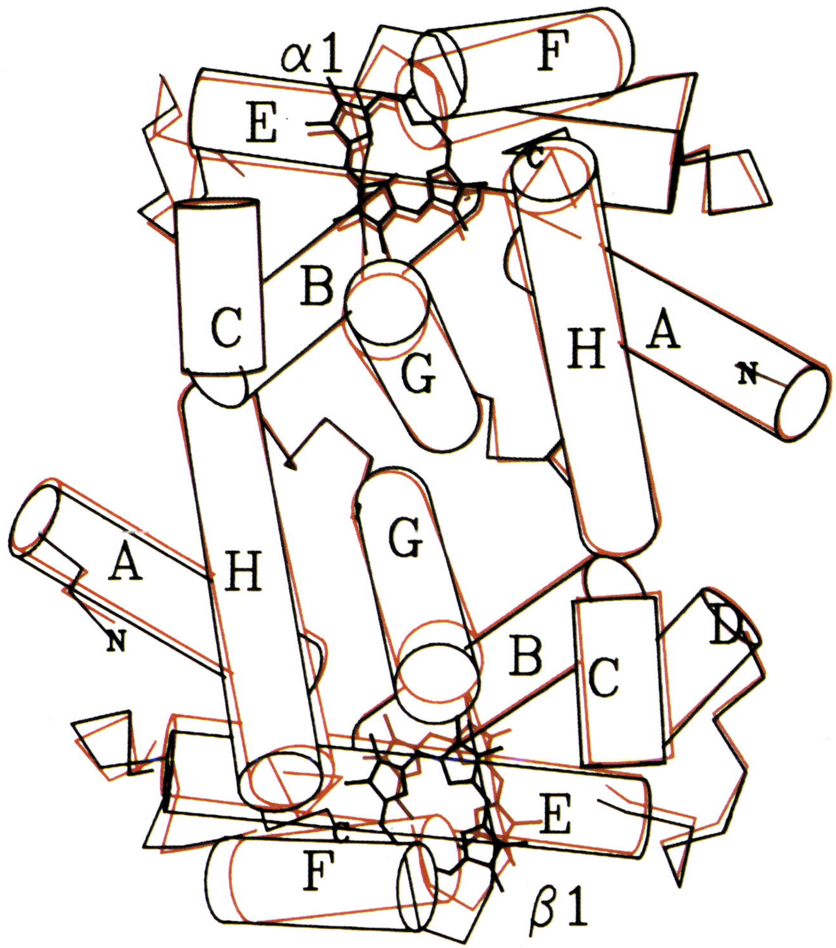

Plate 1 Schematic representation of the difference in tertiary structure between the $\alpha_1\beta_1$ dimers of deoxyhaemoglobin (black) and oxyhaemoglobin (red). The two structures were superimposed by fitting together the α-carbos of the B, G, and H helices. Cylinders, representing helical segments, were fitted to the appropriate α-carbons. Non-helical regions are represented by line segments joining α-carbons; the haems are shown with side-chains omitted.

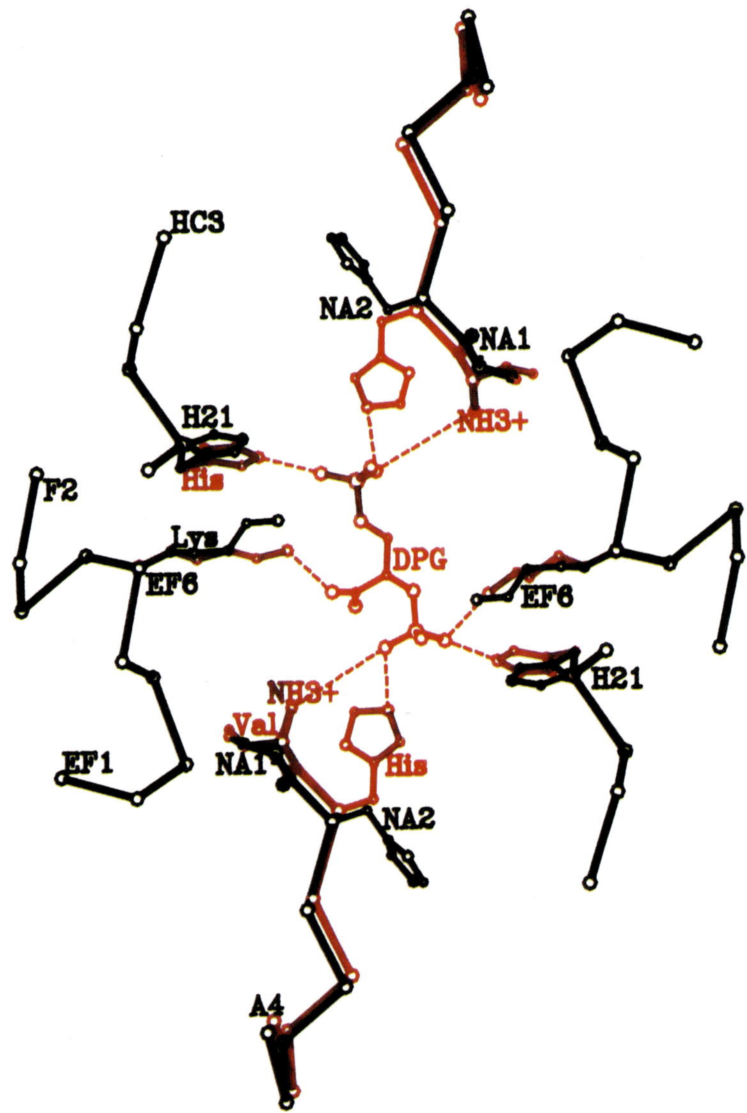

Plate 2 Approximate representation of the binding of 2,3-diphosphoglycerate (DPG) to deoxyhaemoglobin. Blue bonds accurately represent the structure of deoxyhaemoglobin in the absence of DPG (Fermi *et al.* 1984); red bonds show an approximation to the structure in the presence of DPG, based on the description by Arnone (1972); broken single lines indicate likely salt bridges or other ionic interactions. The molecular dyad is perpendicular to the screen, at the centre of the diagram. Strict molecular symmetry is lost when DPG binds, because of lack of symmetry in the DPG molecule. Binding DPG causes changes in the conformation of the N-terminus, His NA2 and His H21; it also causes movements of the A helices into the central cavity (as illustrated) and of the H helices away from the central cavity (not illustrated).

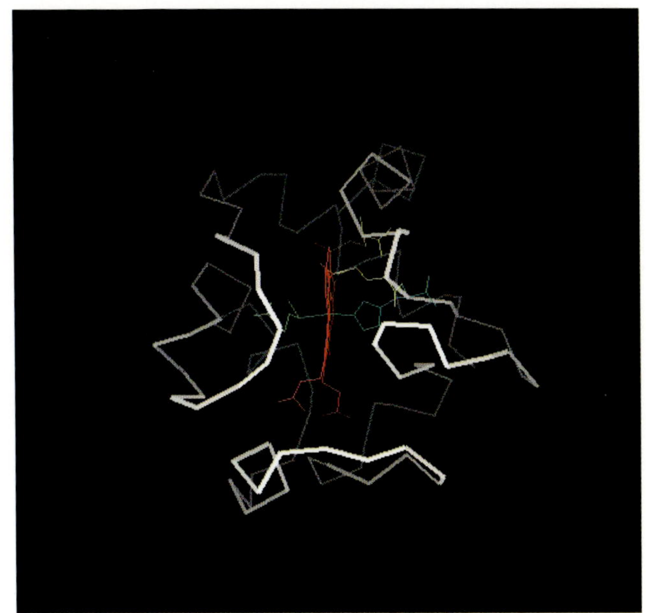

Plate 3 See p. 196

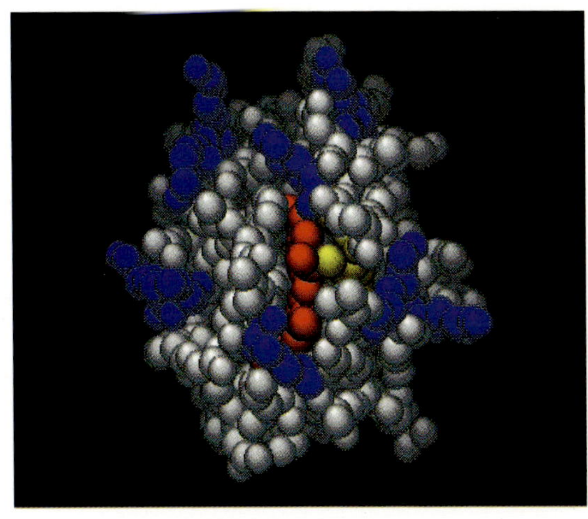

Plate 4 See p. 196

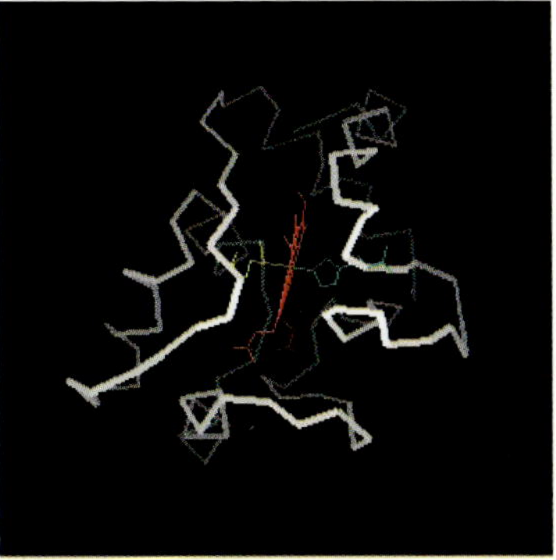

Plate 5 See p. 198

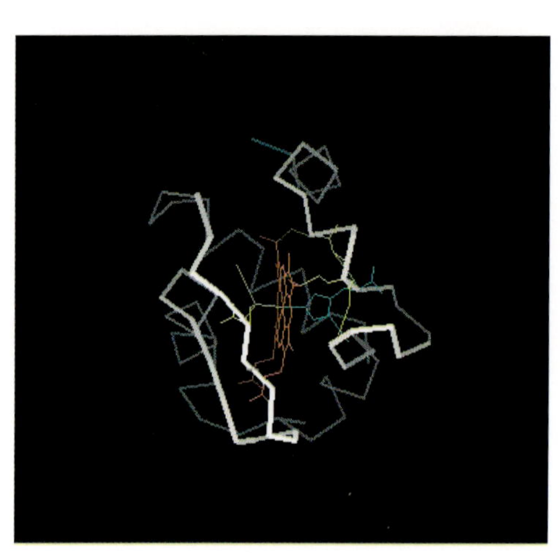

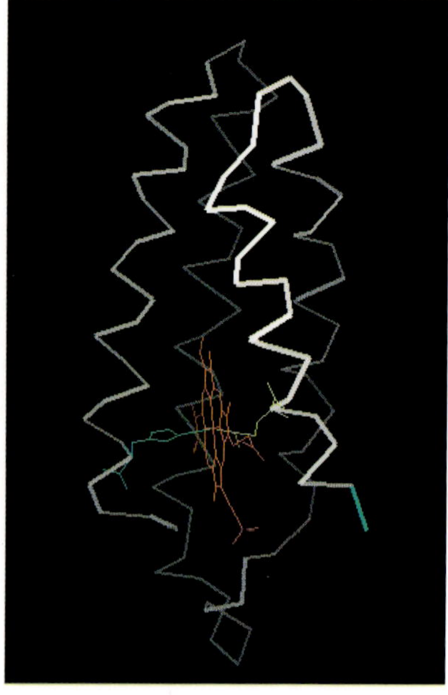

Plate 6 See p. 200   Plate 7 See p. 202

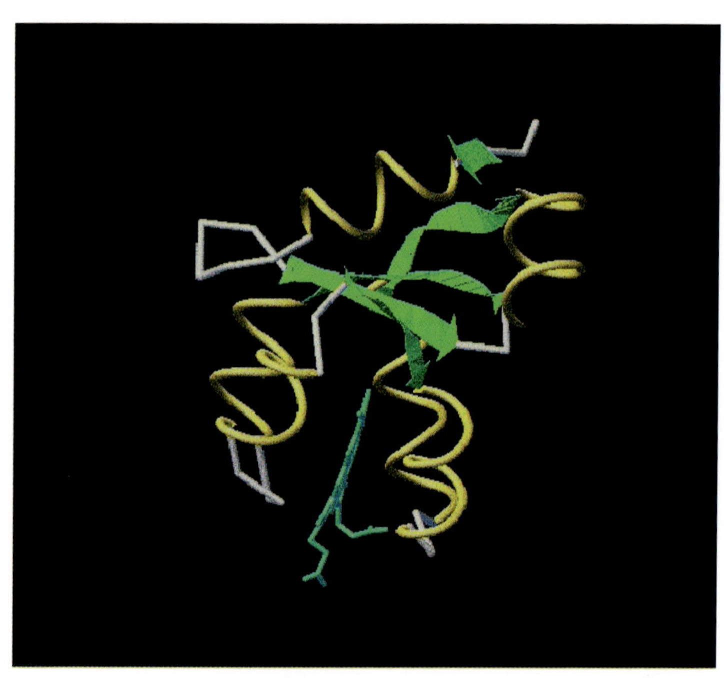

Plate 8 See p. 203

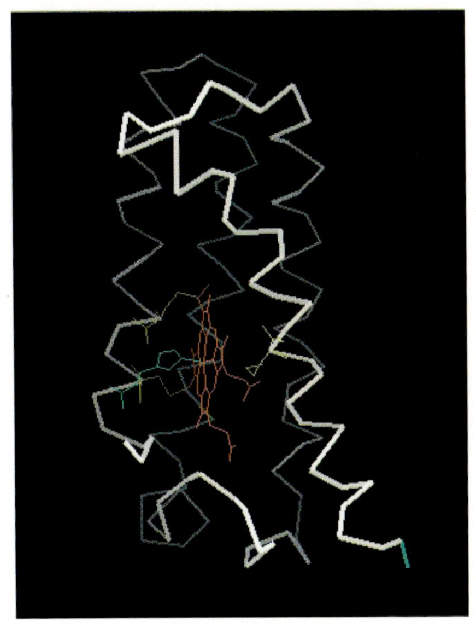

Plate 9 See p. 204

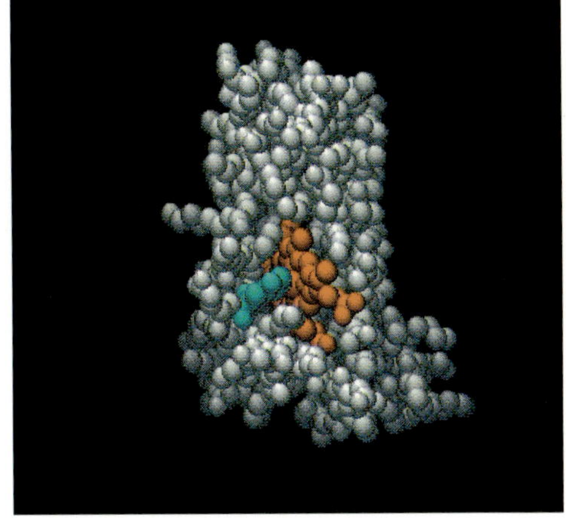

Plate 10 See p. 204

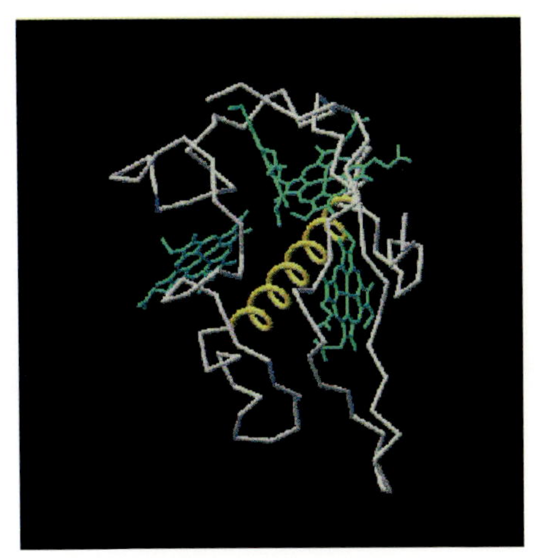

Plate 11 See p. 206

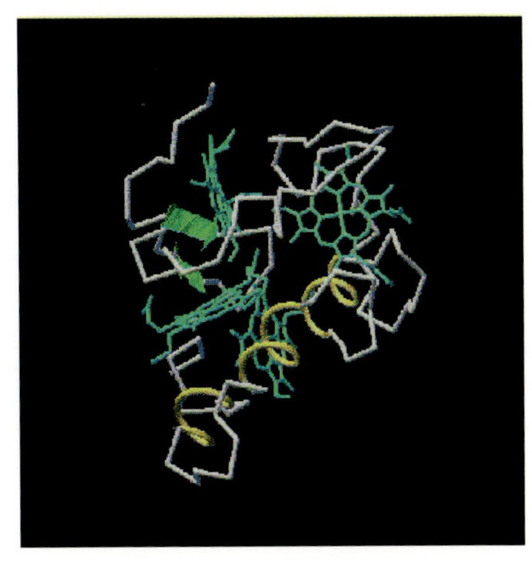

Plate 12 See p. 207

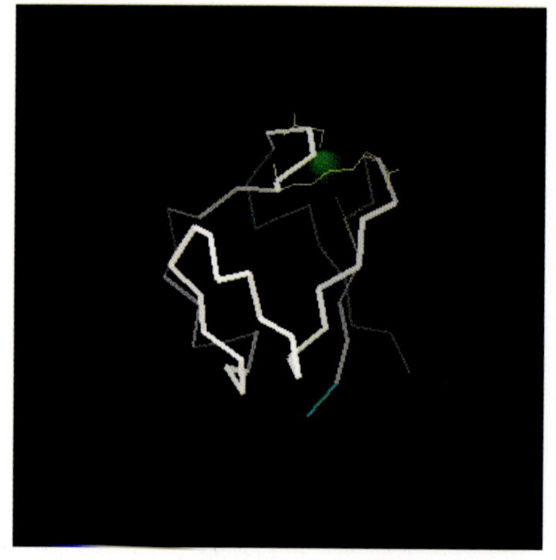

Plate 13 See p. 208

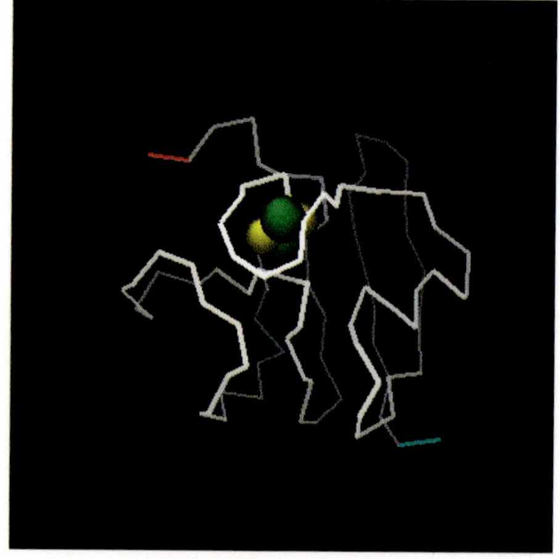

Plate 14 See p. 209

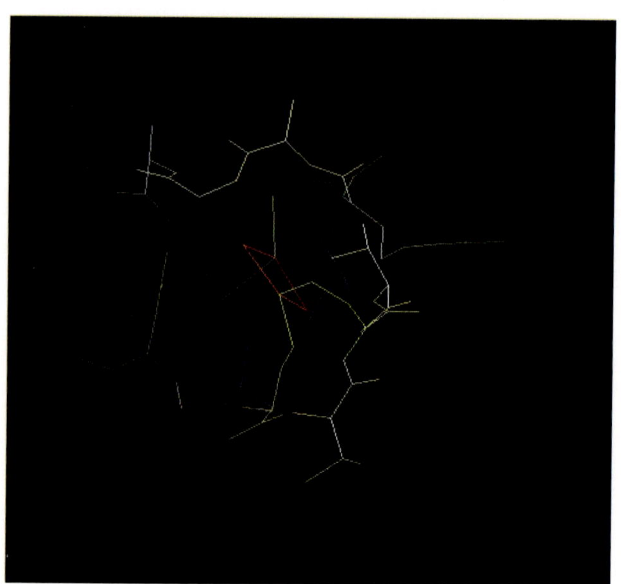

Plate 15 See p. 210

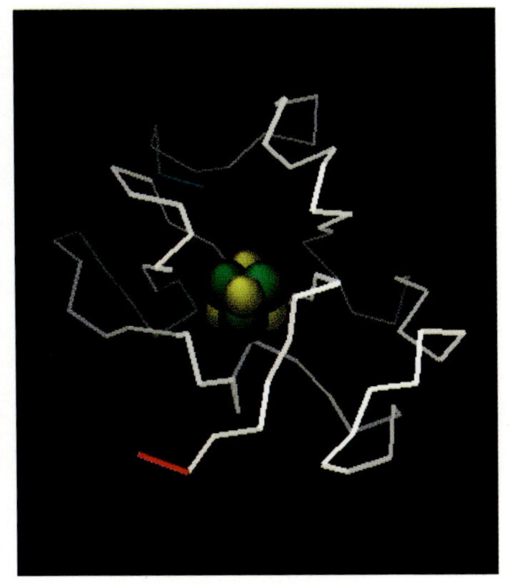

Plate 16 See p. 210

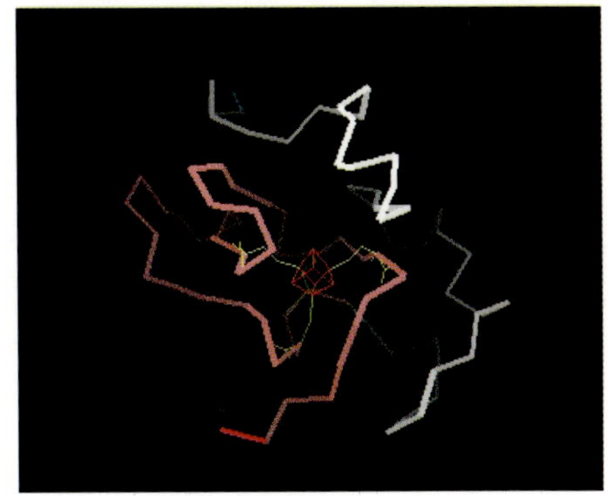

Plate 17 See p. 211

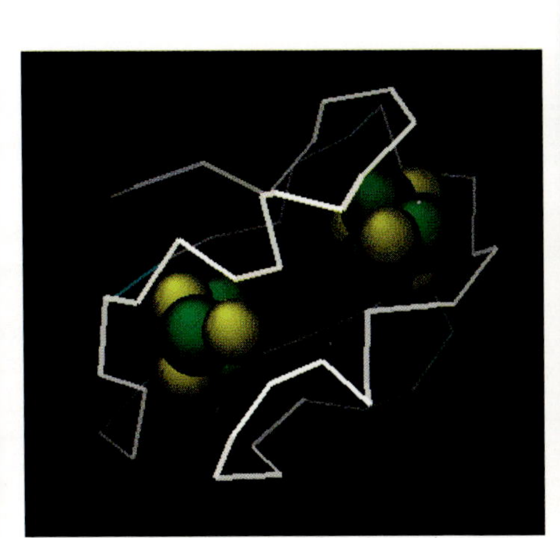

Plate 18 See p. 212

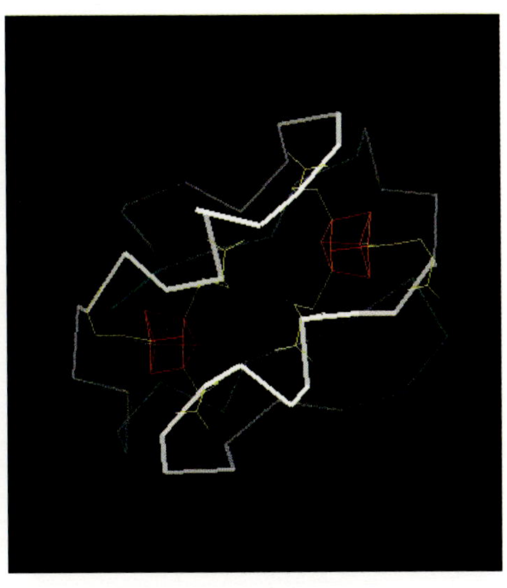

Plate 19 See p. 212

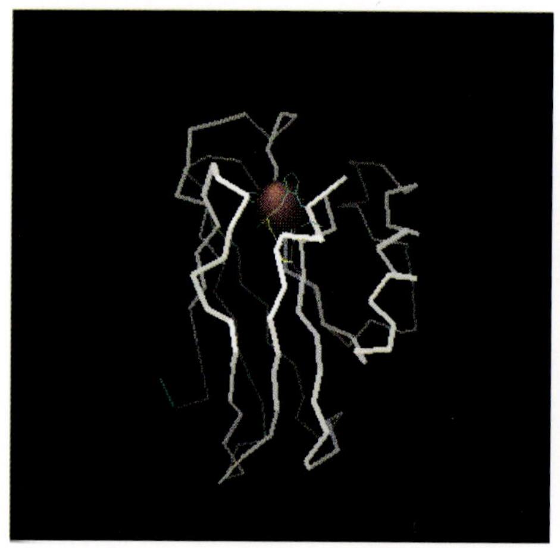

Plate 20 See p. 215

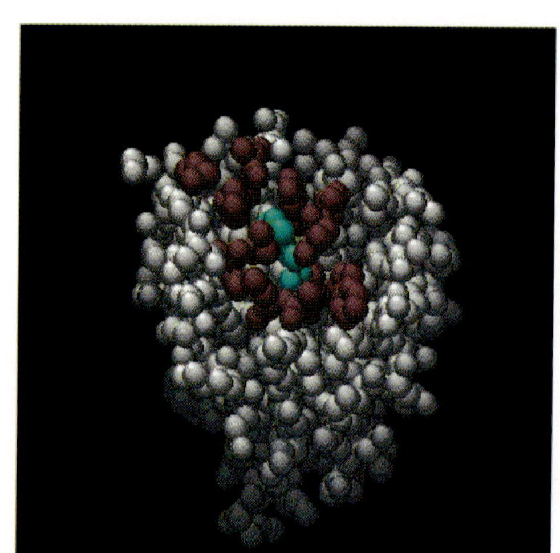

Plate 21 See p. 217

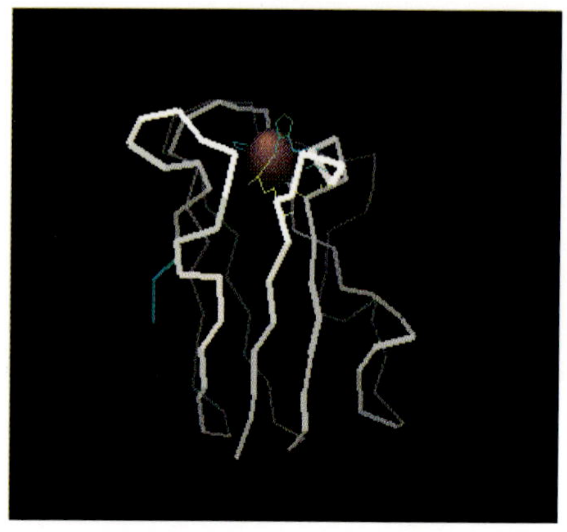

Plate 22 See p. 217

Plate 23 See p. 218

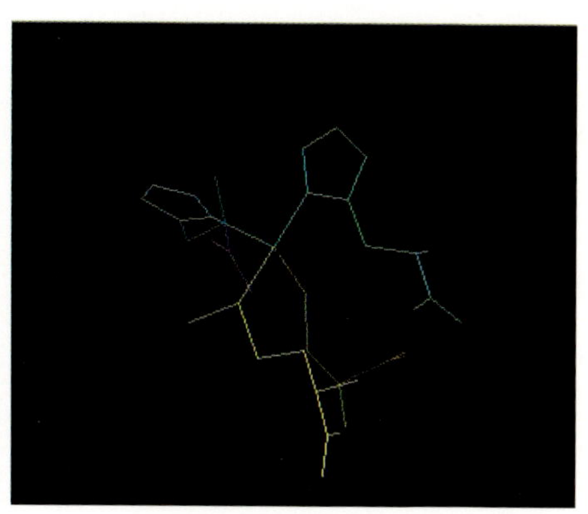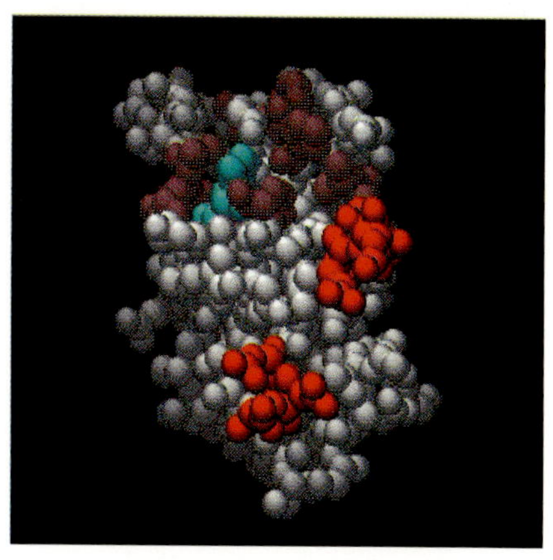

Plate 24 See p. 218

Plate 25 See pp. 218–219

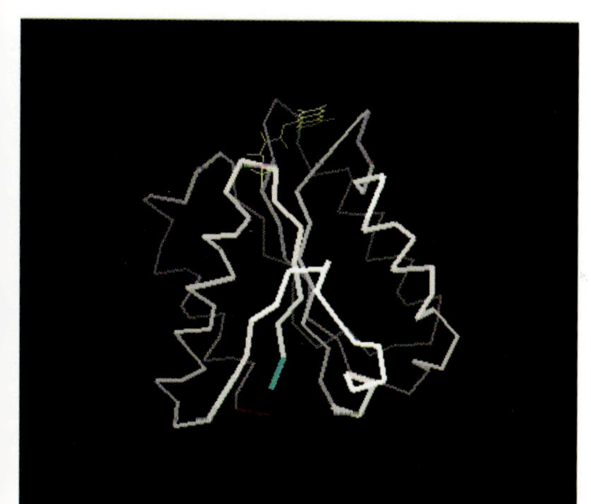

Plate 26 See p. 220

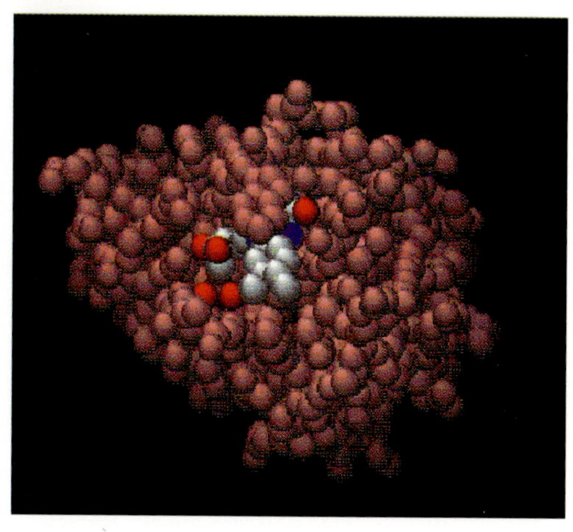

Plate 27 See p. 220

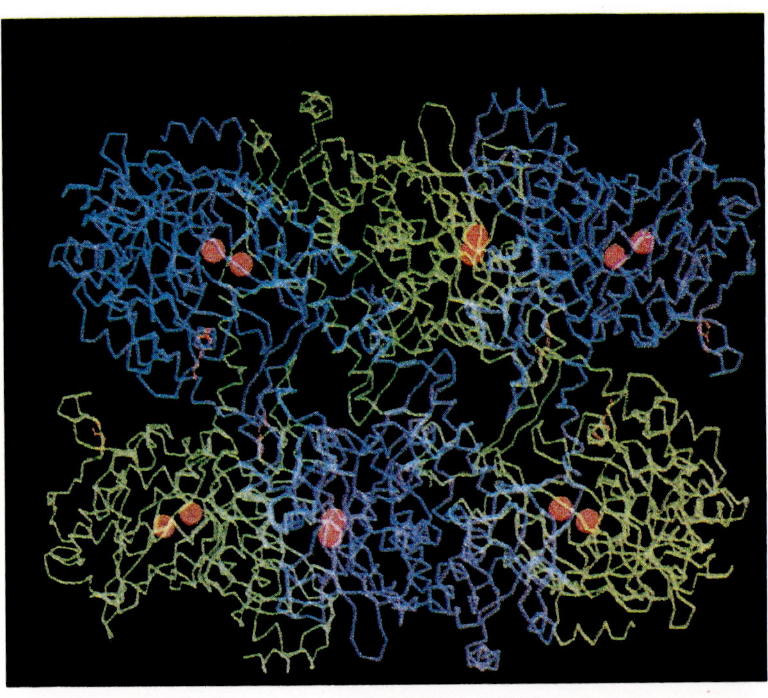

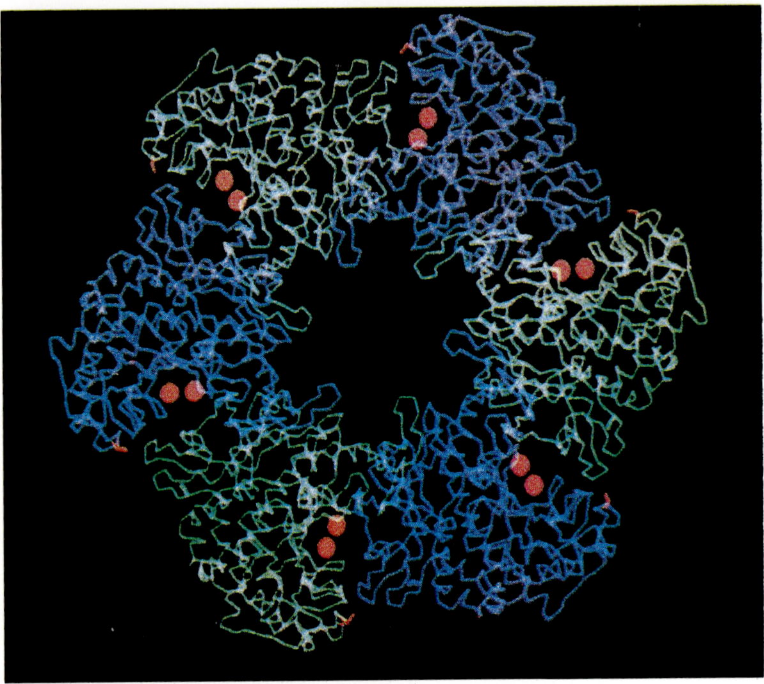

Plate 28 Two views of the dodecameric structure of glutamine synthetase (courtesy of David Eisenberg, University of California at Los Angeles).

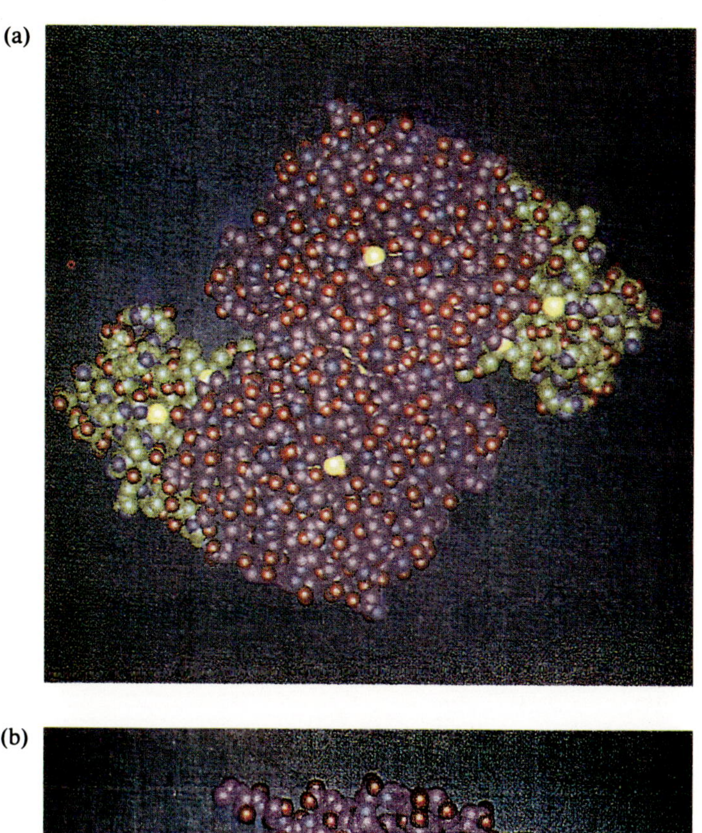

Plate 29 Three-dimensional structure of citrate synthase. (a) Closed conformation after substrate binding; (b) open conformation before substrate binding or after product release. (Reproduced with permission from Remington *et al.* 1982.)

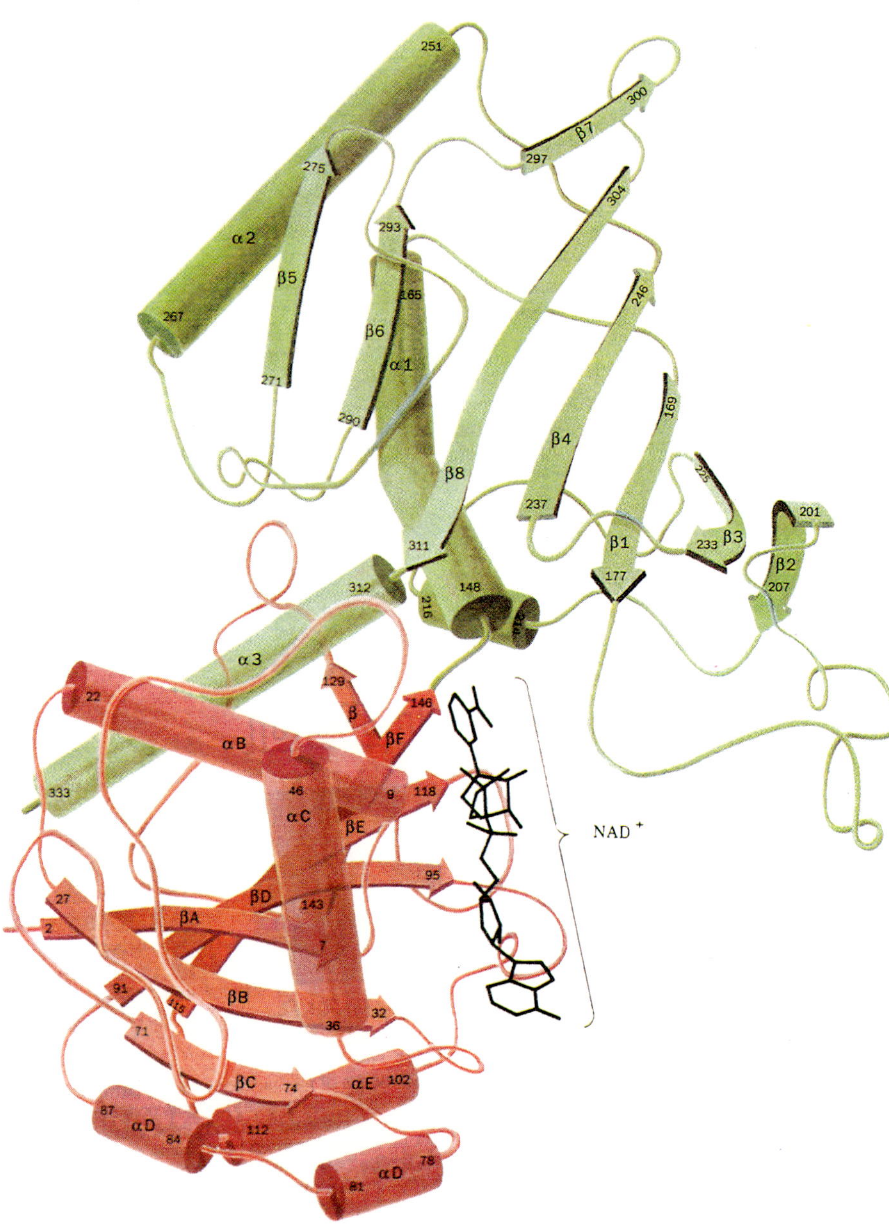

Plate 30 Three-dimensional structure of glyceraldehyde-3-phosphate dehydrogenase (Biesecker *et al.* 1977).

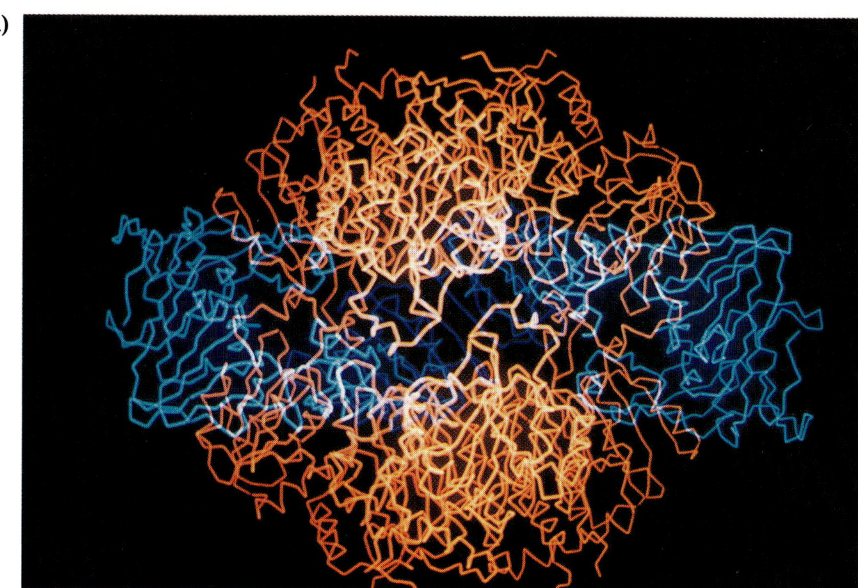

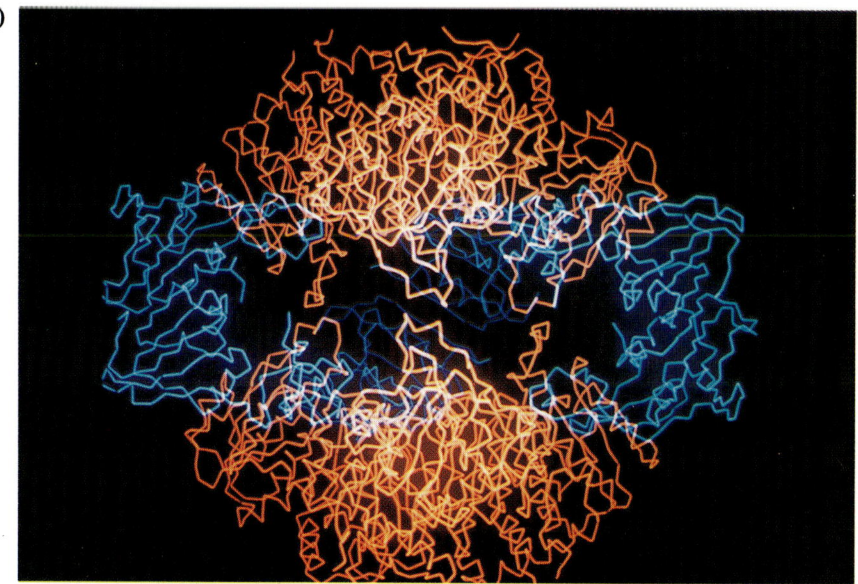

Plate 31 (a) T-state structure of aspartate transcarbamylase. (b) R-state structure of aspartate transcarbamylase. Ligation of substrate analogues to the T state induces an 10.8Å elongation of the molecule along the threefold axis, accompanied by opposite 6° rotations of each catalytic trimer (red) relative to one another. The regulatory dimers (blue) rotate 15° about the non-crystallographic twofold axis.

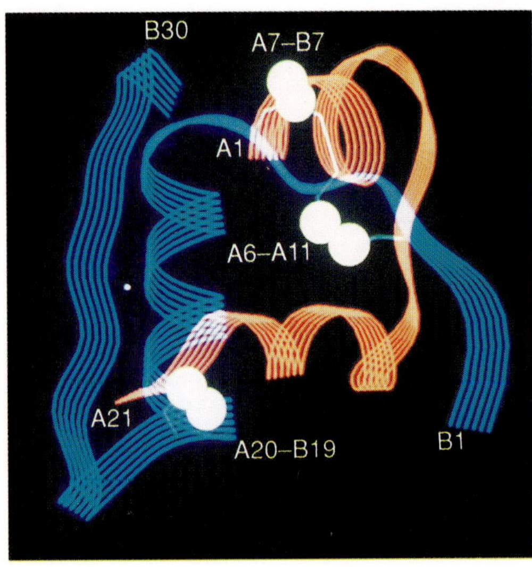

Plate 32 The tertiary structure of the insulin molecule.

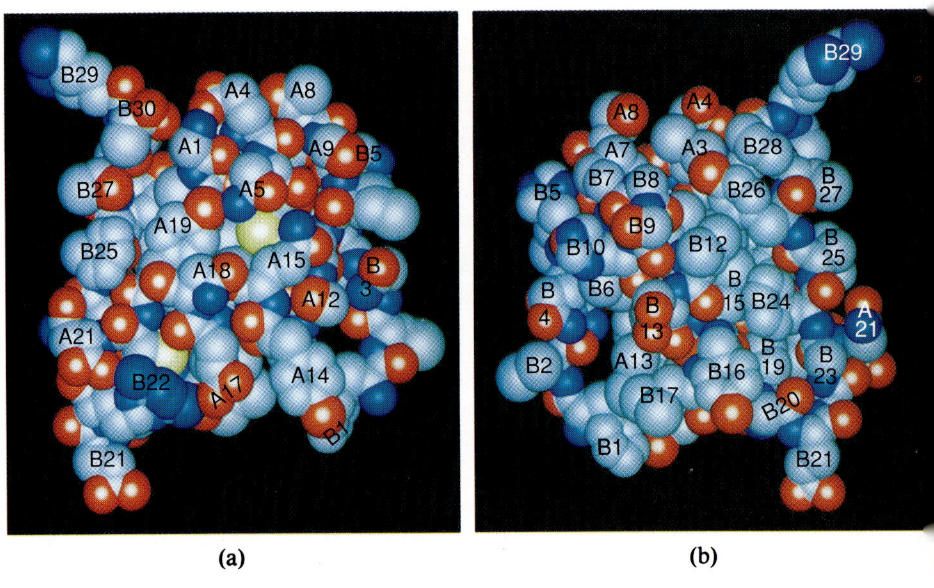

Plate 33 Surface of the insulin molecule (a) View from the A19 site (b) view from the B9-B19 helix site.

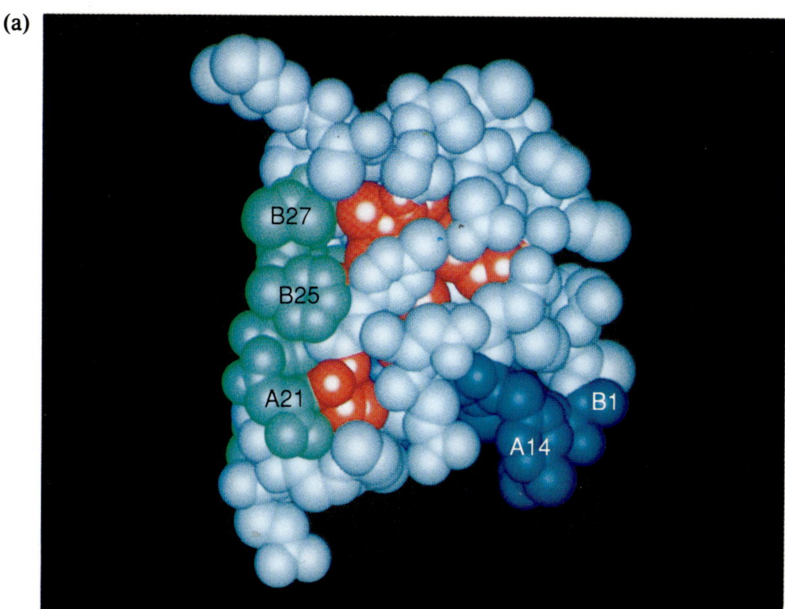

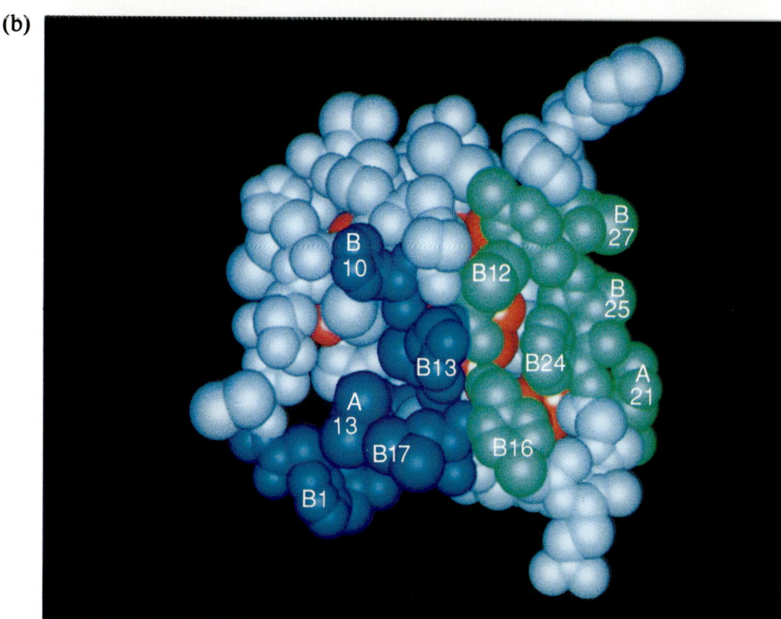

Plate 34 (a) Residues important for association of insulin molecule (a) view from the A19 site (b) view from the B9-B19 helix site. Dimer-forming residues are green; hexamer-forming residues, blue; and monomer interior, red.

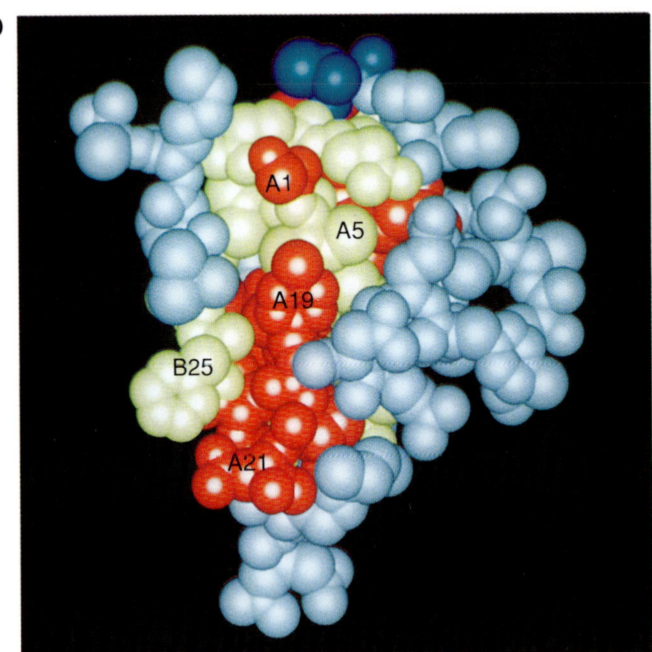

Plate 35 Invariant and semi-variant residues — the A19 Tyr site. The invariant residues in red, the semi-variant residues in yellow. A8 in dark blue.

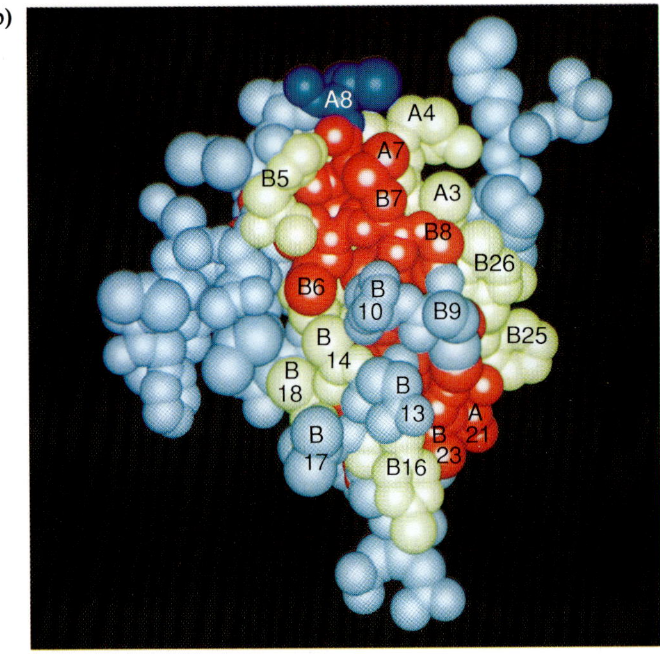

Plate 36 Invariant and semi-variant residues — the B-chain helix site. Red — invariant residues, yellow — semi-variant residues. A8 site dark blue.

and contain a single copper atom which cycles between the $+1$ and $+2$ oxidation states (reviewed by Adman 1985). The redox potential of the protein-bound copper is generally higher than for copper when associated in low molecular weight complexes, and ranges from about $+200\,mV$ to $+700\,mV$. They are called 'blue' because of a strong absorbance at about 600 nm by the oxidized protein, which gives them a deep blue colour. Other types of copper proteins, e.g. superoxide dismutase have a more complicated arrangement of copper atoms and usually have some kind of enzymatic activity.

There are four general categories of blue copper proteins, ranging in size from about 10 kDa to 23 kDa. These are azurins, of bacterial origin; plasto-yanins, from chloroplasts; and two classes of non-photosynthetic, plant blue copper proteins, one glycosylated and the other non-glycosylated. The non-glycosylated copper proteins range in size from 9 kDa to 16 kDa, while the glycosylated proteins can range up to 23 kDa in size.

7.4.1 Azurin

Azurins are bacterial blue copper proteins of M_r about 14 000, with redox potential between 230 mV and 395 mV. Their exact metabolic role is not well known, but one likely function is to transfer electrons from cytochrome c_{551} to cytochrome oxidase.

The azurin from *Alcaligenes denitrificans* contains 129 residues and a single copper atom. The structure of azurin in its 'standard' view is shown in Plate 20. This 'standard' view shows the copper atom bound at the top, or 'northern' end, of the molecule. The molecule is folded as an eight-stranded, mixed β-barrel. The β-barrel is interrupted between strands 4 and 5 by a 28-residue segment which forms a 'flap' on the outside of the molecule. The 'flap' contains the only α-helix in the molecule. There is also a single disulphide bond, between residues Cys-3 and Cys-26, which links the beginnings of strands β-1 and β-3.

The β-barrel of azurin can be better described as a β-sandwich consisting of two twisted β-sheets that pack together with a hydrophobic filling. The folding topology of the β-sandwich and its hydrogen-bonding pattern is shown in Fig. 7.8. The sequential arrangement of strands in the barrel is 1, 3, 6, 5, 4, 7, 8, 2. The two β-sheets of the sandwich are formed by the first half of strand 2 and strands 1, 3, 6 (sheet I) and by strands 5, 4, 7, 8, and the second half of strand 2 (sheet II). There are no hydrogen bonds between strands 6 and 5, but their side-chains are in contact. Thus the two sheets are joined on one side only, by strand 2 which is split between them. The β-sheets are of mixed type, with strands 1 and 3 parallel, strands 2 and 8 parallel, and the remaining adjacent strands antiparallel. This is a 'Greek key' type of antiparallel β-barrel topology.

The flap consists of a three-turn helix, residues 55–67, and an extended

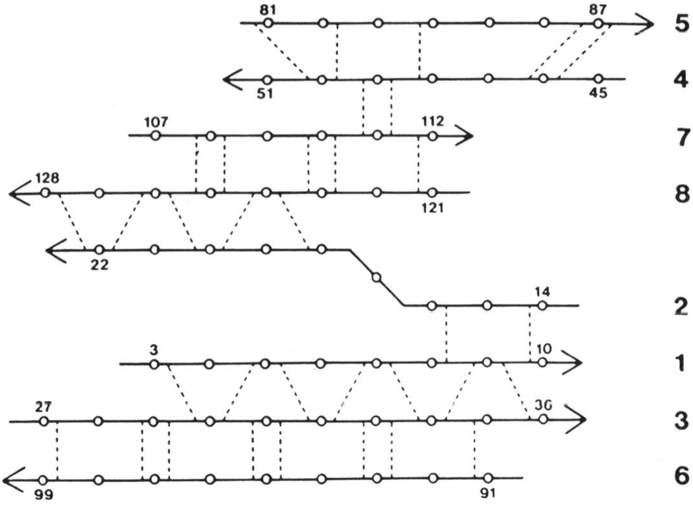

Fig. 7.8 β-sheet topology and hydrogen-bonding pattern for azurin from *Alcaligenes denitrificans* (reproduced by permission from Baker 1988).

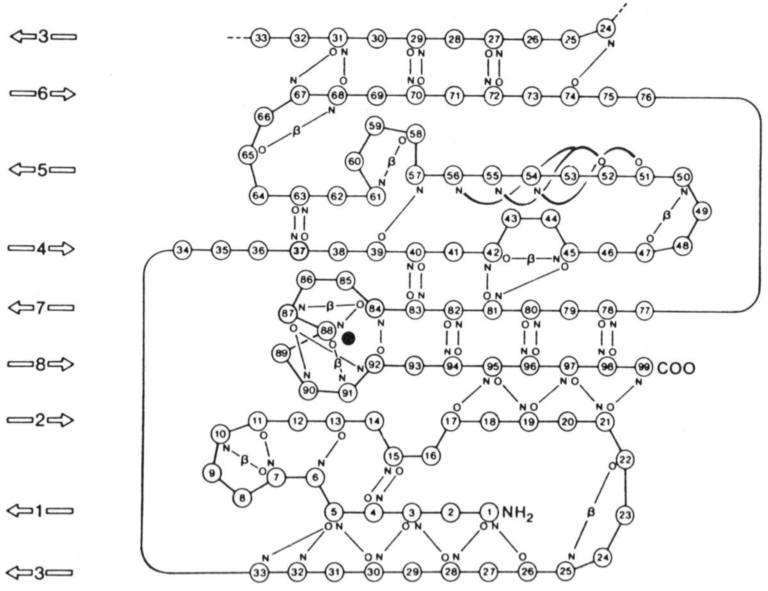

Fig. 7.9 β-sheet topology and hydrogen-bonding pattern for poplar plastocyanin (reproduced by permission from Guss and Freeman 1983).

chain, residues 68–81, which forms irregular loops. The interface between the flap and the barrel is hydrophobic, containing seven aliphatic and aromatic residues. The interior of the β-sandwich is also hydrophobic, containing 19 aliphatic and aromatic side-chains, including the ring portions of several tyrosines.

The copper atom is bound by five ligands, which form a distorted, axially elongated, trigonal bipyramid. This arrangement of the copper ligands is shown in Plate 21. The Met and Cys residues are yellow, Gly green, and His blue. Three of the copper ligands involve short bonds, from the ND atoms of His-46 and His-117 and the SG atom of Cys-112. The Cu-ND distances average 2.04 Å while the Cu-SG distance is 2.14 Å. The coordination geometry of these short ligands about the copper is distorted trigonal.

The other two copper ligands are the SD of Met-121 and the carbonyl oxygen of Gly-45. These are at considerably greater distance from the copper, 3.11 Å and 3.13 Å, respectively. These ligands are arranged above and below the (N_2S) trigonal plane, respectively, to form the distorted (N_2S_2O) trigonal bipyramid.

Three of the ligands, Cys-112, His-117, and Met-121, are located in the loop linking strands 7 and 8 at the 'northern' end of the molecule. The other two ligands, His-46 and the Gly-45 peptide oxygen, are located near the beginning of strand 4.

The orientations of the three strong ligands to the copper, His-46, His-117, and Cys-112, are constrained by elements of the protein structure. His-117 is sandwiched between the side-chains of Phe-114 and Met-13, although the solvent-exposed NE atom is uncomplexed. The NE atom of His-46 forms a hydrogen bond to the peptide oxygen of Asn 10. The SG atom of Cys-112 receives two hydrogen bonds from peptide NH groups, one from the NH of Asn 47 and the other from the NH of Phe-114. The latter hydrogen bond follows the pattern '$n+2$ to n', observed in the NH-SG hydrogen-bonding arrangement found in the iron-sulphur proteins. Thus the loop 112–121, strand 45–47, and the loop 10–13, all on the 'northern' end of the molecule, are connected via copper coordination and hydrogen bonding to form a very rigid structure.

There is a hydrophobic patch on the surface of azurin surrounding His-117, one of the strong copper ligands. This patch and the exposed histidine residue are shown in Plate 22. The patch is quite prominent and is a likely site for protein-protein interaction during electron transfer. His-117 is surrounded on the protein surface by two concentric rings of hydrophobic residues which form a flat surface 14 Å by 18 Å in size. The inner ring contains Met-13, Met-44, Phe-114, Pro-115, and the methylene of Gly-116. The outer ring is formed by Met-39, Ala-43, Trp-118, Ala-119, and Met-120.

There is an approximately equal number of acidic and basic residues (16 and 14, respectively) in azurin. These residues are all located on the protein surface and are distributed uniformly, except for the hydrophobic patch. There is no clustering of the charged side-chains into positively or negatively charged patches as observed in some of the other electron transporters.

7.4.2 Plastocyanin

Plastocyanin is a blue copper protein with an intense absorption at about 600 nm, found in chloroplasts of plants and algae. It serves as the primary electron donor to photosystem I, transferring electrons from the membrane-bound cytochrome b_6/f complex of photosystem II to pigment $P700^+$ of photosystem I. Its redox potential is quite high, ranging from $+350$ to $+400$ mV.

Poplar plastocyanin has an M_r of about 10 000 and contains 99 amino acids and one copper ion. The protein is acidic with a pI below 4. The ratio of acidic to basic side-chains is 15:8, giving the oxidized protein a net charge of approximately -7.

The structure of plastocyanin is shown in Plate 23. The polypeptide chain is folded as an eight-stranded barrel, with the copper atom bound at one end. The molecule is oriented with the copper-containing 'northern' end at the top of the diagram. Seven of the strands, 1–4 and 6–8, are well-formed β-strands, but 'strand 5' is quite irregular, and includes 1.5 turns of helix and a β-turn.

The overall structure of plastocyanin is very similar to that of azurin. Its folding topology and hydrogen-bonding pattern are shown in Fig. 7.8. Like azurin, it forms a β-sandwich. One face is formed by the first half of strand 2 and by strands 1, 3, and 6. The other face is formed by the second half of strand 2 and by strands 8, 7, and 4. 'Strand 5' does not belong to either face of the sandwich and is called a strand only by analogy with azurin. The interior of the sandwich is filled with hydrophobic residues. Included among these are seven aromatic residues, six Phe and one Tyr, which point toward the centre of the molecule.

The copper atom is held in a pocket formed by three loops at the 'northern' end. It is bound by four side-chains, with the atoms of the copper ligands forming a distorted tetrahedron. This arrangement of the copper ligands is shown in Plate 24. The four ligands are His-37 ND (blue), Cys-84 SG (yellow), His-87 ND (blue), and Met–92 SD (yellow). For the first three ligands the bond distances to copper are 2.04 Å, 2.10 Å, and 2.13 Å, respectively. The distance of the fourth ligand, Met-92 SD, to copper, is 2.90 Å. Both of the Cu-S bond distances are unusual, compared with model compounds, with the Cys SG to copper distance shorter than expected and the Met SD to copper distance longer. The observed copper-ligand bond distances are consistent with EXAFS studies on plastocyanin.

The copper-binding pocket is defined by three loops between β-strands, connecting strands 1 with 2, 3 with 4, and 7 with 8. His-37 is located in the N-terminal half of strand 4. The other three ligands are in the loop between strands 7 and 8. There is a high degree of sequence conservation among the various plastocyanins around the latter loop, from residues 82 to 94.

There is one hydrogen bond from a peptide NH (residue Asn 38) to the copper ligand, SG of Cys-84. Thus the '$n+2$ to n' hydrogen-bond pattern for peptide amide groups found in azurin and many iron-sulphur proteins is not observed in plastocyanin. The orientation of His-37 is stabilized by hydrogen bonding between its ND atom and the carbonyl oxygen of Ala-33.

The copper-binding pocket is ringed by hydrophobic residues, which form a rim around the pocket. These residues are Leu-12, Ala-13, Ala-33, Phe-35, Pro-36, Leu-62, Pro-86, and Ala-90. The distribution of these residues on the surface of plastocyanin is shown in Plate 25. All of these side-chains are exposed to solvent. These hydrophobic residues form a relatively planar surface. With the His-87 copper ligand side-chain in its centre, this area seems to be a likely place for electron transfer to a redox partner to occur.

The acidic side-chains, which are present in abundance over the basic side-chains, are unevenly distributed over the protein surface. There are none located in the 'northern' quarter of the molecule. However, there are six acidic groups located in two kinks, residues 42–44 and residues 59–61, which are exposed to solvent and form an elongated patch. This patch is also shown in Plate 25. The acidic patch could serve as another attachment site for a redox cofactor.

7.5 Flavodoxin

Flavodoxins are one-electron carrier proteins found in several types of bacteria and algae. They contain FMN bound non-covalently to a single polypeptide chain, and have molecular weights ranging from about 15 000 to 22 000 Da. (For a review of flavodoxins, see Meyhew and Ludwig 1975.) They will often substitute for ferredoxin in intermolecular electron transfer reactions, particularly when an organism is grown in an iron-deficient medium. Like ferredoxin, they function as low-potential electron carriers ($\approx -400\,\text{mV}$) and are rather acidic.

FMN can exist in three oxidation states: oxidized, semiquinone, and reduced. In the reduced state, neutral FMN contains an additional two electrons and two protons with respect to the oxidized state; while in the semiquinone state, there is only a single additional electron and proton. The single extra electron of the semiquinone is unpaired, causing the FMN to be a free radical. For free FMN, the redox potential for the oxidized/

semiquinone couple, E_1, is $-172\,\text{mV}$, while for the semiquinone/reduced couple the potential, E_2, is $-238\,\text{mV}$.

In flavodoxin, the redox potentials for the bound FMN is considerably different from those of free FMN, indicating that the protein environment has considerable effect in modulating the potential. For the flavodoxin from *Clostridium* MP, the redox potential E_1 is raised to a value of $-92\,\text{mV}$, while the value for E_2 drops to $-400\,\text{mV}$. The physiologically important redox couple of flavodoxin is the latter, between the semiquinone and fully reduced states of FMN, since it can substitute for ferredoxin, which has a comparably low potential.

Flavodoxin from *Clostridium* MP has a molecular weight of about 15 kDa and contains 150 amino-acid residues. Its structure (Plate 26), is very regular, containing four α-helices and a five-stranded parallel β-sheet. The four α-helices are distributed two on each side of the β-sheet, lying approximately antiparallel to the β-strands and parallel to each other.

Flavodoxin is a classic example of a 'nucleotide-binding domain' or 'doubly wound α/β structure'. Its conformation is very similar to that of other nucleotide-binding proteins, particularly the NAD-dependent lactate dehydrogenase. The latter contains a six-stranded parallel β-sheet with four α-helical connections running in the opposite direction. When the five β-strands of flavodoxin are aligned with the first five strands of lactate dehydrogenase, the phosphate groups of FMN and NAD are nearly coincident and the first three helices of flavodoxin lie close to helices α_B, α_E, and α_F of lactate dehydrogenase.

The FMN is attached to flavodoxin with the isoalloxazine ring at the surface of the molecule, near the C-terminal end of the central β-strands, and the ribityl phosphate chain extending toward the protein interior. The dimethyl benzene portion of the isoalloxazine ring is exposed to solvent while the pyrimidine ring portion is buried. The flavin ring is planar, and is sandwiched between two hydrophobic residues, Met-56 and Trp-90. The latter is inclined to the flavin ring by about $20°$.

Three segments of the polypeptide chain interact with the FMN group of flavodoxin, residues 7–12, 55–59, and 89–91. The extent of solvent exposure of the FMN moiety when the FMN is bound to flavodoxin is indicated in Plate 27. The phosphate moiety of FMN is deeply buried in the protein, with little solvent exposure and no nearby compensating, positively charged, amino-acid side-chain. Instead, the phosphate forms hydrogen bonds to the side-chains of several Thr or Ser residues, one water molecule, and the five peptide nitrogen atoms of residues 8–12. The latter peptide segment forms part of the first turn of helix 1, which provides partial compensating positive charge for the phosphate in the form of a helix dipole. The ribityl chain and flavin ring interact exclusively with main-chain amide or carbonyl groups, mostly from residues 55–59 and 89–91, and with water molecules. The exceptions are an Asn and a Thr side-chain, which interact with ribityl hydroxyl groups, and a Glu side-chain, which interacts with atom N3 of the flavin ring. The last interaction is the only one between FMN and a charged amino acid.

The overall charge of the flavodoxin from *Clostridium* MP is quite negative, with a total of 28 Asp and Glu residues distributed over the surface of the molecule, compared with a total of 12 Arg and Lys residues. There are four pairs of well-defined salt bridges, and several other close charge interactions. There are two patches of negative charges on the protein surface, one of which (residues 62–67) is close to the flavin ring.

The structures of the oxidized and semiquinone forms of flavodoxin from *Clostridium* MP have been determined independently. The most significant difference between their structures is a change in conformation of the residue Gly-57 in the peptide chain segment 55–59 which interacts with the flavin ring. In the oxidized form of flavodoxin, the plane of the peptide of Gly-57 is oriented so that the carbonyl group points away from N5 of the flavin ring. In the semiquinone form, the peptide plane flips so that the carbonyl group points toward N5 and can form a hydrogen bond with it. In this process the peptide segment, 56–59, which forms a reverse turn, is transformed from a less energetically favoured Type II 3_{10} bend to a more stable Type I bend.

References

Adman, E.T. (1979). A comparison of the structures of electron transfer proteins. *Biochimica et Biophysica Acta* **549**, 107–44.

Adman, E.T. (1985). Structure and function of blue copper proteins. In *Metalloproteins*, (ed. P. Harrison), Part 1, pp. 1–42. Verlag Chemie, Weinheim, FRG.

Argos, P. and Mathews, F.S. (1975). The structure of ferrocytochrome b5 at 2.8 Å resolution. *Journal of Biological Chemistry* **250**, 747–51.

Baker, E.N., (1988). Structure of azurin from *Alcaligenes denitrificans*. Refinement at 1.8 Å resolution and comparison of the two crystallographically independent molecules. *Journal of Molecular Biology* **203**, 1071–95.

Bushnell, G.W., Louie, G.V., and Brayer, G.D. (1990). High-resolution three-dimensional structure of horse heart cytochrome c. *Journal of Molecular Biology* **214**, 585–95.

Carter, C.W. (1977). X-ray analysis of high-potential iron-sulfur proteins and ferredoxins. In *Iron-sulfur proteins*, (ed. W. Lovenberg), Vol. 3, pp. 157–204. Academic Press, New York.

Dickerson, R.E. and Timkovich, R. (1975). Cytochromes c. In *The enzymes*, (ed. P. Boyer), Vol. 11, pp. 397–547. Academic Press, New York.

Guiard, B. and Lederer, F. (1979). The 'cytochrome b5 fold': structure of a novel protein superfamily. *Journal of Molecular Biology* **135**, 639–50.

Guss, J.M. and Freeman, H.C. (1983). Structure of oxidized plastocyanin at 1.6Å resolution. *Journal of Molecular Biology* **169**, 521–63.

Louie, G.V. and Brayer, G.D. (1990). High-resolution refinement of yeast *iso*-1-cytochrome c and comparisons with other eukaryotic cytochromes c. *Journal of Molecular Biology* **214**, 527–55.

Mathews, F.S. (1985). The structure, function and evolution of cytochromes. *Progress in Biophysics and Molecular Biology* **45**, 1–56.

Meyer, T.E. and Kamen, M.D. (1982). New perspectives on c-type cytochromes. *Advances in Protein Chemistry* **35**, 105–212.

Meyhew, S.G. and Ludwig, M.M. (1975). Flavodoxins and electron-transferring flavoproteins. In *The enzymes* (ed. P. Boyer), Vol. 12, pp. 57–118. Academic Press, New York.

8

Allosteric enzymes

Raymond C. Stevens and William N. Lipscomb

8.1 Allostery

8.1.1 Basic tenets of allosteric behaviour

The term allosteric (Greek, *allos* = other, *steros* = solid or space) is an ill-defined term used to describe the behaviour of numerous proteins. *Webster's dictionary* defines allosteric as 'of or having to do with a protein or structure that is altered reversibly by a small molecule so that its original function is modified'. Biochemistry textbooks define allostery as the behaviour of those proteins which do not kinetically conform to the 'classical' Michaelis-Menten model. That is, a plot of reaction velocity V versus substrate concentration [S],

$$V = \frac{V_{max}[S]}{[S] + K_M}$$

for allosteric proteins does not yield the hyperbolic curves of Michaelis-Menten kinetics but instead displays a sigmoidal curve (Fig. 8.1). This sigmoidal curve suggests co-operativity in the substrate binding of multiple subunit proteins. The binding of the first substrate molecule induces changes in substrate affinity or catalytic efficiency or both at the other, sometimes distant, active sites.

The allosteric phenomena excite interest because of the physiological significance and the challenge offered for a physical interpretation. Classical models of protein co-operativity are based on the concept of multiple equilibria between several conformational states. The states range from the T state with low affinity/low activity for substrates to the R state with high affinity/high activity for substrates. Monod, Wyman, and Changeaux proposed the first model to explain allosteric behaviour (MWC model; Monod *et al.* 1965).

The basic tenets of allostery proposed by Monod, Wyman, and Changeaux (Monod *et al.* 1965) are as follows:

1. Allosteric proteins are oligomers, the protomers of which are associated in such a way that they all occupy equivalent positions. This implies that the molecule possesses at least one axis of symmetry.

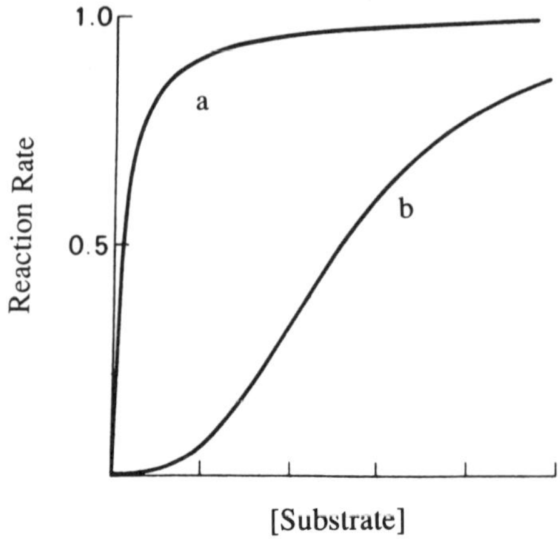

Fig. 8.1 Hyperbolic (a) versus sigmoidal (b) kinetics.

2. To each ligand able to form a *stereospecific* complex with the proteins, there corresponds one, and only one, site on each protomer. In other words, the symmetry of each set of stereospecific receptors is the same as the symmetry of the molecule.

3. The conformation of each protomer is constrained by its association with the other protomers.

4. Two (at least two) states are reversibly accessible to allosteric oligomers. These states differ by the distribution and/or energy of interprotomer bonds, and therefore also by the conformational constraints imposed upon the protomers.

5. As a result, the affinity of one (or several) of the stereospecific sites towards the corresponding ligand is altered when a transition occurs from one to the other state.

6. When the protein goes from one state to another state, its molecular symmetry (including the symmetry of the conformational constraints imposed upon each protomer) is conserved.

The treatise further states:

'By far the most striking and, physically if not physiologically, the most interesting property of allosteric proteins is their capacity to mediate homotropic (see Glossary) co-operative interactions between stereospecific ligands. Although there may be exceptions to this rule, we shall consider that this property characterizes allosteric proteins.'

The assumption of the two states allows the formulation of a mathematical model with three independent variables: K_T and K_R, the association constants of the ligands with the protein in the T and R states, respectively, and L_o = [T]/[R], the concentration ratio of the two structures in the absence of ligands. The reader is cautioned that the MWC model is often referred to as the 'two-state model' and the assumption above is the simplest case of the MWC model. As stated in the basic tenets, *at least two* states must be reversibly accessible to the oligomeric protein. The reader is also cautioned that allosteric behaviour is not a prerequisite for the observation of sigmoidal kinetics. Even though the co-operative model explains sigmoidal data, co-operativity is not essential. In fact, a model that requires a single, independent active site with multiple reaction pathways by which the binding sites are occupied can also explain sigmoidal data.

The allosteric model defines conformational states in terms of substrate affinity and activity. Since the two states can have similar affinities for substrate but different catalytic rates, or conversely, the allosteric model is defined further. In a system where the allosteric states differ in binding affinities and not in activity, the enzyme is defined as a 'K-system'. In a system where the allosteric states differ in catalytic activity but not in substrate affinity, the enzyme is defined as a 'V-system'.

Shortly after the MWC model was proposed, alternative explanations for the co-operative behaviour of multiple subunit proteins emerged. One explanation by Koshland, Némethy, and Filmer (KNF; Koshland *et al.* 1966) suggests that sigmoidal behaviour of oligomeric proteins can be explained by a sequential binding mechanism instead of the concerted model of Monod and co-workers. Because symmetry must be conserved in the MWC model, the allosteric transition must be concerted. The KNF sequential model does not have this restriction and is able to explain negative co-operativity in substrate binding. The debate of symmetrical versus asymmetrical models of allostery continues today. Generally, both models are mentioned in most articles concerning allostery.

Until recently, haemoglobin was the only allosteric protein with detailed X-ray structural evidence available in more than one allosteric state. Presently, four enzymes have been observed via X-ray crystallography in at least two different conformational states. They are: *Escherichia coli* aspartate transcarbamylase, rabbit muscle glycogen phosphorylase, bacterial phosphofructokinase, and pig kidney fructose-1,6-bisphosphatase. Two excellent review articles have recently appeared on allosteric proteins (Perutz 1989) and on allosteric enzymes (Hervé 1989). The X-ray studies provide a description of the low affinity/low activity and high affinity/high activity states. Under certain conditions the equilibrium is perturbed to favour almost exclusively either the T state or the R state. Knowledge of both the T-and R-state structures is essential in elucidating the routes for communication between the sites. X-ray crystallography has not provided the only evidence for the allosteric transition. The molecular conversion from the T to the R state has been monitored in solution by ultraviolet difference spectroscopy, difference sedimentation, circular dichroism, and X-ray solution scattering. The structural studies provide information on the complex interactions that stabilize a particular state. Unfortunately, they provide little information on the pathway between states. Furthermore, if only one conformational state is known, it is difficult to predict the second conformational state.

More than 30 enzymes have been described as allosteric in the scientific literature. In this chapter, we will discuss only those allosteric enzymes that have been observed in more than one conformational state via X-ray crystallography. A few proteins that are not discussed in the main text are mentioned briefly in the following section. These proteins have been observed in at least one conformational state via X-ray crystallography and display certain allosteric properties that deserve mention.

8.1.2 Allosteric and pseudo-allosteric proteins

Although haemoglobin does not catalyse a chemical transformation, it is generally and justifiably mentioned in any discussion on allostery. Haemoglobin was the initial protein used by Monod, Wyman, and Changeaux in developing their allosteric principles. The oxygen affinity is modulated by H^+, Cl^-, CO_2, and DPG (2,3-diphosphoglycerate). At low haemoglobin concentrations, co-operativity effects arise as a result of reversible dissociation of tetrameric deoxy- to dimeric oxyhaemoglobin. At high haemoglobin concentrations, the co-operative effects are due to an equilibrium between two alternative quaternary structures of the tetramer,

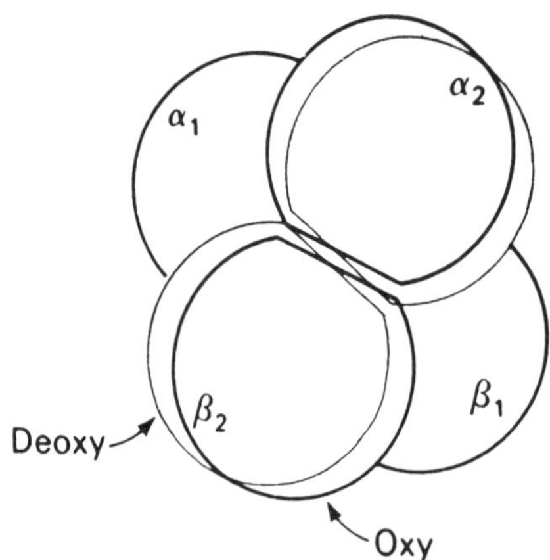

Fig. 8.2 Schematic diagram illustrating the rotation of the $\alpha_2\beta_2$ dimer relative to the $\alpha_1\beta_1$ dimer that occurs in the quaternary structure change from deoxy-(light lines) to oxy-(heavy lines) haemoglobin. (Reproduced with permission from Baldwin and Chothia 1979.)

the deoxy, or T, state and the oxy, or R, state. Human haemoglobin is a tetramer composed of two α-chains and two β-chains, each containing one haem group (Perutz *et al.* 1960). The T and R structures differ in the arrangement of the four subunits (Fig. 8.2; Baldwin and Chothia 1979). The T→R transition consists of a rotation of dimer $\alpha_1\beta_1$ relative to dimer $\alpha_2\beta_2$ by 15° and a translation of one dimer relative to the other by 0.8 Å. The $\alpha\beta$ dimers move relative to each other at the symmetry-related contacts $\alpha_1\beta_2$ and $\alpha_2\beta_1$ and at the contacts $\alpha_1\alpha_2$ and $\beta_1\beta_2$; the contacts $\alpha_1\beta_1$ and $\alpha_2\beta_2$ remain rigid. Haemocyanin and haemerythrin are two more examples of oxygen-carrying allosteric proteins. These three proteins are discussed in Chapter 6.

An excellent example of an enzyme that displays cumulative feedback inhibition is glutamine synthetase (GS) from *Salmonella typhimurium*. GS catalyses the condensation of ammonia and glutamate to yield glutamine, and thus plays an essential role in bacterial nitrogen metabolism. Glutamine is a source of nitrogen in the biosynthesis of many metabolites, eight of which act as feedback inhibitors on GS (Ginsberg 1972). The enzyme's synthesis of glutamine is repressed by ammonia, and its catalytic activity is inhibited cumulatively by alanine, glycine, histidine, tryptophan, cytidine triphosphate, adenosine monophosphate, carbamyl phosphate, and glucosamine 6-phosphate. The synthesis is also repressed by reversible adenylation of a specific tyrosine residue catalysed by an enzyme system that is regulated by other metabolites. David Eisenberg and his colleagues at the University of California at Los Angeles have solved the structure of glutamine synthetase from *Salmonella typhimurium* combined with manganese ions in its unadenylated state at 3.5 Å resolution (Yamashita *et al.* 1989). The enzyme is a dodecamer of identical subunits (52 kD) stacked together in two rings of six, placed face to face, so that the adjacent rings are related by the twofold axis (Plate 28). Each subunit touches four of its neighbours. Structurally, the enzyme shows hydrophobic α-helices from one ring of subunits fitted into hydrophobic sockets of the opposite layer. Because the present resolution level is 3.5 Å, it is impossible to determine more precise details of the structure. What is known is that the manganese ions are approximately 20 Å from tyrosine 397, whose adenylation site sensitizes the enzyme to its inhibitors. To date, this is the only structure of the enzyme. Adenylation of tyrosine 397, or ligation of several feedback inhibitors, does not affect the X-ray diffraction pattern relative to the unadenylated X-ray diffraction pattern. Addition of transition state analogues such as methionine-sulphoxime together with ATP formed a methionine-sulphoxime-ADP complex with the enzyme. The diffraction pattern of the methionine-sulphoxime-ADP complex is different from that of the unligated enzyme and may be due to the second state of glutamine synthetase. The simplest suggested allosteric transition appears to be a rotation of the two rings relative to each other about the sixfold axis and/or a translation along that axis. This transition is similar to the rotational transition found in the allosteric transition of aspartate transcarbamylase. If the enzyme alternated between only two quaternary structures R and T, then different inhibitors could stabilize the same T-state structure by combining with a variety of different sites. This model would account for the cumulative rather than competitive action.

Pyruvate kinase (PKase; EC 2.7.1.40) catalyses the third and last irreversible step in glycolysis. Physiologically, PKase interconverts adenosine diphosphate (ADP) and phosphoenolpyruvate (PEP) to ATP and pyruvate (Boyer 1962). The reaction requires both magnesium and potassium ions as activating cations. ATP allosterically inhibits the isozyme to slow catalysis when the energy charge is high. Alanine, which is synthesized in one step from pyruvate, allosterically inhibits the enzyme, in this case to signal that the building blocks are abundant. Fructose-1,6-bisphosphate, the product of the preceding step in glycolysis, activates pyruvate kinase to enable it to keep pace with the oncoming high flux of intermediates. PKase is a tetramer of 60 kD subunits and 500 amino-acid residues. The crystal structure of pyruvate kinase (1PYK) has been solved to 2.6 Å resolution in the laboratory of Hilary Muirhead at the University of Bristol (Levine *et al.* 1978). Each subunit is folded into three distinct domains, as shown in Fig. 8.4. Domain A is the largest, composed of 220 residues, and contains a cylindrical β-sheet of eight parallel strands. Adjacent strands are connected by α-helices, which form an outer cylinder coaxial with the first. Between the third strand of β-sheet and third helix of domain A the chain folds up into domain B (100 residues) made up of an antiparallel β-sheet. The last 120 residues form domain C, where the first section comprises two long antiparallel α-helices. The rest of the domain folds up into a five-stranded

Domain B   Domain A   Domain C

Fig. 8.4 Three-dimensional structure of pyruvate kinase .

β-sheet flanked by α-helices. Similarities have been observed in the first two domains of PKase with triose phosphate isomerase (TIM), which is also located in the glycolytic pathway. The structural transition of PKase has been monitored by a combination of approaches, including small angle neutron scattering, protein denaturation, partial tryptic digestion, and peptide sequencing. Results from these studies are consistent with the description of a conformational change that involves domain-domain interaction. The sigmoidal nature of the kinetic data suggests that the system exhibits positive co-operativity. Although the alternate state of PKase has not been determined, molecular modelling has been performed by altering the α-carbon coordinates of the known state. The comparison of this model with the structure of the second state in the presence of the substrate, PEP, will be interesting.

A number of enzymes undergo large conformational changes upon substrate binding. These enzymes are able to exist in more than one conformational state and the two states display different affinity/activity behaviour. Because the enzymes do not display co-operative binding character, they are not considered allosteric. One example of this type of enzyme is citrate synthase (EC 4.1.3.7), a component of nearly all living cells, which plays a key role in the central metabolic pathway of aerobic organisms, the citric acid cycle. The citric acid cycle functions as a source of two reducing equivalents for the electron transport chain and as a source of intermediates required for the biosynthesis of amino acids, for ketogenesis, lipogenesis, and gluconeogenesis. The flux throughout the citric acid cycle is maintained by activated acetic acid, originating as acetyl coenzyme A (acetyl CoA) from the degradation of carbohydrate, fatty acids, and amino acids. Citrate synthase catalyses the condensation of acetyl CoA and oxaloacetic acid to form citrate, thus forming a new carbon-carbon bond. The three-dimensional structures of citrate synthase with and without substrates have been determined in the laboratory of Robert Huber at the Max-Planck Institut für Biochemie and Stephen Remington at the University of Oregon, Eugene (Remington *et al.* 1982; Weigand *et al.* 1984). The molecular structure is divided into two domains, a large domain of residues 1–274 and 381–437, and a small domain of residues 275–380. The large domain forms the site of interaction of the two subunits in the dimeric molecule. In the structure of the free enzyme the two domains are separated by a deep cleft. When oxaloacetate binds to the enzyme, the small domain rotates 18° relative to the large domain in each subunit (Plate 29). This movement can be described, to a good approximation, as a rigid body motion around a hinge axis passing through the molecule close to His274. Movements as large as 15 Å are produced by the rotation of α-helices elicited by quite small shifts in side-chains around bound oxaloacetate. The structural rearrangement leads to the creation of a binding site for acetyl

CoA, which binds second. Citrate synthase catalyses the condensation reaction by bringing the substrates into close proximity, orienting them, and polarizing certain bonds. After formation of the intermediate, citryl CoA, additional structural changes are observed while citryl CoA is hydrolysed. CoA then leaves the enzyme, followed by citrate, and the enzyme returns to the open conformation. The conformational transition is similar to the cleft closure in hexokinase induced by the binding of glucose (Steitz *et al.* 1977).

The most abundant form of co-operativity is positive co-operativity, whereas negative co-operativity is less frequently observed. One enzyme that displays negative co-operativity is glyceraldehyde-3-phosphate dehydrogenase (GAPDH). GAPDH catalyses the oxidative phosphorylation of D-glyceraldehyde-3-phosphate to form 1,3-diphosphoglycerate in the presence of NAD^+ and inorganic phosphate (Harris and Waters 1976). It is composed of four chemically identical subunits and has a molecular weight of approximately 145 kD. Medium-and low-resolution studies of GAPDH from various sources have been reported, and the high-resolution model (1.8 Å) has been obtained from *Bacillus stearothermophilus* (Skarzynski and Wonacott 1988). GAPDH undergoes conformational changes upon NAD binding, and early investigations revealed that the changes are sequential (Plate 30). The first molecule of NAD to bind to the apoenzyme (see Glossary) induces a conformational change in the ligand subunit to give an asymmetric tetramer. The GAPDH molecule becomes symmetrical again only when it is fully saturated with NAD, i.e. four molecules of NAD are bound to the tetramer. Since the physiological level of NAD in cells is high and relatively invariant, it is unlikely that the apoenzyme tetramer is of biological significance. However, the apo state of individual subunits plays an important part in catalysis, since one of the reaction steps involves release of the cofactor NADH in the forward reaction or NAD^+ in the reverse reaction. Because apoenzyme prepared from thermophilic organisms is much more stable than its counterpart from mesophiles, crystallographic studies of this form are possible. The main difference between the T and R states of the enzyme is an approximate rigid-body rotation of the coenzyme-binding domain relative to the catalytic domain by 4.3°. The catalytic dimers rotate approximately 2° relative to each other.

An analogy of allosteric behaviour occurs in the crystal structures of hexameric insulin (Smith *et al.* 1984; Derewenda *et al.* 1989). See also Chapter 9. In the bloodstream, insulin is a monomer, while in the stored zinc-containing form, insulin aggregates as a hexamer. Binding and activity data suggest that the insulin molecule undergoes a substantial conformational change when it binds to the receptor (Dodson *et al.* 1983). The 2-Zn (per hexamer) structure is analogous to T_6, the 4-Zn to T_3/R_3, and the 2-Zn (per hexamer)—one phenol (per monomer)—is analogous to R_6. The first six residues of the B-chain are extended in the T form and helical in the R form.

In eukaryotic cells, most control and communication rely on membrane proteins which respond to chemical and electrical signals. The low-resolution structure of the nicotinic acetylcholine receptor (AChR) from *Torpedo californica* was recently reported by Stroud and co-workers at the University of California at San Francisco (Mitra *et al.* 1989). AChR is a 295 kD complex of five homologous transmembrane glycoprotein subunits. Gating of the AChR may be provided by tight ion-binding sites that are located in the transmembrane portion of the resting channel. Antagonists close the ion channel, while agonists open the channel. Thus, careful allosteric control of the influx of ions is allowed into the cell. The gap junction is another membrane-bound protein that displays allosteric behaviour. The gap junction provides a communication pathway between cells. Cryoelectron microscopy has provided the first view of the allosteric changes in the gap junction (Unwin 1987). The gap junction is a hexameric protein that undergoes a symmetrical change in quaternary structure in response to H^+ and Ca^{2+}.

8.2 Allosteric enzymes

8.2.1 *Aspartate transcarbamylase*

8.2.1.1 Introduction   Aspartate transcarbamylase from *Escherichia coli* (also called aspartate carbamoyltransferase, ATCase; EC 2.1.3.2) catalyses the condensation of carbamyl phosphate with L-aspartate to produce *N*-carbamyl-L-aspartate and inorganic phosphate (Jones *et al.* 1955). The reaction is particularly important because once carbamylaspartate is formed, it is committed to the biosynthesis of pyrimidines, a necessary component for nucleic acid biosynthesis. Aspartate transcarbamylase controls the rate of pyrimidine biosynthesis by altering its catalytic velocity in response to cellular levels of both pyrimidines and purines. The end product of the pyrimidine pathway, cytidine triphosphate (CTP), induces a decrease in catalytic velocity, whereas adenosine triphosphate (ATP), the end product of the parallel purine pathway, exerts the opposite effect, stimulating the catalytic activity (Fig. 8.7). In part, the relative amounts of purines and pyrimidines in the cell are kept in balance for nucleic acid synthesis.

8.2.1.2 Crystal structures   Three-dimensional crystal structures of ATCase at 2.5–2.8 Å resolution have been established in W.N. Lipscomb's laboratory at Harvard University. ATCase is a hexamer in both catalytic (310 amino acids; 34 kD) and regulatory (153 amino acids; 17 kD) chains (Plate 31); the holoenzyme (see Glossary) is composed of two catalytic

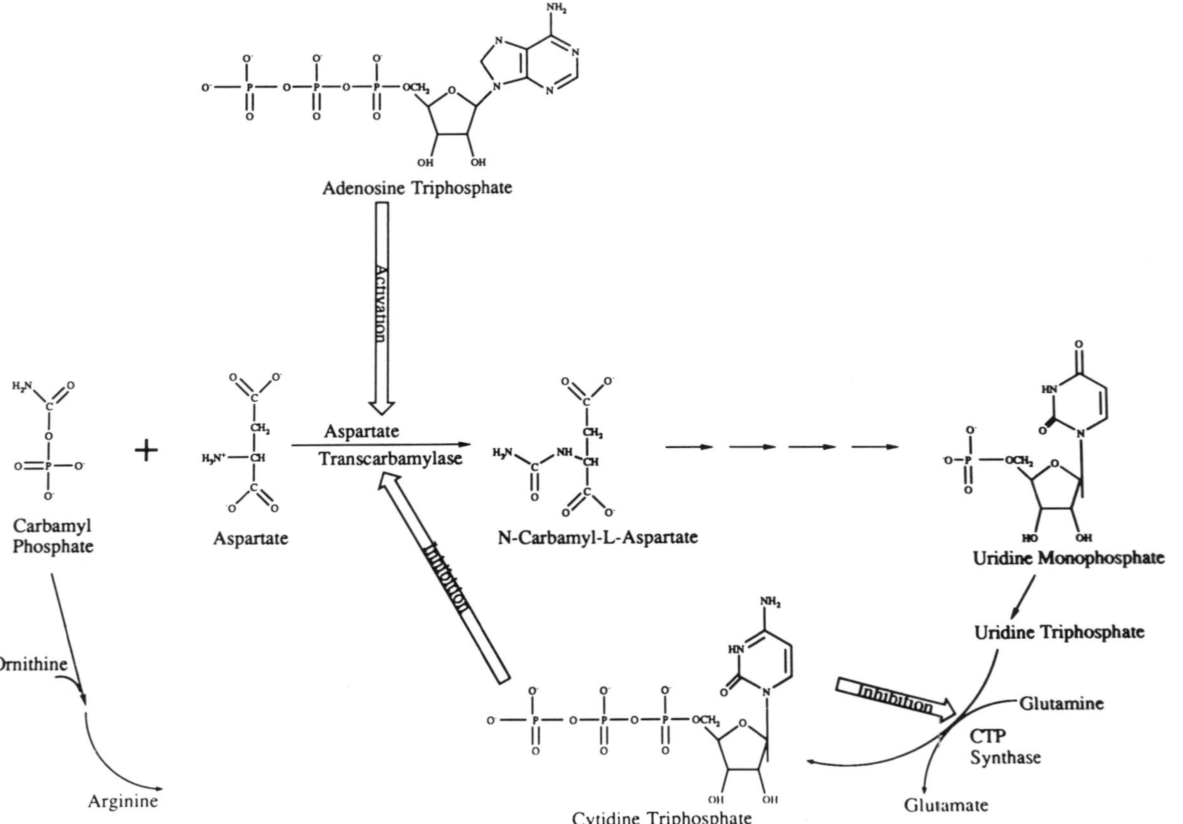

Fig. 8.7 Schematic diagram of catalysis and regulation of aspartate transcarbamylase from *E. coli* in the pyrimidine biosynthetic pathway.

trimers $2c_3$ and three regulatory dimers $3r_2$ (Weber 1968; Wiley and Lipscomb 1968). Functionally, the catalytic trimers and the regulatory dimers are distinct. The catalytic trimer facilitates the carbamoylation of aspartate by carbamoyl phosphate, while the regulatory dimer binds CTP and ATP but is catalytically inert. T-state structures in the absence of heterotropic effectors (see Glossary) have been determined for unligated ATCase (Stevens *et al.* 1990) and phosphonoacetamide (PAM; a carbamoyl phosphate analogue) bound ATCase (Gouaux and Lipscomb 1990). R-state structures in the absence of heterotropic effectors have been determined in the presence of the bisubstrate analogue *N*-(phosphonoacetyl)-L-aspartate (PALA; Ke *et al.* 1988); and the substrate analogues carbamyl phosphate + succinate (Gouaux and Lipscomb 1988) and PAM + malonate (Gouaux and Lipscomb 1990). These three R-state structures are essentially identical. Crystal structures of both the T- and R-state enzymes complexed with the heterotropic effectors ATP, CTP, and CTP + UTP (uridine triphosphate) have also been determined (Gouaux *et al.* 1990; Stevens and Lipscomb 1990; Stevens *et al.* 1990).

8.2.1.3 Allosteric transition   Comparison of the R-state structure to the unligated T-state structure shows a 10.8 Å increase in the separation of the catalytic trimers along the crystallographic threefold axis, a 12° rotation of the trimers relative to one another, and a 15° rotation of the regulatory dimers along each of the three non-crystallographic twofold axes (Ke *et al.* 1988). During the T → R transition, domain closure is sterically hindered until the molecule expands along the threefold axis and allows for the reorientation of the 240s loop. This reorientation occurs symmetrically within the $c_3 \ldots c_3$ interface. The interface interactions between the catalytic trimers and regulatory dimers that stabilize the T state of the enzyme are replaced by new interactions between the substrate analogues and the enzyme and by interactions bridging the carbamyl phosphate and aspartate domain interface.

8.2.1.4 Homotropic effects   The kinetic mechanism for ATCase can be described as 'preferred order', with carbamyl phosphate binding first, aspartate binding second, followed by dissociation of carbamylaspartate and then inorganic phosphate (Hsuanyu and Wedler 1987). When PAM binds to the holoenzyme, the main chain atoms of Glu50c-Thr55c [residues are defined by the amino acid, position in the polypeptide chain, and the catalytic (*c*) or regulatory (*r*) chain] and of His265c-Asp271c move approximately 1.5–2.0 Å, and the side-chains of these residues move as much as 3.5 Å (Gouaux and Lipscomb 1990). Binding of PAM also induces other conformational changes. The guanidinium group of Arg54c moves to bind the phosphonate of PAM. Thr55c moves 1.2 Å so that its hydroxyl group interacts with a phosphonate oxygen and the carbonyl oxygen of PAM. Arg105c moves from a position near Glu50c to bind to the carbonyl oxygen of PAM. Movement of Pro266c-Pro268c by about 0.7 Å appears to form a recognition site for the amino group of PAM. This region is part of the His265c-Asp278c fragment which has contacts with the important 240s loop and has interchain contacts between catalytic chains within c_3 units.

Malonate binds to the PAM-ligated structure in the aspartate domain, as the enzyme is converted from the T- to the R-state conformation (Gouaux and Lipscomb 1990). Relatively large movements of the 80s and 240s loops occur, and the c1 ... r4 (i.e. c4 ... r1) interface disappears. Specific movements towards substrates or analogues are summarized as follows: Arg54c moves 1.5 Å from near Glu86c to bind O-2 of phosphonate; Arg105c moves 1.1 Å from Glu50c to bind to O-3 of phosphonate; Arg167c moves 1.6 Å from Glu50c to bind to O-3 of malonate; and Arg229c moves 2.3 Å from Glu272c to Glu233c and to O-1 of malonate. Substantial movement occurs in the adjacent (c1 ... c2) chain as Ser80c moves 4.6 Å to bind to the phosphonate oxygens in both the PALA and the PAM-malonate complexes; the NH_3^+ group of Lys84c moves 12.1 Å to bind to O-3 of the phosphonate of PALA and to the oxygen atoms of both carboxylates of PALA. However, in the PAM-malonate structure, a similarly large movement of Lys84c brings it within about 4.0 Å of the nearest phosphonate oxygen and 3.8 Å from a carboxylate oxygen. In summary, the T to R transition allows the movement of several important binding residues to the active site, thus increasing the affinity for substrates in the wild-type enzyme.

A plausible mechanism for catalysis involves proton transfer from the amino group of aspartate to the nearest oxygen of the leaving phosphate group of carbamyl phosphate. The lone pair of this amino group attacks the carbonyl carbon of carbamyl phosphate to form a tetrahedral intermediate (Gouaux *et al.* 1987). Formation of the tetrahedral carbon (or an incipient carbonium carbon) is probably facilitated by the binding of Arg105c, His134c, and Thr55c to the oxygen of the carbonyl group of carbamyl phosphate.

8.2.1.5 Heterotropic effects　　How does the binding of ATP and CTP to the same local region on the regulatory chain at sites more than 60 Å from the nearest active site on the catalytic chain produce opposite effects on the enzyme's activity (Fig. 8.9)? And, how is catalysis controlled? Ligation of the allosteric activator, ATP, to the T-state enzyme induces a 0.5 Å increase in the separation of the catalytic trimers, i.e. a movement in the direction of the R state (Stevens *et al.* 1990). In contrast, CTP or CTP + UTP ligation has little effect on the separation of the catalytic trimers in the T state. On the other hand, CTP or CTP + UTP ligation to the R-state enzyme (PAM + malonate complex) causes a decrease of approximately 0.5 Å in the separation of the catalytic trimers, i.e. a movement towards the T

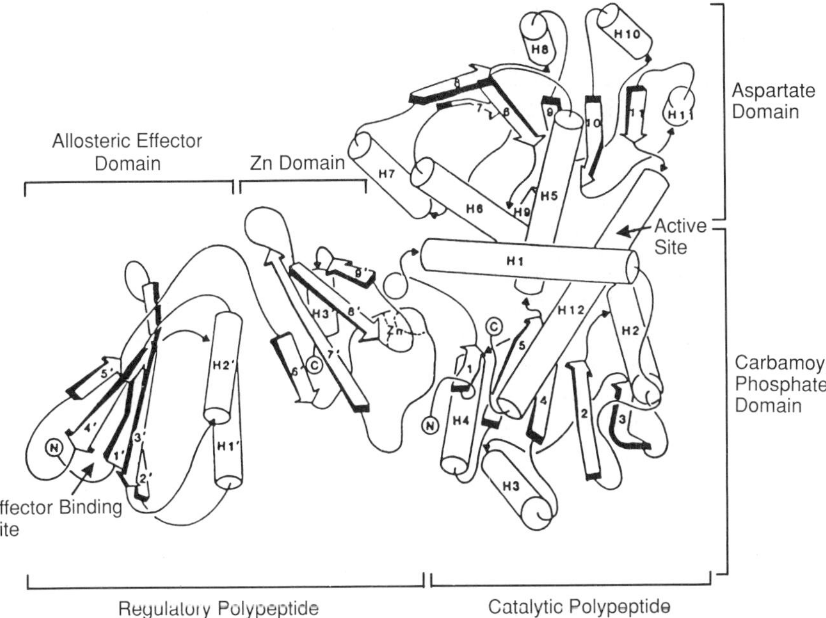

Fig. 8.9 Schematic diagram of the regulatory-catalytic chain of aspartate transcarbamylase. The distance between the allosteric binding site and the active site is approximately 60 Å.

state (Gouaux *et al.* 1990; Stevens and Lipscomb 1990). It has been shown that UTP and CTP synergistically inhibit catalysis in ATCase (Wild *et al.* 1989). Ligation of ATP to the R-state enzyme (PAM + malonate complex) has little effect on the separation of catalytic trimers. In the unligated Glu239→Gln mutant (Ladjimi and Kantrowitz 1988), the T-state c1 ... c4 interface is destabilized so that the enzyme exists in an intermediate quaternary state (Gouaux *et al.* 1989). Ligation of ATP transforms the mutant enzyme to the R state, whereas CTP converts this enzyme to the T state (Stevens and Lipscomb 1990). Thus, this mutant is much more sensitive to heterotropic allosteric control than the native enzyme. It appears from these studies that ATP and CTP transmit signals from the effector site to the c1 ... r1 and c1 ... r4 interfaces which alter the enzyme's catalytic trimer separation or quaternary state. These changes of the quaternary state in the wild-type enzyme control the affinity of substrate binding. Control of substrate binding regulates the T→R transition, thereby regulating catalysis at the active site. Although CTP is not observed to alter the T quaternary state and ATP is not observed to alter the R quaternary state, it is likely that they do stabilize the two allosteric states (as we have noted above, CTP does slightly alter the R state and, ATP does slightly alter the T state).

Locally, ATP and CTP (and most likely UTP) bind in similar positions and to similar residues in both the T and R states (Gouaux *et al.* 1990; Stevens *et al.* 1990). The adenosine or cytidine base moiety interacts with Val9r-Ile12r, Asp19r, His20r, and Tyr89r. The ribose moiety interacts with Lys60r and Asn84r. The triphosphate moiety interacts with His20r and Lys94r. Crystallographic studies have been conducted at pH 5.8 and pH 7.0 and the main difference in structure appears to be the tighter binding of the triphosphate moiety in the pH 7.0 structure relative to the pH 5.8 structure. Near the base moiety region, ATP enlarges the effector-binding cavity slightly more than CTP. The short strand, Val9r-Ile12r, moves approximately 0.5 Å away from the base ring in the ATP structure in comparison to the CTP-ligated structure. The difference in strand movement is most likely propagated to the allosteric . . . zinc interface. Between the effector-binding site and the allosteric . . . zinc domain are two α-helices (H1′ and H2′). Movement of the Val9r-Ile12r strand alters the arrangement of the helices. We propose that alteration in the position of the helices may then transmit a heterotropic signal through the allosteric . . . zinc interface. The allosteric . . . zinc interface of the R state is more open to solvent than in the T state. In the R state, Lys28r makes polar interactions with the N-terminal residue Asn153r. Leu76r and Tyr77r make non-polar interactions with Leu151r. In the T state, the Lys28r . . . Asn153r interaction is replaced by a non-polar interaction between Lys28r and Leu151r. The Leu76r, Tyr77r . . . Leu151r non-polar contacts remain in the T state. It is unclear how the signal is transferred across the allosteric . . . zinc interface but the residues mentioned are likely to be involved. Once the heterotropic signal passes through the allosteric . . . zinc interface, communication continues through the c1 . . . r4 and/or c1 . . . r1 interface before arriving at its final destination, the active site. In the T state, both the c1 . . . r4 and c1 . . . r1 interfaces are present. In the R state, only the c1 . . . r1 interface is present. Several residues are suspected to be of critical importance in transmitting heterotropic effects. For example, the Lys143r . . . Asp236c (c1 . . . r4) and Arg130r . . . Glu204c (c1 . . . r1) interactions of the T form appear to be important in transmitting heterotropic effects based on site-specific mutagenesis (see Glossary) and X-ray crystallographic studies. Clearly, the c1 (c236-c244) . . . r4 (r143-r149) interface involves the 240s loop, which is involved in binding of aspartate.

Further mutagenic and crystallographic studies are required to elucidate heterotropic effects. The loss of CTP effects when the c1 . . . r4 interface is lost, and the loss of ATP effects when mutants are made in the Zn domain near the c1 . . . r1 interface to the carbamoyl phosphate domain (c1–c140 and c280–c310) are indications of some separation of these effects. However, this separation is likely to be an oversimplification. For example, the

Tyr77r → Phe mutant responds to ATP as an inhibitor, not an activator (Van Vliet *et al.*, 1991). Finally, the ATP and CTP effects are still evident in modified forms of the enzyme in which homotropic co-operativity is absent or very small; this result is consistent with propagation of ATP and CTP effects within a given quaternary state. However, this aspect of heterotropic co-operativity does not diminish the importance of the T to R transition in the wild-type enzyme for both homotropic and heterotropic effects. The two-state model is a first approximation and substantial modifications of this model can be expected.

8.2.2 Glycogen phosphorylase

8.2.2.1 Introduction   Glycogen phosphorylase (GP; EC 2.4.1.1) was the first allosteric enzyme to be isolated and studied biochemically (Cori and Cori 1936). An excellent review article on GP, including the extensive structural studies, has appeared recently (Johnson and Barford 1990). GP catalyses the degradative phosphorylation of glycogen to glucose-1-phosphate, the initial step in the generation of metabolic energy in muscle and free glucose in liver (Fig. 8.10). The biological role of GP is to supply the glycolytic pathway with a regulated amount of phosphorylated glucose units derived from tissue glycogen stores. Glycogen is distributed widely in nature and its utilization via GP appears to be related to energy needs in growth, development, or starvation. For example, in muscle, the enzyme is regulated according to the energy needs of contraction. In liver, the enzyme's activity is modulated to provide enough glucose to maintain the fasting blood-sugar level. GP is regulated by allosteric interactions and reversible phosphorylation.

Unlike aspartate transcarbamylase, GP is found to be metabolically interconvertible between two discrete forms, GP *a* and GP *b*, that exhibit different modes of allosteric control. GP *b* requires AMP for activity and is inhibited by ATP and glucose-6-phosphate. The enzyme exhibits homotropic effects for both substrate and AMP. GP *a* is active in the absence of AMP, although substrate binding remains co-operative. Each GP has been observed in two oligomeric forms, dimeric and tetrameric. Each monomer contains 842 amino acids (97 kD). The tetrameric form has only 12–33 per cent of the full phosphorylase activity, but may be dissociated by glycogen or oligosaccharides to give fully active dimers. A substantial amount of GP *in vivo* is bound to glycogen particles, and the dimer is considered to be the physiologically active form of the enzyme. The phosphate groups at Ser14 in GP *a* can be hydrolytically removed by the enzyme phosphorylase phosphatase to yield the *b* form. The relatively less active GP *b* can be converted to GP *a* by the enzyme phosphorylase kinase, which catalyses the phosphorylation of Ser14 at the expense of ATP. In this way, the activity of

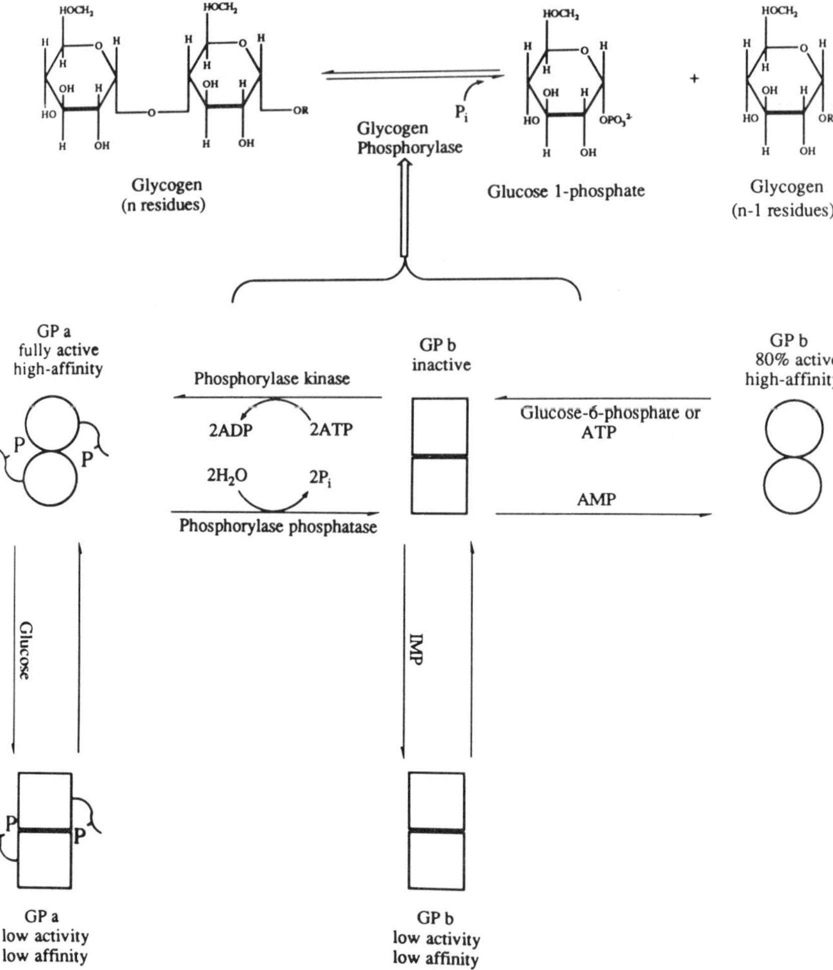

Fig. 8.10 Schematic diagram of catalysis and regulation of glycogen phosphorylase. T and R structures are shown as squares and circles, respectively. (Reproduced with permission from Barford and Johnson 1989.)

GP is regulated by the action of these two enzymes. One molecule of phosphorylase kinase can convert thousands of molecules of the less active GP *b* into the more active GP *a*. GP *a* can then catalyse the production of thousands of molecules of glucose-1-phosphate molecules from glycogen. Phosphorylase kinase and glycogen phosphorylase thus constitute an amplification cascade with two steps.

8.2.2.2 Crystal structures   The structures of rabbit muscle glycogen phosphorylase have been determined in the laboratories of L. N. Johnson at the

University of Oxford (GP b in the T and R states), and R. J. Fletterick at the University of California, San Francisco (GP a in the T state). The T-state structure of GP b has been determined at 1.9 Å resolution in the presence of IMP (Sansom *et al.* 1985). IMP (inosine monophosphate) is a weaker activator in comparison to AMP (Black and Wang 1968). The T-state structure of GP a has been determined at 2.1 Å resolution in the presence of glucose (Sprang and Fletterick 1979), an inhibitor which promotes the T state by binding to and restricting access to the catalytic site (Kasvinsky *et al.* 1978). The structures of GP a and GP b are very similar (rms difference in C_α position is approximately 0.6 Å in the T state), with the exception of the conformational changes resulting from Ser14 phosphorylation. Comparison of the dimeric GP a and GP b show that upon phosphorylation of Ser14, the N-terminus of each subunit assumes an ordered helical conformation and binds to the surface of the dimer. The structural changes at the N- and C-terminal regions lead to the strengthened interactions between subunits and alter the binding site for substrates and allosteric effectors (Sprang *et al.* 1988). The R-state GP b structure in the tetrameric form has been determined in the presence of ammonium sulphate and AMP at 2.9 Å resolution (Barford and Johnson 1989). Association of the two dimers to form a tetramer creates an interface that comprises residues from the catalytic region. The resulting limitation of access to the catalytic site may contribute to the 70–90 per cent reduction in activity in tetrameric GP relative to dimeric GP. In the crystal structure, sulphate simulates phosphate by binding to the catalytic site, the AMP effector site, and very near the Ser14 site, resulting in localized changes in the tertiary structure. Even though the R state of GP b does not contain a phosphorylated serine, the protein conformation near the Ser14 site in the presence of sulphate is almost identical to that of the glucose-inhibited T-state GP a (Barford and Johnson 1989). R-state GP a is identical to R-state GP b with the exception that phosphate is bound to Ser14 in GP a, while sulphate is near to Ser14 in GP b (Barford *et al.* 1991).

Structurally, GP is typical of the α-β class of proteins. Each monomer is composed of two domains, the N-terminal domain of amino-acid residues 1–484 and the C-terminal domain of residues 485–842 (Fig. 8.11). The major domains each consist of a core of twisted β-sheets surrounded by α-helices. The enzyme has five recognition sites:

(1) the binding site for the cofactor pyridoxal phosphate in the centre of the molecule;

(2) the substrate-binding site (approximately 5 Å from the cofactor phosphate site);

(3) the AMP-binding site [located between the two helices α-2 (47–78) and α-8 (289–314); approximately 30 Å from the catalytic site];

(4) the Ser14-phosphate site (approximately 12 Å from the AMP site);

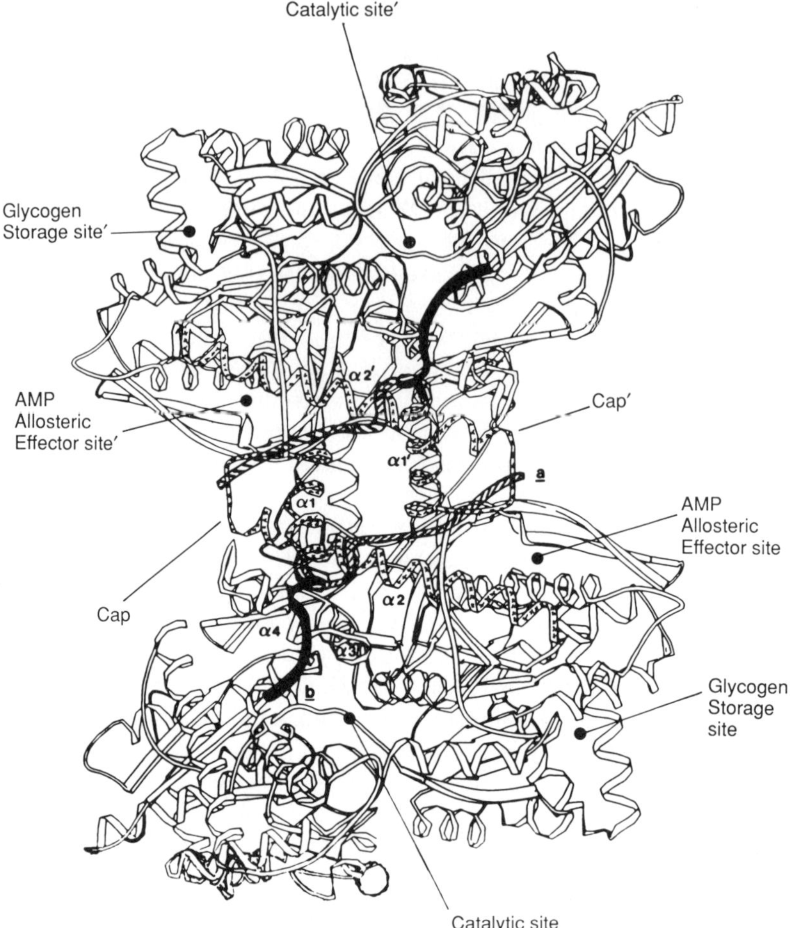

Fig. 8.11 Schematic ribbon diagram of the T-state glycogen phosphorylase *b* viewed down the twofold axis of symmetry (reproduced with permission from Johnson *et al.* 1990).

(5) the glycogen-storage site (Glu433 is a key residue in the binding of glycogen; the site is located on the surface of the enzyme, approximately 30 Å from the catalytic site).

The catalytic, inhibitory, and glycogen-storage sites are all located on one side of the molecule, and the allosteric site is on the opposite side of the molecule.

8.2.2.3 Allosteric transition   The allosteric transition is activated by phosphate or sulphate ions buried at the phosphorylation and catalytic sites (Barford and Johnson 1989). The overall tertiary structural changes between the T and R states are small (the root mean square difference in C_α is 1.3 Å for GP *b*). The transition consists of a rotation of one subunit relative to the other by 10° about an axis at the subunit boundary that is close to the cap' (35'–46')/α-2 (47–78) and is normal to the twofold symmetry axis (Fig. 8.12). This transition exerts a change in structure involving the subunit interface, the tower helices (262–276 and 262'–276'), and the cap'α-2 interface. It is the change in tilt and relative slide of the tower helices which link the subunit interface changes to the changes in the active site. These helices pull apart and change their angle of tilt so that the cap'/α-2 interface is tightened. The alternate packing adopted by the tower helices provides a particularly suitable method by which to couple tertiary and quaternary structural changes. The geometry of one helix is constrained by that of the symmetry-related helix so that molecular symmetry is preserved and a concerted conformational change occurs.

The tower helices terminate in a polypeptide loop (281–287) that restricts access to the catalytic site in the T structure. In the R structure, the loop becomes disordered. The catalytic site is located between the N-terminal and C-terminal domains and is close to the essential cofactor, pyridoxal phosphate. Access to the catalytic site is via a channel some 12 Å in length. In the T state the diameter in one region of this channel is severely restricted. In the T state, oligosaccharides and glycogen analogues do not bind to the catalytic site but bind tightly to the glycogen-storage site on the surface of the enzyme. The glycogen-storage site is approximately 30 Å from the catalytic site and 40 Å from the allosteric effector site. The allosteric site is located on the opposite side of the α-2 helix relative to the Ser14-P site, with a separation of 12 Å. The closest neighbouring catalytic site is roughly 66 Å away. The displacement of glycogen allows substrates as well as additional cationic side-chains to move into the catalytic site. They convert the coenzyme phosphate from the mono-to the dianionic form that is necessary for catalysis. Helix α-8 links the catalytic to the AMP-binding site and appears to be a possible vehicle for communication between the catalytic site and the AMP site. Helix α-8 connects to the 280s loop at its N-terminus and contributes two arginines (residues 309 and 310) to the AMP site at the C-terminus. The helix is part of a rigid internal core in which no changes in tertiary structure between the T and R states are observed. The tertiary structural changes are limited to the binding sites and the subunit-subunit contacts.

Tertiary and quaternary movements are coupled by the alternate packing adopted by the tower helices. Rotation of the two subunits in the T→R transition alters the relative disposition of the two helices. A concerted

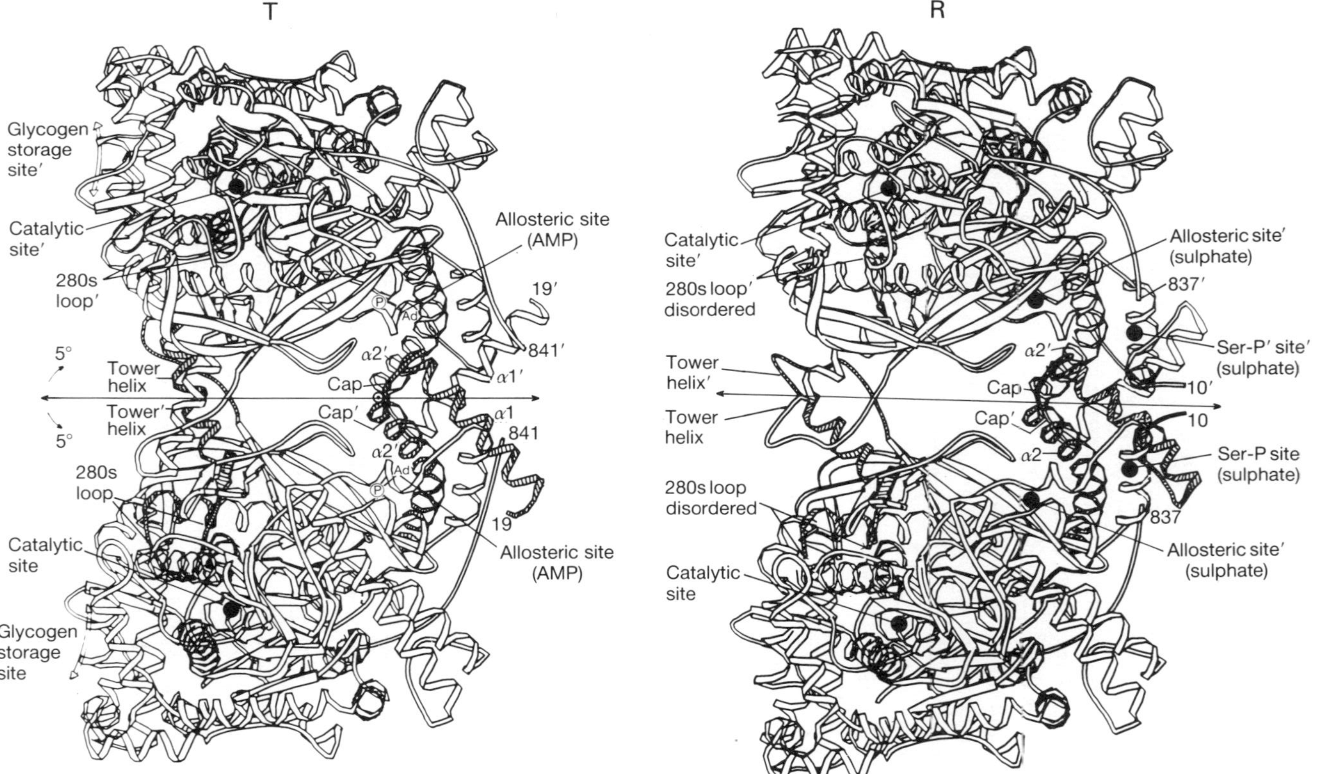

Fig. 8.12 T→R transition of glycogen phosphorylase *b*. On transition to the R state, each subunit rotates by 5° about an axis positioned close to the cap'/α-2 interface. (Reproduced with permission from Johnson *et al.* 1990).

conformational change is observed because the geometry of one helix is constrained by the symmetry of a related helix. A strict coupling of quaternary to tertiary structure is required so that heterotropic allosteric effectors bound at the cap'/α-2 interface, 35 Å from the tower helix interface, can transmit a conformational change to the catalytic site. Homotropic and heterotropic mechanisms therefore involve the same gross structural changes and are mutually independent. The simple two-state model of allosteric response is an oversimplification but does provide a working hypothesis in a number of experiments. Three-, four-, and six-state models have been proposed. It appears that the T→R transition can be achieved in more than one way and in more than one step.

At the catalytic site, the position of Arg569 changes in the T to R transition in order to create a high-affinity phosphate site, which is close to the phosphate of pyridoxal–5'-phosphate. The channel to the catalytic site is opened up as the 280s loop becomes disordered as a result of this transition. Finally, the phosphorylation at Ser14, in addition to the binding of AMP at the cap'/α-2 interface, promotes and stabilizes the R-state quaternary structure.

8.2.3 Phosphofructokinase

8.2.3.1 Introduction   The allosteric enzyme phosphofructokinase (PFK; EC 2.7.1.11) catalyses the key control step of glycolysis, the phosphorylation of fructose-6-phosphate (Fru-6-P) by adenosine triphosphate (ATP) to form fructose-1,6-bisphosphate (Fru-1,6-P$_2$; Blangy *et al.* 1968). Figure 8.13 illustrates the catalytic reaction and allosteric control. PFK from the bacterial sources *Escherichia coli* and *Bacillus stearothermophilus* forms tetramers with a monomeric molecular weight of approximately 35 kD (319 amino acids). The enzyme from the two bacterial sources displays similar kinetic parameters and similar allosteric control. Co-operativity is observed with respect to Fru-6-P; ADP and other diphosphonucleotides are allosteric activators of PFK. Phosphoenolpyruvate (PEP), a later product in the glycolytic pathway, inhibits PFK. The MWC model describes the kinetics parameters of *E. coli* PFK well. *Bacillus stearothermophilus* PFK also displays sigmoidal kinetics but only in the presence of PEP. These two sources of PFK are 54 per cent homologous and the three-dimensional structures are similar. Bacterial PFK behaves approximately as a 'K-system': the two allosteric states differ in their affinity (K_M) for the co-operative substrate Fru-6-P but not in their catalytic rates (V_{max}).

8.2.3.2 Crystal structures   The structures of T- and R-state PFK from *B. stearothermophilus* and T-state PFK from *E. coli* have been determined in the laboratory of Philip Evans at the MRC Laboratory of Molecular

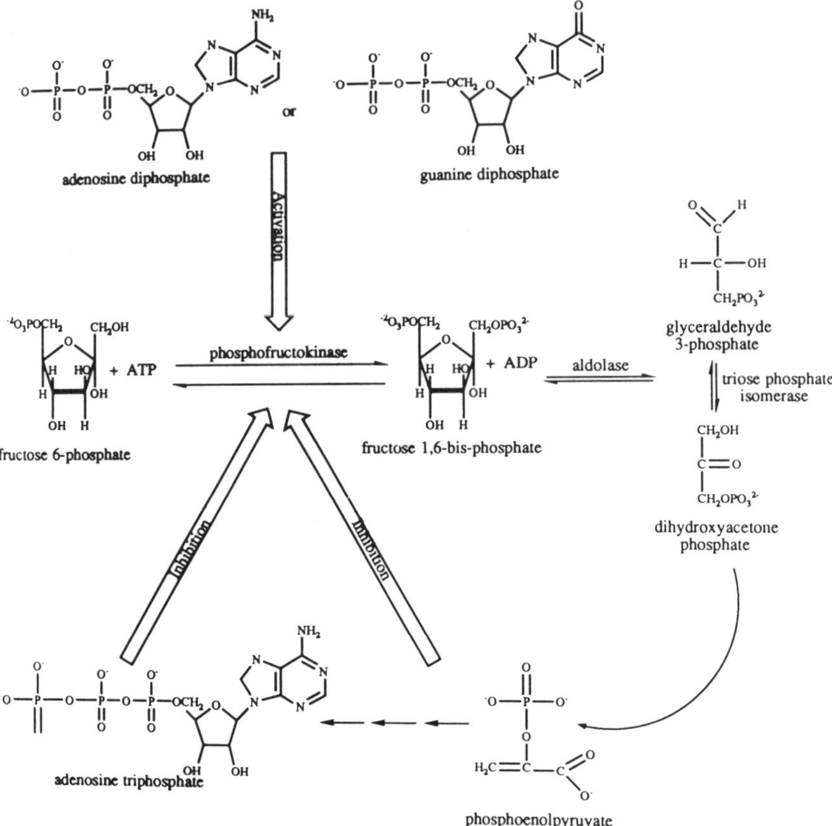

Fig. 8.13 Schematic diagram of catalysis and regulation of phosphofructokinase.

Biology in Cambridge, England. Unlike aspartate transcarbamylase and glycogen phosphorylase, unligated PFK exists in the R conformational state. The crystal structure of unligated R-state PFK from *E. coli* has been determined to a resolution of 2.4 Å (Rypniewski and Evans 1989). PFK from *E. coli* has been complexed with its reaction products Fru-1,6-P_2 and ADP.Mg^{2+}, and the R-state structure has been determined at a resolution of 2.4 Å (Shirakihara and Evans 1988). Two crystal forms of PFK from *B. stearothermophilus* have been examined. Complexation of *B. stearothermophilus* PFK with the substrate Fru-6-P produces the R-state structure which has been determined at 2.4 Å resolution (Evans and Hudson 1979). Ligation of 2-phosphoglycolate (PGC), an analogue of PEP (allosteric inhibitor), transforms the enzyme to the T-state conformation. The PGC-complexed PFK from *B. stearothermophilus* has been determined at 2.5 Å resolution (Schirmer and Evans 1990).

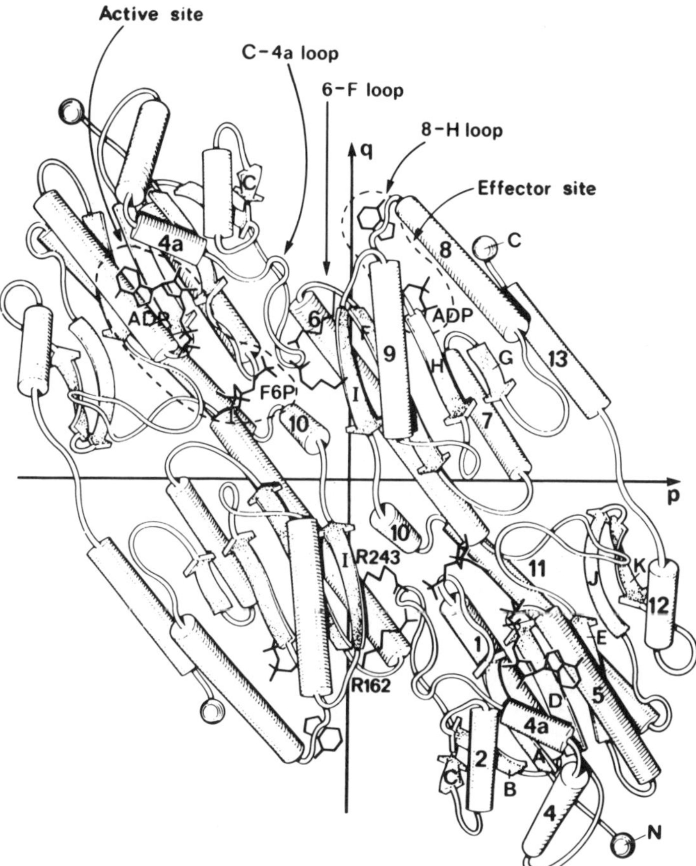

Fig. 8.14 R-state dimeric structure of phosphofructokinase (reproduced with permission from Schirmer and Evans 1990).

The four monomers of PFK form a distorted tetrahedron with a solvent-filled hole ≈ 7 Å in diameter, passing through the middle of the tetramer. This hole separates the pairs of subunits that do not interact. Each monomer is composed of two domains, a larger one (residues 1–137 and 252–306) which binds the substrate ATP, and a smaller one (residues 138–251 and 305–319) which binds Fru-6-P: both domains are three-layered $\alpha\beta\alpha$ sandwich structures (Fig. 8.14). There is a large interface between monomers which form the dimers, and a smaller interface between dimers which form the tetramers. The large interface within the dimer includes the allosteric effector site, which binds both activator and inhibitor. Three binding sites exist on each subunit. Two binding sites are for the substrates Fru-6-P or Fru-1,6-P$_2$ and ATP·Mg^{2+} or ADP·Mg^{2+}. The third binding site is the allosteric effector site that binds both the activator, $ADP \cdot Mg^{2+}$, and the inhibitor, PEP. The ATP-binding site is located almost entirely in the large domain, while Fru-6-P is found to bind in both domains as well as in a neighbouring dimer. This difference in binding sites provides a reasonable explanation for the co-operativity found with respect to Fru-6-P but not ATP. The binding site of Fru-6-P is altered drastically by the 6-F loop (155–162) of the adjacent dimer.

8.2.3.3 Allosteric transition   Comparison of the R- and T-state structures show that the quaternary change is a relative 7° rotation of rigid dimers in the tetramer (Fig. 8.15). This rotation is coupled to the restructuring of the polypeptide loop 6-F, which interchanges the positions of a vital arginine (162) and a glutamate (161) in the substrate-binding site. During the allosteric transition, a layer of water molecules is inserted (or removed) between the pairs of subunits. Only a small amount of movement was observed within the tertiary structure. That is, helices 8 and 9, which form the outer layer of the $\alpha\beta\alpha$ sandwich in the small domain, are observed to move. Their movement does not affect the subunit contacts, nor do they carry any residues that bind ligands. The large domain remains virtually unchanged and it retains its position relative to the rigid core of the small domain. No detectable changes are observed in the large dimer interface, even though the effectors are found to bind there. Due to the 7° rotation of the dimers in

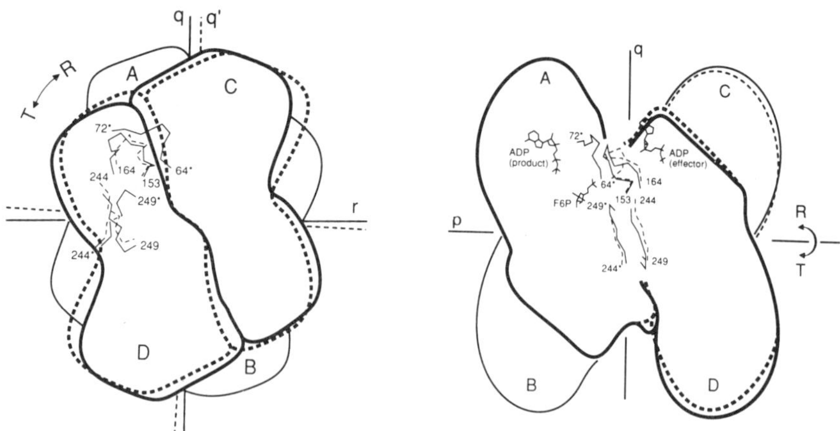

Fig. 8.15 Schematic diagram of the R-state (dashed outline) and T-state (solid outline) PFK tetramer viewed along axis *p*. The three mutual perpendicular twofold symmetry axes are labelled *p* (left), *q*, and *r* (right). The subunits are labelled A–D. Both subunits have been superimposed after fitting the AB dimers. CD rotates ≈ 7° relative to AB about the *p*-axis. (Reproduced with permission from Schirmer and Evans 1990.)

the R→T transition, the interface between the dimers changes slightly. The changes are minimized by the small contact area between dimers and by the preserved interactions close to the quaternary rotation axis. Two main regions are found to change, one passively and the other actively at the interface. The passive change takes place where the quaternary rotation from R to T closes the gap between symmetry-related antiparallel β-sheet strands I (244–248 and 244'–248'). In the R structure, an ordered layer of water molecules bridges these strands, while in the T structure direct H-bonds form across the interface. The active changes take place in contacts between dimers at the end of helix 6 and the 6-F loop. On one side of the contact, the 6-F loop changes its structure to compensate for the quaternary rotation, by maintaining suitable contacts across the dimer interface. If the quaternary rotation to R took place without changing the structure of the loop, the dimer would clash in this region of the interface; conversely, a T-state structure with a loop in the R conformation would leave an open gap. Also, in the T state the last turn of helix 6 and the extended 6-F loop adopt new conformations that cause critical changes in the residues that point into the adjacent Fru-6-P binding site.

8.2.3.4 Homotropic effects   The phosphate that is transferred from ATP to substrate is directly transferred to the substrate (Jarvest *et al.* 1981). Most likely, nucleophilic attack of the 1-OH on the γ-phosphate leads to the formation of a penta-coordinated phosphate transition-state intermediate, followed by hydrolysis of the β-γ phosphate bond of ATP. The small interface between dimers is bridged in the R state by the side-chains of Arg162 and Arg243, which bind the 6-phosphate group of Fru-6-P in the other dimer. This position of the co-operative substrate, Fru-6-P, in a subunit interface is consistent with the change in the dimer-dimer interface in the transition between the T and R states.

8.2.3.5 Heterotropic effects   Comparison of various effectors complexed with PFK displays curious results. Apart from the repositioning of the 8-H loop (211–216), most of the effector-binding site remains unchanged for both activators and inhibitor. When no ligands or phosphate are present, the site is open. When the effectors ADP of the R structure or PGC of the T structure are bound, the cleft closes up to pack tightly around them. The position of the 8-H loop is then controlled by the different sizes of the effectors, the smaller PGC inducing the narrowest cleft. The change in the 8-H loop structure appears to be correlated with the reorganization of the adjacent 6-F loop. ADP (activator) and PGC (inhibitor) bind in identical locations, similar to ATP (activator) and CTP (inhibitor) binding to ATCase (see Section 8.2.1.5). The terminal phosphates are bound in the same way to Arg154, Arg21, and Arg25, deep in the bottom of the effector-binding cleft.

Residues Arg25 and Arg211, which fix the α-phosphate of ADP in the R structure, interact with the bridging oxygen and the carboxylate group of PGC in the T structure. The only major difference in the effector-binding site is the conformation of the side-chain of Glu187, which is co-ordinated to the magnesium ion of ADP in the R structure, and is rotated away from the ligand in the T structure.

Similar to ATCase, site-specific mutagenesis (see Glossary) has provided insight into the role of particular residues. Mutation of a single residue in the effector-binding site, Glu187$\rightarrow$Ala leads to PEP activation instead of inhibition (Lau and Fersht 1987). Mutation of the active-site residues Arg162 or Arg243 to serine results in mutant enzymes with reduced substrate binding and co-operativity but with little effect on the catalytic constant, k_{cat} (Berger and Evans 1990). The reverse reaction displays hyperbolic kinetics with respect to Fru-1,6-P_2; mutation of the active site Arg162$\rightarrow$Ser or Arg243$\rightarrow$Ser, however, yields an enzyme with sigmoidal kinetics. The mutation of Arg72$\rightarrow$Ser reduces the catalytic activity, co-operativity, and binding of Fru-6-P and Fru-1,6-P_2.

8.2.4 Fructose-1,6-bisphosphatase

8.2.4.1 Introduction Fructose 1,6-bisphosphatase (Fru-1,6-Pase; EC 3.1.2.11) catalyses the hydrolysis of D-fructose-1,6-bisphosphate (Fru-1, 6-P_2) to D-fructose-6-phosphate and inorganic phosphate, a key regulatory step in gluconeogenesis (Fig. 8.16). It is now recognized that an important site of regulation in both glycolysis and gluconeogenesis is at the level of fructose bisphosphate formation and hydrolysis. For this reason, Fru-1, 6-Pase is subject to metabolic regulation in a manner complementary to the regulation of the enzyme phosphofructokinase (PFK), which catalyses the opposing step in glycolysis discussed in Section 8.2.3. In the direction of glycolysis, the activity of PFK is inhibited by ATP and citrate; AMP reverses the inhibition by ATP and citrate. Fru-1,6-Pase, on the other hand, is highly sensitive to inhibition by AMP. The choice of pathway might therefore be controlled by the ratio of ATP to AMP; high ratios would prevent glycolysis, and at the same time permit gluconeogenesis to proceed. On the other hand, the inhibition of Fru-1,6-Pase by AMP would protect the cell against the wasteful dephosphorylation of fructose bisphosphate during glycolysis. Fru-1,6-Pase is competitively inhibited by Fru-2,6-P_2, which amplifies the inhibition by enhancing the enzyme's affinity for AMP (Liu and Fromm 1990). The importance of Fru-1,6-Pase can be seen from mutants which lack the enzyme. Various sources that lack the enzyme cannot produce compounds such as glycerol, acetate, or succinate (Fraenkel and Horecker 1968). Because of this effect, children without the enzyme show a tendency towards hypoglycaemia.

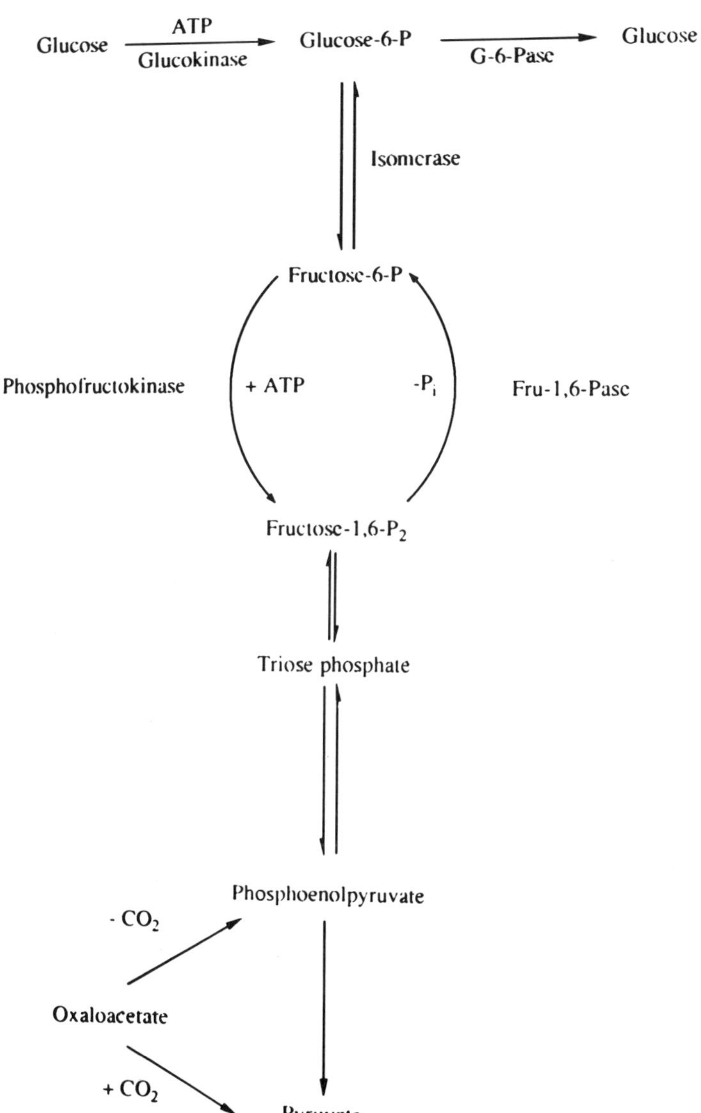

Fig. 8.16 Catalytic/regulatory schematic diagram of the intricate control of glucogenesis by phosphofructokinase and fructose-1,6-bisphosphatase (modified diagram from Ashmore 1964).

Several complete amino-acid sequences are known for Fru-1,6-Pase. The sequence homology is over 85 per cent among the mammalian enzymes (El-Maghrabi *et al.* 1988) and 47 per cent among the mammalian and yeast enzymes (Rogers *et al.* 1988). Two purification schemes of Fru-1,6-Pase yield two enzyme forms with different optimal pH values for catalytic activity. Fru-1,6-Pase with optimal activity at neutral pH has been termed the 'neutral form'. Purification of the enzyme with an optimal activity between pH 8.4 and 9.4 has been termed the 'alkaline form' (Benkovic and deMaine 1982). The alkaline form is known to be subject to cleavage (residues 56–68) by unknown proteases (Marcus 1981). In the neutral form of the enzyme, four AMP molecules bind to the tetramer. When the neutral form is subjected to proteolysis (conversion to the alkaline form), AMP inhibition is decreased or completely lost (Marcus *et al.* 1981). Kinetic evidence suggest that more than two allosteric states exist for Fru-1,6-Pase (Scheffler and Fromm 1986; Kemp and Marcus 1990). So far, two different quaternary states have been observed via X-ray crystallography. Because of the possibility of a further conformational change when both metal cation and substrate are present, we shall refer to the two known structures as R' and T.

8.2.4.2 Crystal structures Crystal structures of the neutral and alkaline forms of pig kidney Fru-1,6-Pase and its complexes have been determined in the laboratory of W. N. Lipscomb at Harvard University. Unligated Fru-1,6-Pase exists in the R' conformation, as does phosphofructokinase. The unligated structure has been determined at 2.8 Å resolution (Ke *et al.* 1989). Ligation of the allosteric inhibitor, Fru-2,6-P_2, also yields an R'-state structure, which has been determined at 2.8 Å resolution; here, the low density for the 2-phosphate suggests disorder or hydrolysis (Ke *et al.* 1990a). The T-state structure was obtained upon ligation of the allosteric inhibitor, AMP; the product, Fru-6-P; and Mg^{2+} at 2.5 Å resolution (Ke *et al.* 1990b). The AMP-complexed alkaline form of the enzyme produces an R'-state structure which has been determined at 2.8 Å resolution (Ke *et al.* 1989).

Pig kidney Fru-1,6-Pase is composed of four monomers (335 amino acids; 36 kD) that aggregate into a relatively flat tetramer with D_2 symmetry (Fig. 8.17). Each monomer is composed of two domains that have an overall hexahedral shape, with two parallel trapezoidal faces having an upper side of about 23 Å, a lower side of approximately 35 Å, and a height of 50 Å. The thickness of the two trapezoids is about 35 Å. Two adjacent monomers in the asymmetric unit resemble a truncated pyramid. The other half of the tetramer interacts at the bottom of the truncated pyramids, so that the whole molecule looks like a pseudohexagonal disc with a thickness of 35 Å. Each monomer is composed of two domains (Fig. 8.18), one that binds the

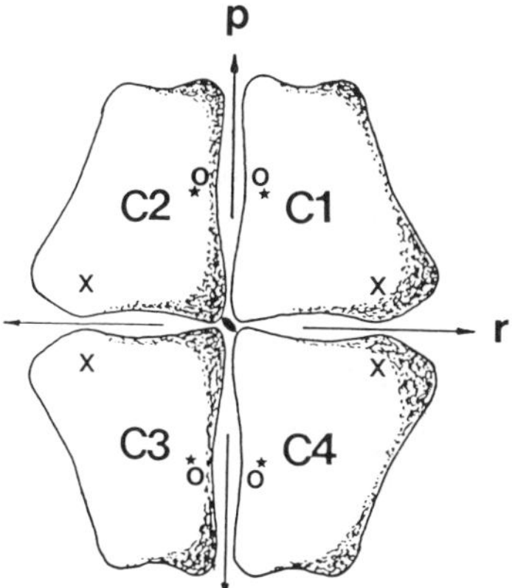

Fig. 8.17 Schematic diagram of fructose-1,6-bisphosphatase, looking down the molecular twofold axis. The *r*-axis is the crystallographic twofold axis and the *p*-axis is the non-crystallographic twofold axis. 'X' marks the AMP-binding site; 'O', the Fru-6-P-or Fru-2,6-P_2-binding site; and '*', the metal-binding site.

allosteric inhibitor, AMP, (AMP domain; residues 6–200) and one that contains the fructose-bisphosphate-binding site (FBP domain; residues 201–335). AMP binds near the bottom of the truncated pyramid of each monomer and interacts with residues from two different monomers. In the initial study of the alkaline form of Fru-1,6-Pase, two AMP molecules were observed to bind to the tetrameric structure (Ke *et al.* 1989). However, further refinement of the alkaline form of Fru-1,6-Pase in the presence of AMP has reduced the occupancy of AMP in the two sites of the tetramer to zero. In the study of the AMP-ligated neutral form of Fru-1,6-Pase (Ke *et al.* 1990b), four AMP molecules were found in the tetrameric structure. Comparison of the alkaline form to the neutral form shows a cleavage of two loop regions (one major cut between residues 54–68 and one minor cut between residues 235–237). Substrate, product, and the divalent metal cation bind at the active site. The active site is an 18 x 12 Å cave, approximately 10 Å deep. The active sites of each dimer open in the opposite direction to one another, and exchange the Arg243 side-chains, indicating potential for communication between the two sites. A binding site for Mg^{2+} has been located about 5 Å away from the position that would be expected for the 1-phosphate, on the basis of a diffraction study at 2.1 Å resolution of the complex of Fru-6-P with the enzyme.

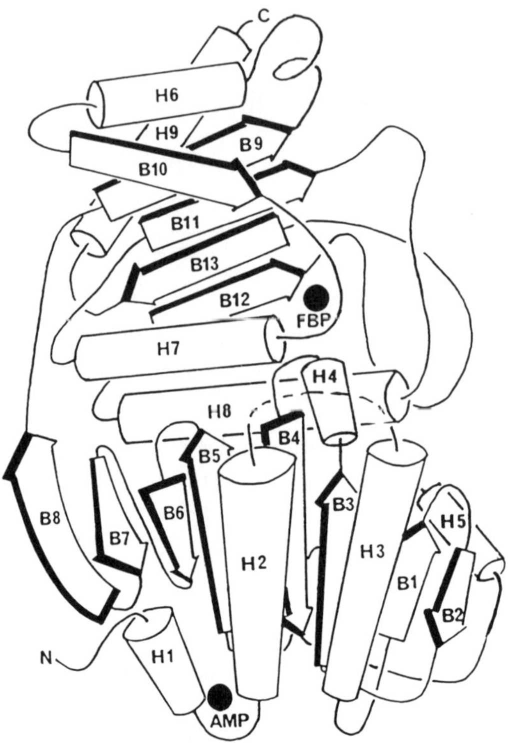

Fig. 8.18 Monomeric structure of fructose-1,6-bisphosphatase. The letters N and C represent the N- and C-terminus, respectively. The two circles labelled FBP and AMP indicate the catalytic and allosteric binding site, respectively.

8.2.4.3 Allosteric transition   Comparison of the R′-and T-state structures shows a 17° twist of dimer C1-C2 relative to dimer C3-C4 and a 1 Å translation of the AMP domains of C1-C2 relative to C3-C4 in the direction opposing the 17° twist, so that many of the interface contacts are conserved (Fig. 8.19). Similar to the other allosteric enzymes discussed, the R′→T transition in Fru-1,6-Pase undergoes significant quaternary structural changes but the secondary structure changes very little. Substantial structural changes were observed near the AMP(domain)-AMP(domain) interdimer interface (C1-C4), which is composed of helices H1 (12–24), H2 (28–50), and H3 (72–88). Significant structural changes were observed in the metal-and AMP-binding sites but not in the part of the catalytic site where Fru-6-P binds.

Two principal intradimer (C1-C2) interfaces exist in Fru-1,6-Pase, the FBP-FBP domain interface and the AMP-AMP domain interface. The

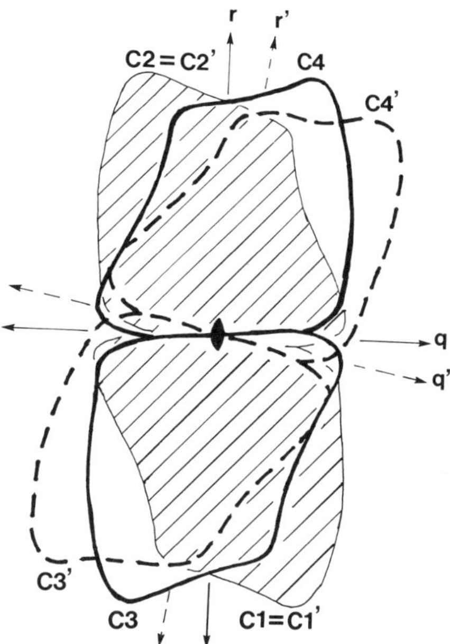

Fig. 8.19 T→R transition of fructose-1,6-bisphosphatase, viewed down the molecular twofold axis, p. Dimer C1-C2 is superimposed and drawn as a hatched region. The heavy solid lines represent dimer C3-C4 of the R state (enzyme-F2,6P$_2$ complex), while the dashed lines are those of the T state (enzyme-F6P-AMP-Mg^{2+} complex). Dimer C3-C4 of the T state is twisted $\approx 19°$ about the molecular twofold axis, p, relative to the R-state structure.

location of the active site near the FBP-FBP domain interface suggests a mechanism for the enzyme's co-operativity. The sharing of active-site ligands between subunits most likely influences the binding of substrate. In all four of the allosteric enzymes discussed, the active site is located at or near an interface region. A possible model for allosteric inhibition has AMP binding near the AMP-AMP interdimer interface in a manner that allows for a shared binding site between subunits. Once AMP binds, the enzyme is held in the T-state conformation and rearrangement of the AMP-AMP domain interface occurs. Rearrangement of the FBP-FBP domain interface may change the position of residues critical for substrate or metal binding.

Insight into the molecular architecture of Fru-1,6-Pase has been obtained via X-ray crystallography. The next step in understanding the mechanisms of the allosteric enzyme is well under way and will come from the structures of the complexed enzyme. Several questions regarding the kinetics of the enzyme have been answered by the determination of the three-dimensional structure of a number of complexed forms of Fru-1,6-Pase:

1. In the unligated form, Fru-1,6-Pase exists in the R′-state conformation similar to phosphofructokinase. Only the allosteric inhibitor, AMP, is able to induce the allosteric transition to the T state. In phosphofructokinase, the T state can only be obtained in the presence of the allosteric inhibitor, phosphoenolpyruvate (PEP).

2. In the allosteric transition, displacements of about 1.2 Å are seen in helices H1 and H2 near the allosteric AMP site. Also, changes of more than 1 Å are noted for the C_α locations for Glu97, Glu98, and Asp121, at or very near the metal-binding site. At the C1-C4 interface large changes are observed in the packing of H1, H2, and H3 of one AMP domain against helices H3, H2, and H1 of the adjacent AMP domain. The inhibitory AMP site is connected to this interface partly through Arg22, which undergoes a bonding switch in the transition. Finally, this interface connects to the active site via the three helices H1, H2, and H3, and also via a five (or so)-stranded β-structure, against which these helices pack.

Acknowledgements

This work was supported by National Institute of Health Grant GM06920 (WNL) and a National Institute of Health Postdoctoral Fellowship (RCS). We thank Professor Louise Johnson, Dr Philip Evans, Dr Hengming Ke, and Karin Reinisch for comments on this manuscript.

References

Ashmore, J. (1964). *Fructose 1,6-diphosphatase and its role in gluconeogenesis*, (ed. R.W. McGilvery and B.M. Pogell), p. 43. American Institute of Biological Sciences, Washington, DC.

Baldwin, J.M. and Chothia, C. (1979). Haemoglobin: the structural changes related to ligand binding and its allosteric transition. *Journal of Molecular Biology* **129**, 183–220.

Barford, D. and Johnson, L.N. (1989). The allosteric transition of glycogen phosphorylase. *Nature* **340**, 609–16.

Barford, D., Hu, S.-H., and Johnson, L.N. (1991). The structural mechanism for glycogen phosphorylase control by phosphorylation and AMP. *Journal of Molecular Biology*, in press.

Benkovic, S.J. and deMaine, M.M. (1982). Mechanism of action of fructose 1,6-bisphosphatase. *Advances in Enzymology* **53**, 45–82.

Berger, S.A. and Evans, P.R. (1990). Active-site mutants altering the cooperativity of *E. coli* phosphofructokinase. *Nature* **343**, 575–6.

Biesecker, G., Harris, J.I., Thierry, J.C., Walker, J.E., and Wonacott, A. (1977). Sequence and structure of D-glyceraldehyde 3-phosphate dehydrogenase from *Bacillus stearothermophilus*. *Nature* **266**, 328–33.

Black, W.J. and Wang, J.H. (1968). Allosteric activation of glycogen phosphorylase *b* by nucleotides. I. Activation of phosphorylase *b* by inosine monophosphate. *Journal of Biological Chemistry* **243**, 5892–8.

Blangy, D., Buc, H., and Monod, J. (1968). Kinetics of the allosteric interactions of phosphofructokinase from *Escherichia coli*. *Journal of Molecular Biology* **31**, 13–35.

Boyer, P.D. (1962). *The enzymes*, (2nd edn), Vol. 6. Academic Press, New York.

Cori, C.F. and Cori, G.T. (1936). Mechanism of formation of hexose monophosphate in muscle and isolation of a new phosphate ester. *Proceedings of the Society for Experimental Biological Sciences* **34**, 702–15.

Derewenda, U., *et al.* (1989). Phenol stabilizes more helix in a new symmetrical zinc insulin hexamer. *Nature* **338**, 594–6.

Dodson, E.J., Dodson, G.G., Hubbard, R.E., and Reynolds, C.D. (1983). Insulin's structural behavior and its relation to activity. *Biopolymers* **22**, 281–91.

El-Maghrabi, M.R. *et al.* (1988). cDNA sequence of rat liver fructose–1, 6-bisphosphatase and evidence for down-regulation of its mRNA by insulin. *Proceedings of the National Academy of Sciences USA* **85**, 8430–4.

Evans, P.R. and Hudson, P.J. (1979). Structure and control of phosphofructokinase from *Bacillus stearothermophilus*. *Nature* **279**, 500–4.

Fraenkel, D.G. and Horecker, B.L. (1968). Fructose-1,6-diphosphatase and acid hexose phosphatase of *Escherichia coli*. *Journal of Bacteriology* **90**, 837–41.

Ginsberg, A. (1972). Glutamine synthetase of *Escherichia coli*: Some physical and chemical properties. *Advances in Protein Chemistry* **26**, 1–79.

Gouaux, J.E. and Lipscomb, W.N. (1988). Three-dimensional structure of carbamoyl phosphate and succinate bound to aspartate carbamoyltransferase. *Proceedings of the National Academy of Sciences USA* **85**, 4205–8.

Gouaux, J.E. and Lipscomb, W.N. (1990). Crystal structures of phosphonoacetamide ligated T and phosphonoacetamide and malonate ligated R states of aspartate carbamoyltransferase at 2.8 Å resolution and neutral pH. *Biochemistry* **29**, 389–402.

Gouaux, J.E., Krause, K.L., and Lipscomb, W.N. (1987). The catalytic mechanism of *Escherichia coli* aspartate carbamoyltransferase: a molecular modelling study. *Biochemical and Biophysical Research Communications* **142**, 893–7.

Gouaux, J.E., Stevens, R.C., Ke, H.M., and Lipscomb, W.N. (1989). Crystal structure of the Glu-239→Gln mutant of aspartate carbamoyltransferase at 3.1 Å resolution: an intermediate quaternary structure. *Proceedings of the National Academy of Sciences USA* **86**, 8212–16.

Gouaux, J.E., Stevens, R.C., and Lipscomb, W.N. (1990). Crystal structures of aspartate carbamoyltransferase ligated with phosphonoacetamide, malonate, and CTP or ATP at 2.8 Å resolution and neutral pH. *Biochemistry* **29**, 7702–15.

Harris, J.I. and Waters, M. (1976). Glyceraldehyde 3-phosphate dehydrogenase. In *The enzymes*, (ed. P. D. Boyer), (3rd edn), Vol. 13C, pp. 1–49. Academic Press, New York.

Hervé, G. (1989). *Allosteric enzymes*. CRC Press, Boca Raton, Florida.

Hsuanyu, Y. and Wedler, F.C. (1987). Kinetic mechanism of native *Escherichia coli* aspartate transcarbamylase. *Archives Biochemistry and Biophysics* **259**, 316–30.

Jarvest, R.L., Lowe, G, and Potter, B.V.L. (1981). The stereochemical course of phosphoryl transfer catalysed by *Bacillus stearothermophilus* and rabbit skeletal-muscle phosphofructokinase with a chiral [16O, 17O, 18O] phosphate ester. *Biochemical Journal* **199**, 427–32.

Johnson, L.N. and Barford, D. (1990). Glycogen phosphorylase. *Journal of Biological Chemistry* **265**, 2409–12.

Jones, M.E., Spector, L., and Lipmann, F. (1955). Carbamoyl phosphate, the carbamyl donor in enzymatic citrulline synthesis. *Journal of the American Chemical Society* **77**, 819–20.

Kantrowitz, E.R. and Lipscomb, W.N. (1988). *Escherichia coli* aspartate transcarbamylase: the relations between structure and function. *Science* **241**, 669–74.

Kantrowitz, E.R. and Lipscomb, W.N. (1990). *Escherichia coli* aspartate transcarbamylase: the molecular basis for a concerted allosteric transition. *Trends in Biochemical Sciences* **15**, 53–9.

Kasvinsky, P.J., Shechosky, S., and Fletterick, P.J. (1978). Synergistic regulation of phosphorylase *a* by glucose and caffeine. *Journal of Biological Chemistry* **253**, 9102–6.

Ke, H., Lipscomb, W.N., Cho, Y., and Honzatko, R.B. (1988). Complex of *N*-phosphonoacetyl-L-aspartate with aspartate carbamoyltransferase. *Journal of Molecular Biology* **204**, 725–47.

Ke, H., Thorpe, C.M., Seaton, B.A., Marcus, F., and Lipscomb, W.N. (1989). Molecular structure of fructose-1,6-bisphosphatase at 2.8 Å resolution. *Proceedings of the National Academy of Sciences USA* **86**, 1475–9.

Ke, H., Thorpe, C.M., Seaton, B.A., Lipscomb, W.N., and Marcus, F. (1990a). Structure refinement of fructose-1,6-bisphosphatase and its fructose 2,6-bisphosphate complex at 2.8 Å resolution. *Journal of Molecular Biology* **212**, 513–39.

Ke, H., Zhang, Y., and Lipscomb, W.N. (1990b). Crystal structure of fructose-1,6-bisphosphatase complexed with fructose 6-phosphate, AMP, and magnesium. *Proceedings of the National Academy of Sciences USA* **87**, 5243–7.

Kemp, R.G. and Marcus, F. (1990). Effects of fructose-2,6-bisphosphate on 6-phosphofructo-1-kinase and fructose-1,6-bisphosphatase. In *Fructose-2,6-bisphosphate*, (ed. S. J. Pilkis), p. 27. CRC Press, Boca Raton, Florida.

Koshland, D.E., Némethy, G., and Filmer, D. (1966). Comparison of experimental binding data and theoretical models in proteins containing subunits. *Biochemistry* **5**, 365–85.

Ladjimi, M.M. and Kantrowitz, E.R. (1988). A possible model for the concerted allosteric transition in *Escherichia coli* aspartate transcarbamylase as deduced from site-directed mutagenesis studies. *Biochemistry* **27**, 276–83.

Lau, F.T.-K. and Fersht, A.R. (1987). Conversion of allosteric inhibition to activation in phosphofructokinase by protein engineering. *Nature* **326**, 811–12.

Levine, M., Muirhead, H., Stammers, D.K., and Stuart, D.I. (1978). Stucture of pyruvate kinase and similarities with other enzymes: possible implications for protein taxonomy and evolution. *Nature* **271**, 626–30.

Liu, F. and Fromm, H.J. (1990). Kinetic studies of the mechanism and regulation of rabbit liver fructose-1,6-bisphosphatase. *Journal of Biological Chemistry* **265**, 7401–6.

Marcus, F. (1981). Structure and regulation of fructose-1,6-bisphosphatase. In *The regulation of carbohydrate formation and utilization in mammals* (ed. C. M. Veneziale), pp. 269–90. University Park Press, Baltimore, MD.

Marcus, F, Edelstein, I, Saidel, L.J., Keim, P.S., and Heinrikson, R.L. (1981). The covalent structure of pig kidney fructose-1,6-bisphosphatase: sequence of the 60-residue NH$_2$-terminal peptide produced by digestion with subtilisin. *Archives of Biochemistry and Biophysics* **209**, 687–96.

Mitra, A.K., McCarthy, M.P., and Stroud, R.M. (1989). Three-dimensional structure of the nicotinic acetylcholine receptor and location of the major associated 43-kD cytoskeletal protein, determined at 22 Å by low dose electron microscopy and X-ray diffraction to 12.5 Å. *Journal of Cell Biology* **109**, 755–74.

Monod, J., Wyman, J., and Changeaux, J.-P. (1965). On the nature of allosteric transitions: a plausible model. *Journal of Molecular Biology* **12**, 88–118.

Perutz, M.F. (1989). Mechanisms of cooperativity and allosteric regulation in proteins. *Quarterly Review of Biophysics* **22**, 139–236.

Perutz, M.F., Rossman, M.G., Cullis, A.F., Muirhead, H., Will, G., and North, A.C.T. (1960). Structure of haemoglobin: A three-dimensional Fourier synthesis at 5.5 Å resolution obtained by X-ray analysis. *Nature* **185**, 416–22.

Remington, S., Weigand, G., and Huber, R. (1982). Crystallographic refinement and atomic models of two different forms of citrate synthase at 2.7 and 1.7 Å resolution. *Journal of Molecular Biology* **158**, 111–52.

Rogers, D.T., Hiller, E., Mitsock, L, and Orr, E. (1988). Characterization of the gene for fructose-1,6-biphosphatase from *Saccharomyces cerevisiae* and *Schizosaccharomyces pombe*. Sequence, protein homology, and expression during growth on glucose. *Journal of Biological Chemistry* **263**, 6051–7.

Rypniewski, W.R. and Evans, P.R. (1989). Crystal structure of unliganded phosphofructokinase from *Escherichia coli*. *Journal of Molecular Biology* **207**, 805–21.

Sansom, M.S.P., Stuart, D.I., Acharya, K.R., Hajdu, J., McLaughlin, P.J., and Johnson, L.N. (1985). Glycogen phosphorylase *b* – the molecular anatomy of a large regulatory enzyme. *Journal of Molecular Structure* **123**, 3–25.

Scheffler, J.E. and Fromm, H.J. (1986). Regulation of rabbit liver fructose-1,6-bisphosphatase by metals, nucleotides, and fructose 2,6-bisphosphate as determined from fluorescence studies. *Biochemistry* **25**, 6659–65.

Schirmer, T. and Evans, P.R. (1990). Structural basis of the allosteric behaviour of phosphofructokinase. *Nature* **343**, 140–5.

Shirakihara, Y. and Evans, P.R. (1988). Crystal structure of the complex of phosphofructokinase from *Escherichia coli* with its reaction products. *Journal of Molecular Biology* **204**, 973–94.

Skarzynski, T. and Wonacott, A.J. (1988). Coenzyme-induced conformational change in glyceraldehyde-3-phosphate dehydrogenase from *Bacillus stearothermophilus*. *Journal of Molecular Biology* **203**, 1097–118.

Smith, G.D., Swenson, D.C., Dodson, G.C., and Reynolds, C.D. (1984). Structural stability in the 4-zinc human insulin hexamer. *Proceedings of the National Academy of Sciences USA* **81**, 7093–7.

Sprang, S.R. and Fletterick, R.J. (1979). The structure of glycogen phosphorylase *a* at 2.5 Å resolution. *Journal of Molecular Biology* **131**, 523–51.

Sprang, S.R., *et al.* (1988). Structural changes in glycogen phosphorylase induced by phosphorylation. *Nature* **336**, 215–21.

Steitz, T.A., Anderson, W.F., Fletterick, R.J., and Anderson, C.M. (1977).

High resolution crystal structures of yeast hexokinase complexes with substrates, activators, and inhibitors. *Journal of Biological Chemistry* **252**, 4494–500.

Stevens, R.C. and Lipscomb, W.N. (1990). Allosteric control of quaternary states in *E. coli* aspartate transcarbamylase. *Biochemical and Biophysical Research Communications* **171**, 1312–8.

Stevens, R.C., Gouaux, J.E., and Lipscomb, W.N. (1990). Structural consequences of effector binding to the T state of aspartate carbamoyltransferase: crystal structures of the unligated, ATP, and CTP complexed enzymes at 2.6 Å resolution. *Biochemistry* **29**, 7691–701.

Unwin, P.N.T. (1987). Gap junction structure and the control of cell to cell communication. *Ciba Foundation Symposium* **129** 78–91.

Van Vliet, F., Xi, X.-G., De Staercke, C., De Wannemaeker, B., Jacobs, A., Cherfils, J., Ladjimi, M.M., Herve, G., and Cunin, R. (1991). Heterotropic interactions in aspartate transcarbamoylase: Turning allosteric ATP activation into inhibition as a consequence of a single tyrosine to phenylalanine mutation. *Proc. Natl. Acad. Sci. USA* **88**, 9180–3.

Weber, K. (1968). New structural model of *E. coli* aspartate transcarbamylase and the amino-acid sequence of the regulatory polypeptide chain. *Nature* **218**, 1116–19.

Weigand, G., Remington, S., Deisenhofer, J., and Huber, R. (1984). Crystal structure analysis and molecular model of a complex of citrate synthase with oxaloacetate and *S*-acetonyl-coenzyme A. *Journal of Molecular Biology* **174**, 205–19.

Wild, J.R., Loughrey-Chen, S.J., and Corder, T.S. (1989). In the presence of CTP, UTP becomes an allosteric inhibitor of aspartate transcarbamoylase. *Proceedings of the National Academy of Sciences USA* **86**, 46–50.

Wiley, D.C. and Lipscomb, W.N. (1968). Crystallographic determination of symmetry of aspartate transcarbamylase. *Nature* **218**, 1119–21.

Yamashita, M.M., Almassy, R.J., Janson, C.A., Cascio, D., and Eisenberg, D. (1989). Refined atomic model of glutamine synthetase at 3.5 Å resolution. *Journal of Biological Chemistry* **264**, 17681–90.

9

The structure and sequence of insulin

U. Derewenda and G.G. Dodson

9.1 Some early history of insulin

In 1922 Banting and Best (Banting and Best 1922) finally extracted a hormone factor from the dog pancreas and showed that it was responsible for controlling the level of blood sugar. This hormone was named insulin. It was one of the first proteins to be obtained in a sufficiently pure form for pharmaceutical purposes and was also the first protein for which the full amino-acid sequence was determined (Sanger *et al.* 1955; Sanger 1959).

The first crystals of insulin were obtained by J. J. Abel in 1926 (Abel 1926). Due to the impurities of the first insulin preparations, these early crystals were not always reproducible. The discovery of zinc in crystalline insulin was made by J. Brown in 1931 (Jukes 1979) and then by Scott (Scott 1934). The first X-ray diffraction photographs were taken from zinc-containing crystals. The first crystal structure of insulin to be solved by X-ray crystallography was also that of zinc insulin. The first electron density map was calculated at 2.8 Å resolution (Adams *et al.* 1969). Today the molecular structure of insulin is known to 1.5 Å resolution (Baker *et al.* 1988), which allows one to locate individual atoms with 0.1-0.2 Å accuracy.

Insulin has been studied extensively in different crystalline forms, at different stages of association; many chemically modified insulin molecules have been prepared and some of these have been studied crystallographically.The progress in biological sciences has made it possible to study the physical, chemical, and biological properties of insulin even more thoroughly and accurately. Much of this knowledge comes from the studies of mutants obtained by site-directed mutagenesis.

9.2 The functions of insulin

The principle metabolic effects of insulin on various tissues are shown in Table 9.1 (for extensive reviews on insulin function see Wallis *et al.* 1985; Kono 1988). In addition to the metabolic functions, insulin also exhibits weak growth-promoting activity—it is able to stimulate thymidine incor-

Table 9.1 Metabolic effects of insulin (after Wallis *et al.* 1985)

Metabolic effects	Tissue
Short-term	
increases sugar transport	muscle, adipose tissue
increases protein synthesis	muscle, liver
inhibits lypolysis	adipose tissue
activates enzymes	
glycogen synthetase	liver, muscle
lipoprotein-lipase	blood vessel wall
pyruvate dehydrogenase	adipose tissue
acetyl-CoA carboxylase	
low K_m diesterase	
inactivates enzymes	
phosphorylase	liver
hormone-sensitive lipase	adipose tissue
Long-term	
activates glucokinase	liver
depresses controlling enzymes of gluconeogenesis	liver

poration into DNA. It has been shown that the metabolic effects are mediated through the insulin receptor, whereas the growth-promoting effects are stimulated by binding to the insulin-like growth factor (IGF) receptors.

9.3 The crystal structure of insulin

9.3.1 The monomer

A monomer of insulin is made up of two polypeptide chains designated A and B. Plate 32 shows the arrangement of the molecule. Chain A contains 21 amino acids. The first eight residues in the A-chain (A1-A8) form a slightly distorted α-helix which is followed by an extended turn formed with residues A9-A13. This stretch connects the N-terminal helix and the nearly antiparallel C-terminal 3_{10} helix (A14-A20). Cysteines A6 and A11 form a disulphide link closing an intrachain loop. Both ends of the A-chain are close together and A19 Tyr and A2 Ile side-chains are in van der Waals contact.

The B-chain (30 amino acids) is joined to the A-chain by two disulphide bonds, A7-B7 and A20-B19, situated at the ends of a central α-helix which runs from B9 to B19. B20 Gly breaks the helix and, following a sharp turn, the B-chain forms an extended stretch of mostly hydrophobic residues, B24-B30. The two chains are further secured together by a number of hydrogen bonds. The interior of the molecule is composed of hydrophobic residues: isoleucine A2 and three leucines, A16, B11, and B15. On both sides of the central core are the two disulphide bonds, A6-A11 and A20-B19. The central core is surrounded by non-polar valines (A3, B12, B18) and phenylalanine B24. The molecule is small, so the last group of residues is partly exposed.

The surface of the molecule is about 50 per cent polar. There are two non-polar regions on the surface of the molecule: one includes residues lying on the B-chain α-helix, B8 Gly, B12 Val, B14 Ala, B16 Tyr, and B17 Leu, and the C-terminal B24 Phe, B25 Tyr, B26 Tyr, and B28 Pro. The other non-polar surface contains α-helical B18 Val, N-terminal B1 Phe, B2 Val, and two residues from the A-chain (A13 Leu and A14 Tyr), (see Plate 34).

9.3.2 Aggregation of the insulin molecule

In concentrated solution insulin occurs as a mixture of different states of association: monomers, dimers, tetramers, and hexamers. Only in very dilute ($\approx 10^{-6}$ M) solutions does insulin exist mostly as a monomer. All crystals of native insulin are grown from solutions in which the hormone forms dimers or hexamers. The extended structure of the B-chain at the C-terminus (B21-B30) plays a very important role in the association of insulin molecules into dimers. Together with the corresponding structure in another molecule, it forms an antiparallel β-sheet, with hydrogen bonds between two backbone threads holding the two monomers together. The B-chain helices are packed close together. In Plate 35 all the residues involved in dimer formation are outlined.

In the presence of zinc ions, insulin dimers associate into hexamers. The three dimers are related by a central threefold axis of symmetry with two zinc ions lying on the axis and keeping the dimers together by an interaction between imidazole rings of B10 His and the zinc ions. Through the centre of the hexamer and between the dimers run channels of water.

The dimer-dimer contacts are made mostly from B-chain residues B1, B2, B4, B14, B17, B18, B19, and B20 and the A-chain residues A13, A14, and A17. The surface of the hexamer contains most of the hydrophilic residues. The hexamer has the shape of a torus with two zinc atoms bound in the central cylindrical channel.

9.3.3 Active surface

Identifying the insulin active surface has presented complex problems. Most insulins exhibit binding affinity to the receptor parallel to the biological activity. It is therefore generally assumed that the region associated with metabolic activation is indistinguishable from the receptor-binding region of insulin. On the basis of very strong evidence (despenta peptide insulin, DPI), the monomer is assumed to be the active molecule.

The determination of residues responsible for biological activity was initially based on the distribution of invariant residues and biological activity in different animal species. Blundell *et al.* (1972) first identified residues that might be involved in biological activity. The invariant residues on the monomer surface comprise the A-chain residues: A1 Gly, A4 (conserved as an acid group), A5 Gln, A19 Tyr, and A21 Asn; and the B-chain residues: B10 His, B12 Val, B13 Glu, B16 Tyr, B22 Arg, B24 Phe, B25 Phe (conserved as an aromatic residue), and B26 Tyr (Plates 36 and 37). The importance of these residues has been confirmed by studies of modified insulins; for example, the positively charged group NH_3^+ in A1 glycine is important for affinity -substitutions with different uncharged groups decreased activity to 10–40 per cent (Waelbroeck *et al.* 1979). A19 Phe has activity of $\approx$ 22 per cent (Danho *et al.* 1980). Elimination of A21 Asn reduces the insulin binding potency to about 6 per cent (Chu *et al.* 1987), substitution of B12 Val $\rightarrow$ Ile reduces potency to about 30 per cent (Brange *et al.* 1988). B10 His and B26 Tyr are replaced in guinea-pig and coypu insulin and, as the potency of the hormone in these species is low, these residues are thought to be involved in activity. Insulin activity is drastically diminished after substitution of B22 Arg $\rightarrow$ Lys and disappears after changing B22 Arg $\rightarrow$ Ala (Weitzel *et al.* 1975).

Studies of mutant insulins suggest that many residues other than the invariant ones are relevant for insulin activity: changes of A2 Ile $\rightarrow$ Gly or A2 Ile $\rightarrow$ Ala have almost abolished activity (Kitagawa *et al.* 1984) and mutation of A3 Val $\rightarrow$ Leu reduces activity (Nanjo *et al.* 1986), whereas A8 Thr $\rightarrow$ His markedly activates insulin (Marki *et al.* 1979).

The early assumptions that the β-pleated sheet connecting monomers into dimers may be a pattern for an interaction between the hormone and its receptor (Pullen *et al.* 1976) has not been supported by later experiments. Dimers in which monomers are covalently bound (e.g. tyrosines B16 and B26, which are very close to each other at a dimer interface, are cross-linked) retain some biological activity (*c.* 10 per cent) (S. M. Cutfield *et al.* 1986). This means that the surface involved in aggregation of monomers into dimers cannot be responsible for binding to the receptor. Consistent with this, DPI (des B30-B25) insulin, which has lost the dimer-forming surface, retains at least 40 per cent activity (Bi *et al.* 1983).

The DPI studies have revealed a new surface which is exposed by removing the last five residues from the B-chain. This surface contains residues A1, A2, A3, A16, B11, and B15. Furthermore, the structures of the A2sb (see glossary) insulin with low potency, structure A1-B29 peptide cross-link (with no activity), are essentially identical to the native molecule. This confirms the opinion that the movement of C-terminal B-chain residues exposing the surface seen in the DPI structure is necessary for biological activity of the hormone. The studies of different substitutions at the B25-B30 residues are very numerous, complex, and sometimes very confusing. What they seem to suggest is that B25-B30 are not necessary for receptor binding; however, when present, the movement of these residues and their precise conformation during the binding are important.

A different approach to the receptor-binding region was proposed by Yeung *et al.* (1980). They suggested that the receptor-binding region comprises the five N-terminal residues in the B-chain. Studies with N-terminal residues removed from the insulin monomer point to the involvement of those residues in receptor binding. Although detachment of B1 does not influence the biological activity (Schwartz and Katsoyannis 1978), removing residues B1-B3 decreases activity (to 70 per cent). Identical results were obtained with des-B1-B4 and des-B1-B5 insulins (*c*. 54 per cent and 32 per cent potency, respectively). Analogues that have substitutions at B1 (Ala, Val, Tyr) or B2 (Phe) have biological activities of 34–61 per cent (Saunders and Offord 1977).

Very interesting studies on some aspects of insulin activity were carried out by King and Kahn (1981). They observed striking differences in potency of 23 different insulins and insulin analogues on thymidine incorporation into DNA (growth-promoting activity) and glucose oxidation (metabolic activity). Generally, all insulins examined (both native and modified) retain their growth activity to a higher degree than their metabolic effects. The data indicated that growth effects can withstand amino-acid substitutions better than metabolic effects, and that the growth-promoting activity of insulin is distinct from metabolic activity. In view of this study, the surface exposed in DPI may not be vital for insulin growth effects, whereas B10, B13, B26, and A4 may be important for growth-promoting effects. From the structural point of view it is an important observation. The invariant and semi-invariant residues lie on both sides of the insulin monomer. One part is on the A19 and β-sheet side, the other part is on the opposite side — α-helix in the B-chain. These two sites are connected only by residue A8 — which seems to be important for both metabolic and growth effects.

9.4 Evolution of insulin and related peptides

9.4.1 Evolution of insulin

Insulin was the first protein for which the full amino-acid sequence was determined chemically (see Table 9.2). Up to date, at least 50 amino-acid sequences of insulin have been reported in phylogenetically very diverse vertebrates, ranging from mammals through birds to primitive jawless hagfish.

Insulin and other related peptides have been considered by some as occurring only in the vertebrates, and their existence in invertebrates was not readily accepted. However, there is evidence that invertebrates do contain molecules having functions analogous to vertebrate insulin and insulin receptor.

Table 9.2 Comparison of the amino-acid sequences of polypeptides from the insulin superfamily (residues invariant in the insulin molecule are outlined)

A-chain		1	2	3	4	5	6	7	8	9	10	11	12	13	14	15	16	17	18	19	20	21	
human insulin		G	I	V	E	Q	C	C	T	S	I	C	S	L	Y	Q	L	E	N	Y	C	N	
Lymnea MIP	Q G T T N	N	I	V	C	Q	C	C	M	K	P	C	T	L	S	E	L	R	Q	Y	C	P	
Bombyx PTTH		G	I	V	D	E	C	C	L	R	P	C	S	V	D	V	L	L	S	Y	C	-	
IGF I		G	I	V	D	E	C	C	F	R	I	C	D	L	R	R	L	E	M	Y	C	A	P L K P A K S A
IGF II		G	I	V	E	E	C	C	F	R	I	C	D	L	A	L	L	E	T	Y	C	A	T - - P A K S E
relaxin	R M	T	L	S	E	K	C	C	Q	V	G	C	I	R	K	D	I	A	R	L	C	-	
amylin DAP		-	-	-	-	K	C	N	T	A	T	C	A	T	Q	R	L	A	N	F	L	V	H S S N N F G A I L S S T N V G S N T Y

B-chain	−2	−1	1	2	3	4	5	6	7	8	9	10	11	12	13	14	15	16	17	18	19	20	21	22	23	24	25	26	27	28	29	30	+1	+2	+3	+4	+5
human insulin			F	V	N	Q	H	L	C	G	S	H	L	V	E	A	L	Y	L	V	C	G	E	R	G	F	F	Y	T	P	K	T					
Lymnea MIP			P	H	R	R	G	V	C	G	S	A	L	A	D	L	V	D	F	A	C	S	S	S	N	Q	P	A	M	V	-	-					
Bombyx PTTH			Q	A	V	H	T	T	C	G	R	H	L	A	R	T	L	A	D	L	C	W	E	A	G	V	D	-	-	-	-	-					
IGF I			-	G	P	E	T	L	C	G	A	E	L	V	D	A	L	Q	F	V	C	G	D	R	G	F	Y	N	K	P	T	G					
IGF II			R	P	S	E	T	L	C	G	G	E	L	V	D	T	L	Q	F	V	C	G	D	R	G	F	Y	S	R	P	A	S					
relaxin			N	D	F	I	K	A	C	G	R	E	L	V	R	L	W	V	E	I	C	V	W	S	-	-	-	-	-	-	-	-					

As early as 1964, extracts from larval food of the honey-bee had been found to exhibit insulin-like activity. These findings were recognized as inconclusive because the tests were carried out on the epididymal fat pad, a system that responds to substances other than insulin. In 1965, experiments on digestive tissue of the purple starfish, an echinoderm, suggested that insulin-like activity was present. Low concentrations of mercaptoethanol destroyed hormonal activity of starfish extracts, suggesting that (as in vertebrates) activity depends on the three-dimensional structure of the protein and the existence of the disulphide bonds (Wilson and Falkmer 1965). Insulin production occurs in the gastrointestinal tracts of species of shore crab, whelk, scallop, and octopus, and in the intestine of the sea squirt (Davidson *et al.* 1971) but has not been found in species of Cnidaria (jellyfish and sea anemone).

It has been now established that insulin-like activity can be detected in the gastrointestinal duct, haemolymph, and neuroendocrine system of insects. The hormone has been found in blowfly brains and displayed insulin-like activity when tested on isolated rat fat cells (Duve and Thorpe 1979; Duve *et al.* 1979). The blowflies became hyperglycaemic after removal of the median neurosecretory system (Duve 1978).

In 1981 the presence of a substance very similar to insulin was reported in fruitfly (*Drosophila melanogaster*) heads (LeRoith *et al.* 1981). The material had insulin bioactivity, as measured by glucose oxidation or lipogenesis by isolated rat adipocytes. The activity was neutralized by anti-insulin antibodies. Also, extracts from the skin and interior organs of earthworms contained a compound similar to insulin (LeRoith *et al.* 1981).

The localization of insulin-like material in insect brains supports the opinion that the two major systems of intercellular communication (the endocrine and the nervous systems) must have common origins. It also suggests that the nervous system of primitive multicellular organisms was the original source of regulatory peptides (Thorpe and Duve 1988).

All these findings provide evidence that the evolutionary origins of insulin are far more ancient than the history of vertebrates. They might even go back as far as unicellular organisms, as immunologically and biologically active insulin-like material has been found in fungi (LeRoith *et al.* 1980).

9.4.2 *Evolution of insulin-related molecules*

Insulin is structurally and functionally related to the insulin-like growth factors (IGF I and II). The ratio of the growth-promoting effects compared with the metabolic stimulation is higher for growth factors than for insulin; on the other hand, insulin does exhibit low levels of growth-promoting activity (Blundell and Humbel 1980; Dodson *et al.* 1980; Bajaj *et al.* 1983).

Relaxin is yet another polypeptide with a three-dimensional structure similar to that of insulin (Isaacs *et al.* 1978; Blundell and Humbel 1980;

Dodson *et al.* 1980; Bajaj *et al.* 1983). Relaxin does not show insulin-like activity, neither does it bind to the insulin receptor. The hormone can be found in the corpus luteum and in the placenta. It stimulates dilation of the symphysis pubis prior to parturition.

More recent studies (Nagasawa *et al.* 1986) on material isolated from the brain of the male adult silkworm (*Bombyx mori*) resulted in the determination of the amino-acid sequence of 4K-PTTH-II, a prothoracicotropic hormone responsible for the production of ecdysone by the prothoracic glands. The primary structure of the PTTH has 40 per cent sequence homology with insulin (Table 9.2) and possesses the same distribution of six cysteine residues. Residues composing the interior of the insulin molecule are also preserved in PTTH (A2 Ile, A16 Leu, B11 Leu, and B15 Leu).

Sequence data of an insulin-related peptide precursor (MIP) from the mollusc *Lymnaea stagnalis* (Smit *et al.* 1988) shows that it can assume insulin-like tertiary structure. All cysteine residues correspond to their positions in insulin, and the hydrophobic core (A2, A16, B11) is conserved except that B15 is Val instead of Leu. It is worth noticing that in *Lymnaea*, as well as in the silkworm, B8 Gly and A3 Val (invariant in all known true insulins) are also conserved. The B23 Gly, which helps to define the particular conformation of the B-chain C-terminal residues, is conserved in all known insulin species. The presence of the side-chain at B23 in *Lymnaea* suggests that B-chain C-termini have a somewhat different conformation and function.

No homology with insulin was found in the C-terminal B-chain of PTTH, MIP, or relaxin. As is the case in relaxin, the PTTH and MIP peptides have no detectable insulin-like biological activity. This may perhaps be explained by the different sequence at B23-B26, which results in the loss of the receptor-binding ability (Steiner and Chan 1988).

The sequence analysis of a peptide (diabetes-associated peptide, DAP; or islet amyloid polypeptide, IAPP) which is a major component of deposits of amyloid in pancreatic islets (in human type 2, non-insulin-dependent diabetes) has shown that there are some weak similarities between the 16-residue segment of the peptide and the insulin A-chain. The DAP contains 37 amino acids (Cooper 1987). It has 46 per cent sequence homology with calcitonin gene-related peptide (CGRP). Both human DAP and rat CGRP-1 act as inhibitors of glycogen synthesis in rat muscle *in vitro* (Leighton and Cooper 1988).

The observation that amylin has 30 per cent sequence identity with glyceraldehyde-3-phosphate dehydrogenase (GPDH) from *Bacillus stearothermophilus* has led to a proposed tertiary structure of amylin, based on secondary structure prediction methods. Part of the structure of GPDH (143–178 sequence) shows similarities in organization to the insulin B-chain. The structure of amylin has been predicted to be the same as the insulin B-chain (Soldanha *et al.* 1989), even though the amino-acid sequences of the peptides are not homologous.

Table 9.3 Sequence of the insulin A-chain

A-chain	1	2	3	4	5	6	7	8	9	10	11	12	13	14	15	16	17	18	19	20	21		References
human	G	I	V	E	Q	C	C	T	S	I	C	S	L	Y	Q	L	E	N	Y	C	N		Bell *et al.* (1980)
macaque	G	I	V	E	Q	C	C	T	S	I	C	S	L	Y	Q	L	E	N	Y	C	N		Wetekam *et al.* (1982)
chimpanzee	G	I	V	E	Q	C	C	T	S	I	C	S	L	Y	Q	L	E	N	Y	C	N		Seino *et al.* (1987)
African green monkey	G	I	V	E	Q	C	C	T	S	I	C	S	L	Y	Q	L	E	N	Y	C	N		Peterson *et al.* (1972)
hamster	G	I	V	E	Q	C	C	T	S	I	C	S	L	Y	Q	L	E	N	Y	C	N		Neelson *et al.* (1973)
finback whale	G	I	V	E	Q	C	C	T	S	I	C	S	L	Y	Q	L	E	N	Y	C	N		Hama *et al.* (1964)
sperm whale	G	I	V	E	Q	C	C	T	S	I	C	S	L	Y	Q	L	E	N	Y	C	N		Ishihara *et al.* (1958)
rabbit	G	I	V	E	Q	C	C	T	S	I	C	S	L	Y	Q	L	E	N	Y	C	N		Smith (1966)
pig	G	I	V	E	Q	C	C	T	S	I	C	S	L	Y	Q	L	E	N	Y	C	N		Chance *et al.* (1968)
dog	G	I	V	E	Q	C	C	T	S	I	C	S	L	Y	Q	L	E	N	Y	C	N		Kwok *et al.* (1983)
owl monkey	G	V	V	D	Q	C	C	T	S	I	C	S	L	Y	Q	L	Q	N	Y	C	N		Seino *et al.* (1987)
sei whale	G	I	V	E	Q	C	C	A	S	T	C	S	L	Y	Q	L	E	N	Y	C	N		Ishihara *et al.* (1958)
bovine	G	I	V	E	Q	C	C	A	S	V	C	S	L	Y	Q	L	E	N	Y	C	N		Nolan *et al.* (1971)
cat	G	I	V	E	Q	C	C	A	S	V	C	S	L	Y	Q	L	E	H	Y	C	N		Hallden *et al.* (1986)
Arabian camel	G	I	V	E	Q	C	C	A	S	V	C	S	L	Y	Q	L	E	N	Y	C	N		Dahno (1972)
horse	G	I	V	E	Q	C	C	T	G	I	C	S	L	Y	Q	L	E	N	Y	C	N		Harris *et al.* (1956)
elephant	G	I	V	E	Q	C	C	T	G	V	C	S	L	Y	Q	L	E	N	Y	C	N		Smith (1966)
goat & Arabian camel	G	I	V	E	Q	C	C	A	G	V	C	S	L	Y	Q	L	E	N	Y	C	N		Smith L (1966)
sheep	G	I	V	E	Q	C	C	A	G	V	C	S	L	Y	Q	L	E	N	Y	C	N		Nolan *et al.* (1971)
crested porcupine	G	I	V	D	Q	C	C	T	G	V	C	S	L	Y	Q	L	Q	N	Y	C	N		Horuk *et al.* (1980)
egp. spiny mouse	G	I	V	D	Q	C	C	T	S	I	C	S	L	Y	Q	L	E	N	Y	C	N		Buenzli *et al.* (1972)
mouse 1	G	I	V	D	Q	C	C	T	S	I	C	S	L	Y	Q	L	E	N	Y	C	N		Buenzli *et al.* (1972)
mouse 2	G	I	V	D	Q	C	C	T	S	I	C	S	L	Y	Q	L	E	N	Y	C	N		Buenzli *et al.* (1972)
rat 1	G	I	V	D	Q	C	C	T	S	I	C	S	L	Y	Q	L	E	N	Y	C	N		Clark and Steiner (1969)
rat 2	G	I	V	D	Q	C	C	T	S	I	C	S	L	Y	Q	L	E	N	Y	C	N		Lomedico *et al.* (1979)
chinchilla	G	I	V	D	Q	C	C	T	S	I	C	T	L	Y	Q	L	E	N	Y	C	N		Wood *et al.* (1975)
cuis	G	I	V	D	Q	C	C	T	R	I	C	T	S	Y	Q	L	R	N	Y	C	N		Bajaj *et al.* (1986)
guinea-pig	G	I	V	D	Q	C	C	T	G	T	C	T	R	H	Q	L	Q	S	Y	C	N		Smith (1966)
nutria, coypu	G	I	V	D	Q	C	C	T	N	I	C	S	R	N	Q	L	M	S	Y	C	N	D	Bajaj *et al.* (1986)
casiragua	G	I	V	D	Q	C	C	T	N	I	C	S	R	N	Q	L	L	T	Y	C	N		Horuk *et al.* (1979)

chicken	G	I	V	E	Q	C	C	H	N	T	C	S	L	Y	Q	L	E	N	Y	C	N	Perler *et al.* (1980)
ostrich	G	I	V	E	Q	C	C	H	N	T	C	S	L	Y	Q	L	E	N	Y	C	N	Evans *et al.* (1988)
goose	G	I	V	E	Q	C	C	E	N	P	C	S	L	Y	Q	L	E	N	Y	C	N	Xu *et al.* (1983) and Sundby
duck	G	I	V	E	Q	C	C	E	N	P	C	S	L	Y	Q	L	E	N	Y	C	N	Markussen (1973)
rattle snake	G	I	V	E	Q	C	C	E	N	T	C	S	L	Y	Q	L	E	N	Y	C	N	Kimmel *et al.* (1976)
culubrid snake	G	I	V	E	Q	C	C	E	N	T	C	S	L	Y	E	L	E	N	Y	C	N	Zang *et al.* (1981)
alligator	G	I	V	E	Q	C	C	H	N	T	C	S	L	Y	Q	L	E	N	Y	C	N	Lance *et al.* (1984)
Afr. clawed frog 1	G	I	V	E	Q	C	C	H	S	T	C	S	L	F	Q	L	E	N	Y	C	N	Shuldiner *et al.* (1989)
Afr. clawed frog 2	G	I	V	E	Q	C	C	H	S	T	C	S	L	F	Q	L	E	N	Y	C	N	Shuldiner *et al.* (1989)
alligator gar	G	I	V	E	Q	C	C	H	K	P	C	T	I	Y	E	L	E	N	Y	C	N	Pollock *et al.* (1987)
tuna II	G	I	V	E	Q	C	C	H	K	P	C	N	I	F	D	L	Q	N	Y	C	N	Neumann and Humbel (1969)
tuna skipjack	G	I	H	Z	Z	C	C	H	K	P	C	B	I	F	Z	L	Z	B	Y	C	N	Kotaki (1963)
bonito	G	I	H	E	Q	C	C	H	K	P	C	D	I	F	Q	L	E	N	Y	C	N	Neumann and Humbel (1972)
carp	G	I	V	E	Q	C	C	H	K	P	C	S	I	F	E	L	Q	N	Y	C	N	Makower *et al.* (1982)
salmon (coho)	G	I	V	E	Q	C	C	H	K	P	C	N	I	F	D	L	Q	N	Y	C	N	Plisetskaya *et al.* (1985)
flounder	G	I	V	E	Q	C	C	H	K	P	C	N	I	F	D	L	Q	N	Y	C	N	Conlon *et al.* (1987)
chum salmon	G	I	V	E	Q	C	C	H	R	P	C	N	I	F	D	L	Q	N	Y	C	N	Sorokin *et al.* (1982)
anglerfish	G	I	V	E	Q	C	C	H	R	P	C	N	I	F	D	L	Q	N	Y	C	N	Hobart *et al.* (1980)
goosfish	G	I	V	E	Q	C	C	H	R	P	C	N	I	F	D	L	Q	N	Y	C	N	Hobart *et al.* (1980)
cod	G	I	V	D	Q	C	C	H	R	P	C	D	I	F	D	L	Q	N	Y	C	N	Reid *et al.* (1968)
toadfish 1	G	I	V	E	Q	C	C	H	R	P	C	D	I	F	D	L	Q	S	Y	C	N	Smith (1966)
toadfish 2	G	I	V	E	Q	C	C	H	R	P	C	D	K	F	D	L	Q	S	Y	C	N	Smith (1966)
hagfish	G	I	V	E	Q	C	C	H	K	R	C	S	I	Y	D	L	E	N	Y	C	N	Chan *et al.* (1981)
daddy sculpin	G	I	V	E	Q	C	C	H	R	P	C	N	I	R	V	L	E	N	Y	C	N	J. Cutfield *et al.* (1986)
spotted ratfish	G	I	V	E	Q	C	C	H	N	T	C	S	L	A	N	L	E	G	Y	C	N	Conlon *et al.* (1986)
rabbit fish	G	I	V	E	Q	C	C	H	N	T	C	S	L	A	N	L	E	G	Y	C	N	Conlon *et al.* (1988)
elephantfish	G	I	V	E	Q	C	C	H	N	T	C	S	L	V	N	L	E	G	Y	C	N	Berks *et al.* (1989)
ray, torpedo	G	I	V	E	H	C	C	H	N	T	C	S	L	F	D	L	E	G	Y	C	N	Conlon *et al.* (1986)
shark, spiny dogfish	G	I	V	E	H	C	C	H	N	T	C	S	L	Y	D	L	E	G	Y	C	N Q	Bajaj *et al.* (1983)
lamprey	G	I	V	E	Q	C	C	H	R	K	C	S	I	Y	D	M	E	N	Y	C	N	Plisetskaya *et al.* (1988)

egp., Egyptian; Afr., African.

Table 9.4 Sequence of the insulin B-chain

B-chain	−2	−1	1	2	3	4	5	6	7	8	9	10	11	12	13	14	15	16
human			F	V	N	Q	H	L	C	G	S	H	L	V	E	A	L	Y
Afr. gr. monkey			F	V	N	Q	H	L	C	G	S	H	L	V	E	A	L	Y
chimpanzee			F	V	N	Q	H	L	C	G	S	H	L	V	E	A	L	Y
macaque			F	V	N	Q	H	L	C	G	S	H	L	V	E	A	L	Y
dog			F	V	N	Q	H	L	C	G	S	H	L	V	E	A	L	Y
finback whale			F	V	N	Q	H	L	C	G	S	H	L	V	E	A	L	Y
hamster			F	V	N	Q	H	L	C	G	S	H	L	V	E	A	L	Y
pig			F	V	N	Q	H	L	C	G	S	H	L	V	E	A	L	Y
rabbit			F	V	N	Q	H	L	C	G	S	H	L	V	E	A	L	Y
sperm whale			F	V	N	Q	H	L	C	G	S	H	L	V	E	A	L	Y
elephant			F	V	N	Q	H	L	C	G	S	H	L	V	E	A	L	Y
horse			F	V	N	Q	H	L	C	G	S	H	L	V	E	A	L	Y
bovine			F	V	N	Q	H	L	C	G	S	H	L	V	E	A	L	Y
egp. spiny mouse			F	V	B	Q	H	L	C	G	S	H	L	V	E	A	L	Y
sei whale			F	V	N	Q	H	L	C	G	S	H	L	V	E	A	L	Y
Arabian camel			F	A	N	Q	H	L	C	G	S	H	L	V	E	A	L	Y
cat			F	V	N	Q	H	L	C	G	S	H	L	V	E	A	L	Y
goat & ar. camel			F	V	N	Q	H	L	C	G	S	H	L	V	E	A	L	Y
mouse 1			F	V	K	Q	H	L	C	G	P	H	L	V	E	A	L	Y
owl monkey			F	V	N	Q	H	L	C	G	P	H	L	V	E	A	L	Y
rat 1			F	V	K	Q	H	L	C	G	P	H	L	V	E	A	L	Y
rat 2			F	V	K	Q	H	L	C	G	S	H	L	V	E	A	L	Y
sheep			F	V	N	Q	H	L	C	G	S	H	L	V	E	A	L	Y
mouse 2			F	V	K	Q	H	L	C	G	P	H	L	V	E	A	L	Y
chinchilla			F	V	N	K	H	L	C	G	S	H	L	V	D	A	L	Y
cr. porcupine			F	V	N	Q	H	L	C	G	S	H	L	V	E	A	L	Y
cuis			F	F	N	R	H	L	C	G	S	N	L	V	D	A	L	Y
guinea-pig			F	V	S	R	H	L	C	G	S	N	L	V	E	T	L	Y
casiragua			Y	V	G	Q	R	L	C	G	S	Q	L	V	D	T	L	Y
putria, coypu			Y	V	S	Q	R	L	C	G	S	Q	L	V	D	T	L	Y
duck			A	A	N	Q	H	L	C	G	S	H	L	V	E	A	L	Y
goose			A	A	N	Q	H	L	C	G	S	H	L	V	E	A	L	Y
turkey			A	A	N	Q	H	L	C	G	S	H	L	V	E	A	L	Y
chicken			A	A	N	Q	H	L	C	G	S	H	L	V	E	A	L	Y
ostrich			A	A	N	Q	H	L	C	G	S	H	L	V	E	A	L	Y
alligator			A	A	N	Q	R	L	C	G	S	H	L	V	D	A	L	Y
rattle snake			A	P	N	Q	R	L	C	G	S	H	L	V	E	A	L	F
culubrid snake			A	P	N	Q	R	L	C	G	S	H	L	V	E	A	L	F
Afr. claw. frog 2			L	A	N	Q	H	L	C	G	S	H	L	V	E	A	L	Y
Afr. claw. frog 1			L	V	N	Q	H	L	C	G	S	H	L	V	E	A	L	Y
alligator gar			A	A	N	Q	H	L	C	G	S	H	L	V	E	A	L	Y
bonito			A	A	N	P	H	L	C	G	S	H	L	V	D	A	L	Y
daddy sculpin			A	D	P	Q	H	L	C	G	S	H	L	V	D	A	L	Y
salmon (coho)			A	A	A	Q	H	L	C	G	S	H	L	V	E	A	L	Y
chum salmon			A	A	A	Q	H	L	C	G	L	H	L	V	D	A	L	Y
elephantfish			V	P	T	Q	R	L	C	G	S	H	L	V	D	A	L	Y
flounder		V	V	P	P	Q	H	L	C	G	A	H	L	V	D	A	L	Y
goosfish			A	P	A	Q	H	L	C	G	S	D	L	V	E	A	L	Y
anglerfish		V	A	P	A	Q	H	L	C	G	S	H	L	V	D	A	L	Y
tuna II		V	A	P	P	Q	H	L	C	G	S	H	L	V	D	A	L	Y

17	18	19	20	21	22	23	24	25	26	27	28	29	30	1+2+3+4+5
L	V	C	G	E	R	G	F	F	Y	T	P	K	T	
L	V	C	G	E	R	G	F	F	Y	T	P	K	T	
L	V	C	G	E	R	G	F	F	Y	T	P	K	T	
L	V	C	G	E	R	G	F	F	Y	T	P	K	T	
L	V	C	G	E	R	G	F	F	Y	T	P	K	A	
L	V	C	G	E	R	G	F	F	Y	T	P	K	S	
L	V	C	G	E	R	G	F	F	Y	T	P	K	S	
L	V	C	G	E	R	G	F	F	Y	T	P	K	A	
L	V	C	G	E	R	G	F	F	Y	T	P	K	S	
L	V	C	G	E	R	G	F	F	Y	T	P	K	S	
L	V	C	G	E	R	G	F	F	Y	T	P	K	T	
L	V	C	G	E	R	G	F	F	Y	T	P	K	A	
L	V	C	G	E	R	G	F	F	Y	T	P	K	A	
L	V	C	G	E	R	G	F	F	Y	T	P	K	S	
L	V	C	G	E	R	G	F	F	Y	T	P	K	A	
L	V	C	G	E	R	G	F	F	Y	T	P	K	A	
L	V	C	G	E	R	G	F	F	Y	T	P	K	A	
L	V	C	G	E	R	G	F	F	Y	T	P	K	A	
L	V	C	G	E	R	G	F	F	Y	T	P	K	S	
L	V	C	G	E	R	G	F	F	Y	A	P	K	T	
L	V	C	G	E	R	G	F	F	Y	T	P	K	S	
L	V	C	G	E	R	G	F	F	Y	T	P	M	S	
L	V	C	G	E	R	G	F	F	Y	T	P	K	A	
L	V	C	G	E	R	G	F	F	Y	T	P	M	S	
L	V	C	G	D	R	G	F	F	Y	T	P	M	A	
L	V	C	G	N	D	G	F	F	Y	R	P	K	A	
V	V	C	K	D	K	G	F	F	S	R	P	D	-	
S	V	C	Q	D	D	G	F	F	Y	I	P	K	D	
S	V	C	K	H	R	G	F	Y	R	P	S	E	-	
S	V	C	R	H	R	G	F	Y	R	P	N	D	-	
L	V	C	G	E	R	G	F	F	Y	S	P	K	T	
L	V	C	G	E	R	G	F	F	Y	S	P	K	T	
L	V	C	G	E	R	G	F	F	Y	S	P	K	A	
L	V	C	G	E	R	G	F	F	Y	S	P	K	A	
L	V	C	G	E	R	G	F	F	Y	S	P	K	A	
L	V	C	G	E	R	G	F	F	Y	S	P	K	G	
L	I	C	G	E	R	G	F	Y	Y	S	P	R	S	
L	I	C	G	E	R	G	F	Y	Y	S	P	R	T	
L	V	C	G	E	R	G	F	F	Y	Y	P	K	I	
L	V	C	G	E	R	G	F	F	Y	Y	P	K	V	
L	V	C	G	E	K	G	F	F	Y	N	P	K	-	
L	V	C	G	E	R	G	F	F	Y	Q	P		-	
L	V	C	G	D	R	G	F	F	Y	N	P	K	-	
L	V	C	G	E	K	G	F	F	Y	N	P	K	-	
L	V	C	G	E	K	G	F	F	Y	T	P	K	R	
F	V	C	G	E	R	G	F	F	Y	S	P	K	Q	I
L	V	C	G	E	R	G	F	F	Y	T	P	K		
L	V	C	G	D	R	G	F	F	Y	N	P	K	R	
L	V	C	G	D	R	G	F	F	Y	N	P	K		
L	V	C	G	D	R	G	F	F	Y	N	P	K	-	

Table 9.4 *Continued*

B-chain	−2	−1	1	2	3	4	5	6	7	8	9	10	11	12	13	14	15	16
carp	N	A	G	A	P	Q	H	L	C	G	S	H	L	V	D	A	L	Y
cod		M	A	P	P	Q	H	L	C	G	S	H	L	V	D	A	L	Y
ray, torpedo			L	P	S	Q	H	L	C	G	S	H	L	V	E	A	L	Y
toadfish 1		M	A	P	P	Q	H	L	C	G	S	H	L	V	D	A	L	Y
tuna skipjack			A	A	N	P	H	L	C	G	S	H	L	V	E	A	L	Y
shark, dogfish			L	P	S	Q	H	L	C	G	S	H	L	V	E	A	L	Y
toadfish 2		M	A	P	P	Q	H	L	C	G	S	H	L	V	D	A	L	Y
hagfish			R	T	T	G	H	L	C	G	K	D	L	V	N	A	L	Y
lamprey		SALTG	A	G	G	T	H	L	C	G	S	H	L	V	E	A	L	Y
spiny ratfish			V	P	T	Q	R	L	C	G	S	H	L	V	D	A	L	Y
rabbit fish			V	P	T	Q	R	L	C	G	S	H	L	V	D	A	L	Y

Afr. gr., African green; egp., Egyptian; ar., Arabian; cr., crested; Afr. claw., African clawed.

Table 9.2 shows amino-acid sequence data for all the above-mentioned polypeptides. Interestingly, for all residues invariant in the insulin molecule (in boxes) only cysteines A7, A11, A20, B7, and B19, and additionally B8 Gly and B11 Leu, are completely invariant in all hormones (excluding amylin). It seems that this molecule is capable of maintaining its tertiary structure and biological functions despite very extensive changes in the amino-acid sequence.

9.4.3 *Vertebrate insulin*

Details of the insulin sequences from different species are given in Tables 9.3 and 9.4. Several features emerge from the comparison.

The substitutions occurred at a rate of *c.* 1×10^{-9} sites y^{-1} (Blundell and Wood 1975). This is a typical figure for highly conserved proteins, indicating strict functional roles for most of the regions. Special features of the molecule have been conserved throughout evolution:

1. Residues that are responsible for the structure of the molecule: the precise position of the disulphide bridges; invariant B11 Leu, B15 Leu, and semi-invariant A2 Ile, A3 Val, and A16 Leu, which form the internal core of the molecule.

2. Residues that are responsible for dimerization: B23 Gly, B24 Phe, and B25 Phe (which is conserved as an aromatic site); N-and C-terminal residues in the A-chain.

There are frequent changes in the mammalian insulins within the intra-chain disulphide ring between A8 and A10. However, A8 as histidine in the insulin of all fishes and some birds and reptiles is regarded as responsible for hyperactivity of the molecule (the activities of turkey and chicken insulins are *c.* 300 per cent higher than the activity of porcine insulin). No mammalian insulin has A8 His.

17	18	19	20	21	22	23	24	25	26	27	28	29	30	1+2+3+4+5
L	V	C	G	P	T	G	F	F	Y	N	P	K	-	
L	V	C	G	D	R	G	F	F	Y	N	P	K	-	
F	V	C	G	P	K	G	F	Y	Y	L	P	K	A	T
L	V	C	G	D	R	G	F	F	Y	N	P	K	-	
L	V	C	G	E	R	G	F	F	Y	Q	P	K	-	
F	V	C	G	P	K	G	F	Y	Y	L	P	K	B	Z V
L	V	C	G	D	R	G	F	F	Y	N	S	-	-	
I	A	C	G	V	R	G	F	F	Y	D	P	T	K	M
V	V	C	G	D	R	G	F	F	Y	T	P	S	K	T
F	V	C	G	E	R	G	F	F	Y	S	P	K	P	I R E L E P L
F	V	C	G	E	R	G	F	F	Y	S	P	K	P	I R E L E P L L

Some hystricomorph rodents, such as casiragua, coypu, guinea-pig, and cuis, lack histidine in the B10 position of insulin. This deprives the molecule of the ability to self-associate. Hagfish insulin also lacks B10 His: it cannot form zinc hexamers but associates to dimers. Residues B14, B17, and B20 (which, in other mammalian insulins, create a hydrophobic surface involved in contacts within hexamers) in coypu, guinea-pig, and casiragua have undergone changes into more hydrophilic residues, making the insulin monomer more stable (Blundell *et al.* 1978). Some similarities with IGFs exist in the insulin sequences of hystricomorph rodents (B10 His → Glu, Gln, Asp; B13 Glu → Asp; A4 Glu → Asp). The similarities have been confirmed by the higher potency of these insulins in the growth assay than in their metabolic effects (King and Kahn 1981).

Especially intriguing are two substitutions: A2 Ile → Val in owl monkey insulin and B16 Leu → Met in lamprey insulin, as these residues are conserved in all other species. On the basis of other work done on substitution in A2 (Kitagawa *et al.* 1984), this substitution may be responsible for the reduced activity of owl monkey insulin (*c.* 20 per cent potency).

References

Abel, J. J. (1926). Crystalline insulin. *Proceedings of the National Academy of Sciences, USA* **12**, 132–6.

Adams, M.J., *et al.* (1969). Structure of rhombohedral 2zinc insulin crystals. *Nature* **224**, 491–5.

Bajaj, M., *et al.* (1983). *European Journal of Biochemistry* **135**, 535–42.

Bajaj, M., *et al.* (1986). *Biochemical Journal* **238**, 345–51.

Baker, E.N., *et al.* (1988). The structure of 2Zn pig insulin crystals at 1.5 Å resolution. *Philosophical Transactions of the Royal Society of London* **319**, 369–456.

Banting, F.G. and Best, C.H. (1922). Pancreatic extracts. *Journal of Laboratory and Clinical Medicine* **7**, 464–72.

Bell, G.I., Pictet, R.L., Rutter, W.J., Cordell, B., Tischer, E., and Goodman, H.M. (1980). *Nature* **284**, 26–32.

Berks, B.C., Marshall, C.J., Carne, A., Galloway, S.M., and Cutfield, J.F. (1989). *Biochemical Journal* **263**, 261–6.

Bi, R.C., *et al.* (1983). *Proceedings of the Indian Academy of Sciences* **92**, 473–83.

Blundell, T.L., Bedarkar, S., Rinderkrecht, E., and Humbel, R.E. (1978). *Proceedings of the National Academy of Sciences USA* **75**, 180–4.

Blundell, T.L. and Humbel, R.E. (1980). Hormone families – pancreatic hormones and homologous growth factors. *Nature* **87**, 781–7.

Blundell, T.L. and Wood, S.P. (1975). Is the evolution of insulin Darwinian or due to selectively neutral mutation? *Nature* **257**, 197–203.

Blundell, T.L., Dodson, G. G., Hodgkin, D. C., and Mercola, D. (1972). *Advances in Protein Chemistry* **26**, 280–402.

Brange, J., *et al.* (1988). *Nature* **333**, 679–82.

Buenzli, H.F., Glatthaar, B., Kunz, P., Muelhaupt, E., and Humbel, R.E. (1972). *Hoppe-Seyler's Zeitschrift Physiologische Chemie* **353**, 444–50.

Chan, S.J., Emdin, S.O., Kwok, S.C.M., Kramer, J.M., Folkmer, S., and Steiner. D.F. (1981). *Journal of Biological Chemistry* **256**, 7595–602.

Chance, R.E. Ellis, R.M., and Bromer, W.W. (1968). *Science* **161**, 165–7.

Chu, Y.-Ch., Wang, R.-Y., Burke, G.T., Chanley, J.D., and Katsoyannis, P.G. (1987). *Biochemistry* **26**, 6975–9.

Clark, J.L. and Steiner, D.F. (1969). *Proceedings of the National Academy of Sciences, USA* **62**, 298–302.

Conlon, J.M., Dafgard, E., Falkmer, S., and Thim, L. (1986). *FEBS Letters* **208**, 445–50.

Conlon, J.M., Davis, M.S., and Thim, L. (1987). *General and Comparative Endocrinology* **66**, 203–9.

Conlon, J.M., Andrews, P.C., Falkmer, S., and Thim, L. (1988). *General and Comparative Endocrinology* **72**, 154–60.

Cooper, G.J.S. (1987). Purification and characterization of a peptide from amyloid-rich pancreases of type 2 diabetic patients. *Proceedings of the National Academy of Sciences, USA* **84**, 8628–32.

Cutfield, J.F., Cutfield, S.M., Carne, A., Emdin, S.O., and Falkmer, S. (1986). *European Journal of Biochemistry* **158**, 117–23.

Cutfield, S.M., Dodson, G.G., Ronco, N., and Cutfield, J.F. (1986). Preparation and activity of nitrated insulin dimer. *International Journal of Peptide and Protein Research* **27**, 335–43.

Danho, W.A., Sasaki, A., Bullesbach, E., Fohles, J., and Gattner, H.-G.

(1980). [A19-Phe] insulin: a new synthetic analogue. *Hoppe-Seyler's Zeitschrift Physiologische Chemie* **361**, 735–46.

Danho, W.O.J. (1972). *Faculty of Medicine, Baghdad* **14**, 16–18.

Davidson, J.K., Falkmer, S., Mehrotra, B.K., and Wilson, S. (1971). *General and Comparative Endocrinology* **17**, 388–401.

Dayhoff, M.O. (1972). In *Atlas of protein sequence and structure* and subsequent supplements (ed. M.O. Dayhoff), Washington DC National Biomedical Research Foundation.

Dodson, G.G., Cutfield, S., Hoenjet, E., Wollmer, A., and Brandenburg, D. (1980). In *Insulin: chemistry, structure and function of insulin and related hormones* (ed. D. Brandenburg and A. Wollmer), de Gruyter, Berlin.

Duve, H. (1978). *General and Comparative Endocrinology* **36**, 102–10.

Duve, H. and Thorpe, A. (1979). *Cell and Tissue Research* **200**, 189–91.

Duve, H., Thorpe, A., and Lazarus, N.R. (1979). *Biochemical Journal* **184**, 221–7.

Evans, T.K., Litthauer, D., and Oelofsen, W. (1988). *International Journal of Peptide and Protein Research* **31**, 454–62.

Hallden, G., Gafvelin, G., Mutt, V., and Jornvall, H. (1986). *Archives of Biochemistry and Biophysics* **247**, 20–7.

Hama, H., Titani, K., Sakaki, S., and Narita, K.J. (1964). *Journal of Biochemistry* **56**, 285–93.

Harris, J.I., Sanger, F., and Naughton, M.A. (1956). *Archives of Biochemistry and Biophysics* **65**, 427–8.

Hobart, P.M., Lu-Ping Shen, Crawford, R., Pictet, L.W., and Rutter, J.W. (1980). *Science* **210**, 1360–3.

Horuk, R., Goodwin, P., O,Connor, K., Neville, R.W.J., Lazarus, N.R., and Stone, D. (1979). *Nature* **279**, 439–40.

Horuk, R., Blundell, T.L., Lazarus, N.R., Neville, R.W.J., Stone, D., and Wollmer, A. (1980). *Nature* **286**, 822–4.

Isaacs, N. *et al.*(1978). Relaxin and its structural relationship to insulin. *Nature* **271**, 278–81.

Ishihara, Y., Saito, T., Ito, Y., and Fujino, M. (1958). *Nature* **181**, 1468–9.

Jukes, T.H. (1979). Dr. Best, insulin, and molecular evolution. *Canadian Journal of Biochemistry* **57**, 455–8.

Kimmel, J.R., Maher, M.J., Pollock, H.G., and Vensel, W.H. (1976). *General and Comparative Endocrinology* **28**, 320–33.

King, G.L. and Kahn, C.R. (1981). Non-parallel evolution of metabolic and growth-promoting functions of insulin. *Nature* **292**, 644–6.

Kitagawa, K., Ogawa, H., Burke, G.T., Chanley, J.D., and Katsoyanis, P.G. (1984). *Biochemistry* **23**, 1405–13.

Kono, T. (1988). Insulin-sensitive glucose transport. *Vitamins and Hormones* **44**, 103–54.

Kotaki, A.J. (1963). *Biochemistry* **53**, 61–70.

Kwok, S.C.M., Chan, S.J., and Steiner, D.F. (1983). *Journal of Biological Chemistry* **258**, 2357–63.

Lance, V., Hamilton, J.W., Rouse, J.B., Kimmel, J.R., and Pollock, H.G. (1984). *General and Comparative Endocrinology* **55**, 112–24.

Leighton, B. and Cooper, G.J.S. (1988). Pancreatic amylin and calcitonin gene-related peptide cause resistance to insulin in skeletal muscle *in vitro*. *Nature* **335**, 632–5.

LeRoith, D., Shiloack, J., and Lesniak, M.A. (1980). Evolutionary origins of vertebrate hormones. *Proceedings of the National Academy of Sciences, USA* **10**, 6184–8.

LeRoith, D., Lesniak, M.A., and Roth, J. (1981). Insulin in insects and annelids. *Diabetes* **30**, 70–6.

Lomedico, P., Rosenthal, N., Efstratiadis, A., Gilbert, W., Kolodner, R., and Tizard, R. (1979). *Cell* **18**, 545–58.

Makower, A., *et al.* (1982). *European Journal of Biochemistry* **122**, 339–45.

Marki, F., *et al.* (1979). *Hoppe-Seyler's Zeitschrift Physiologische Chemie* **360**, 1619–32.

Markussen, J. and Sundby, F. (1973). *International Journal of Peptide and Protein Research* **5**, 37–48.

Nagasawa, H., *et al.* (1986). Amino acid sequence of a prothoracico-tropic hormone of the silkworm *Bombyx mori*. *Proceedings of the National Academy of Sciences, USA* **83**, 5840–3.

Nanjo, K., *et al.* (1986). *Journal of Clinical Investigation* **77**, 514–19.

Neelson, F.A., Delcher, H.K., Steinman, H., and Lebovitz, H.E. (1973). *Federation Proceedings* **32**, 300.

Neumann, P. and Humbel, R.E. (1969). *International Journal of Peptide and Protein Research* **1**, 125–40.

Nolan, C., Margoliash, E., Peterson, J.D., and Steiner, D.F. (1971). *Journal of Biological Chemistry* **246**, 2780–95.

Perler, F., Efstratiadis, A., Lomedico, P., Gilbert, W., Kolodner, R., and Dodgson, J. (1980). *Cell* **20**, 555–66.

Peterson, J.D., Nehrlich, S., Oyer, P.E., and Steiner, D.F. (1972). *Journal of Biological Chemistry* **247**, 4866–71.

Plisetskaya, E.M., Pollock, H.G., Rouse, J.B., Hamilton, J.W., Kimmel, J.R., and Gorbman, A. (1985). *Regulatory Peptides* **11**, 105–16.

Plisetskaya, E.M., Pollock, G.H., Elliott, W.M., Youson, J.H., and Andrews, P.C. (1988). *General and Comparative Endocrinology* **69**, 46–55.

Pollock, H.G., *et al.* (1987). *General and Comparative Endocrinology* **67**, 375–82.

Pullen, R.A., *et al.* (1976). *Nature* **259**, 369–73.

Reid, K.B.M., Grant, P.T., and Youngson, A. (1968). *Biochemical Journal* **110**, 289–96.

Sanger, F. (1959). Chemistry of insulin. *Science* **129**, 1340–4.

Sanger, F., Thompson, O.P., and Kitai, R. (1955). The amide groups of insulin. *Biochemical Journal* **59**, 509–18.

Saunders, D.J. and Offord, R. (1977). *Biochemical Journal* **165**, 479–86.

Schwartz, G. and Katsoyannis, P.G. (1978). *Biochemistry* **21**, 4550–6.

Scott, D.A. (1934). Crystalline insulin. *Biochemical Journal* **28**, 1592–1602.

Seino, S., Steiner, D.F., and Bell, G.I. (1987). *Proceedings of the National Academy of Sciences, USA* **84**, 7423–7.

Shuldiner, A.R., Bennett, C., Robinson, E.A., and Roth, J. (1989). *Endocrinology* **125**, 469–77.

Smit, A.B., Vreugdenhil, E., Ebberink, R.M.M., Geraerts, W.P.M, Klootwijk, J., and Joosse, J. (1988). Growth-controlling molluscan neurons produce the precursor of an insulin-related peptide. *Nature* **331**, 535–8.

Smith, L.F. (1966). *American Journal of Medicine* **40**, 662–6.

Soldanha, J., Banga, P., Hutton, J.C., and Mahadevan, D. (1989). The tertiary structure of human amylin found in amyloid deposits of type II non-insulin dependent diabetics has a similar tertiary structure to insulin. *Conference on Insulin, York, August 1989*, (poster).

Sorokin, A.V., Petrenko, O.I., Kavsan, V.M., Kozlov, Y.I., Debarov, V.G., and Zlochevskij, M.L. (1982). *Gene* **20**, 367–76.

Steiner, D.F. and Chan S.J. (1988). *Hormone and Metabolic Research* **20**, 443–44.

Thorpe, A. and Duve, H. (1988). Insulin found at last. *Nature* **331**, 483.

Waelbroeck, M. (1982). *Journal of Biological Chemistry* **257**, 8284–91.

Waelbroeck, M., van Obbergen, F., and de Meyts, P. (1979). *Journal of Biological Chemistry* **254**, 7736–40.

Wallis, M., Howell, S.L., and Taylor, K.W. (1985). *The biochemistry of the polypeptide hormones*. Wiley, New York.

Weitzel, G., Eisele, K., and Stock, W. (1975). *Hoppe-Seyler's Zeitschrift Physiologische Chemie* **356**, 583–90.

Wetekam, W., Groneberg, J., Leineweber, M.M., Wengenmayer, F., and Winnacker, E.-L. (1982). *Gene* **19**, 179–83.

Wilson, S. and Falkmer, S. (1965). *Canadian Journal of Biochemistry* **43**, 1615–24.

Wood, S.P., Blundell, T.L., Wollmer, A., Lazarus, N.R., and Neville, R.W.J. (1975). *European Journal of Biochemistry* **55**, 531–42.

Xu, Y., Lin, N., Zhang, Y., and Zhang, Y. (1983). *Kexue Tongbao* **28**, 966–8.

Yeung, C.W.-T., Moule, M.L., and Yip, C.C. (1980). Functional role of the N-terminal region of the B-chain of insulin. In *Insulin: chemistry, structure and function of insulin and related hormones*, (ed. D. Brandenburg and A. Wollmer), pp. 417–23. de Gruyter, Berlin.

Zhang, Y., Cao, Q., and Zhang, Y. (1981). *Scientia Sinica* **24**, 1585–9.

10

Oligonucleotides

Olga Kennard, William N. Hunter, and Madeleine H. Moore

10.1 Single-crystal X-ray structure determination of oligonucleotides containing four or more bases

Tables 10.1–10.5 list the principal references to structure determinations published between 1979 and mid-1989.

There are three main families of double-helical oligonucleotide structures. These are the right-handed A- and B-forms and the left-handed Z-form. Figures 10.1–10.3 show examples of each family, based on a single-crystal structure determination using coordinates deposited in the Brookhaven Protein Data Bank (Bernstein *et al.* 1977). Useful references to the poly-

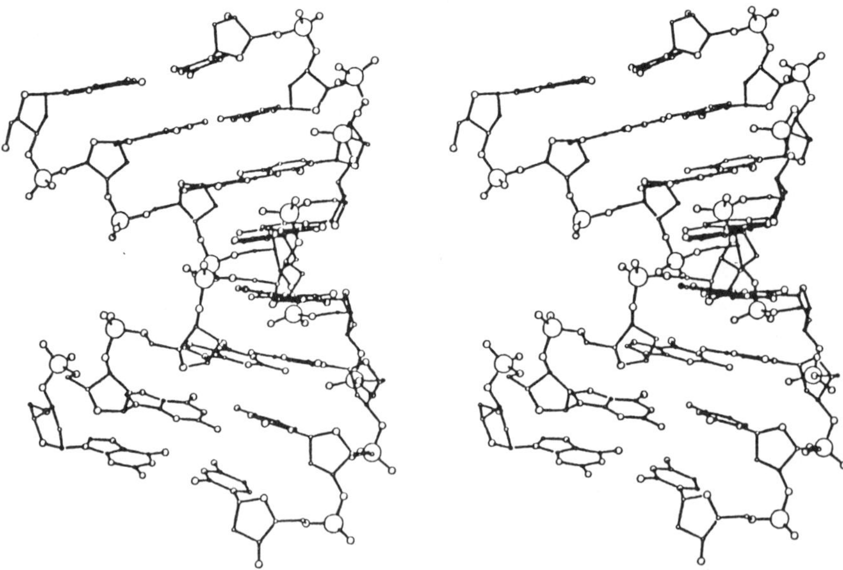

Fig. 10.1 Stereoview showing A-form DNA as represented by the octamer d(GGGGCCCC) (McCall *et al.* 1985). This view shows the shallow but wide minor groove at the bottom, and the narrow, deep major groove above.

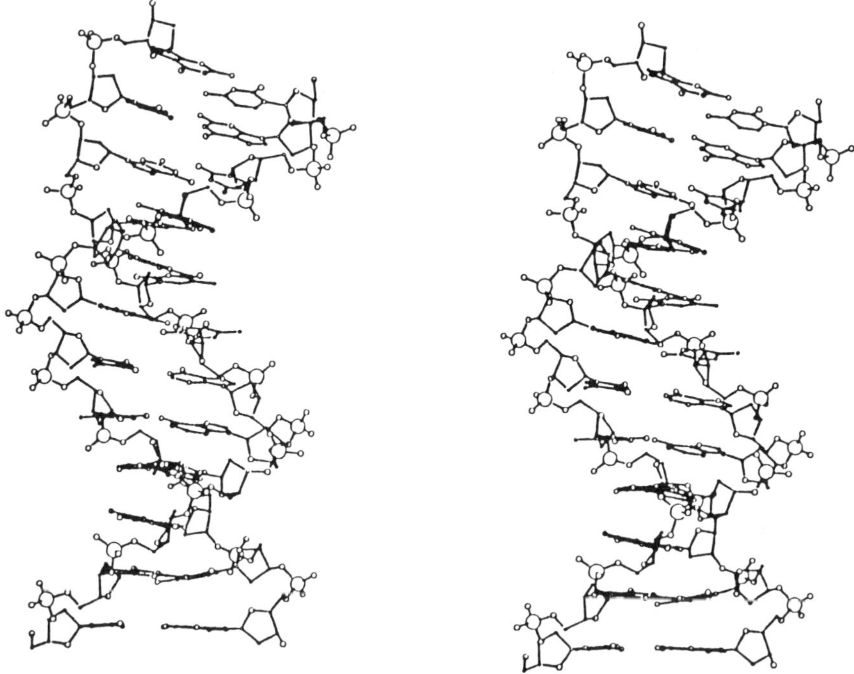

Fig. 10.2 Stereoview of B-form DNA as depicted by the Drew-Dickerson dodecamer, d(CGCGAATTCGCG). Numerous references of this structure are given in the text.

morphic variants of DNA are Rich *et al.* (1984), Shakked and Kennard (1984), Jovin *et al.* (1987) and Drew *et al.* (1988). Detailed numeric data on oligonucleotide studies can be found in the Landolt-Bornstein Tables (Kennard and Hunter 1989a) and a recent review in *Quarterly Reviews of Biophysics* (Kennard and Hunter 1989b).

The number of distinct structures that have been successfully crystallized and analysed during the past decade is still relatively limited. There are six types of B-DNA structures: represented by two dodecamers, d(CGCGAATTCGCG) (Drew *et al.* 1981) and d(ACCGGCGCCACA) (Timsit *et al.* 1989), a hexamer, d(GpsCpGpsCpGpsC) (Cruse *et al.* 1986), decamers, d(CCAGGCCTGG) (Heinemann *et al.* 1989) and d(CGATCGATCG) (Grzeskowiak *et al.* 1989), and two sequences with extra adenines (Miller *et al.* 1988; Joshua-Tor *et al.* 1988). These six structure types make use of different interactions between adjacent double helices to create the regular crystal lattice. In contrast, A-DNA helices which have been crystallized utilize a very similar packing motif, involving hydrogen bonding and van der Waals contacts between minor grooves of adjacent molecules.

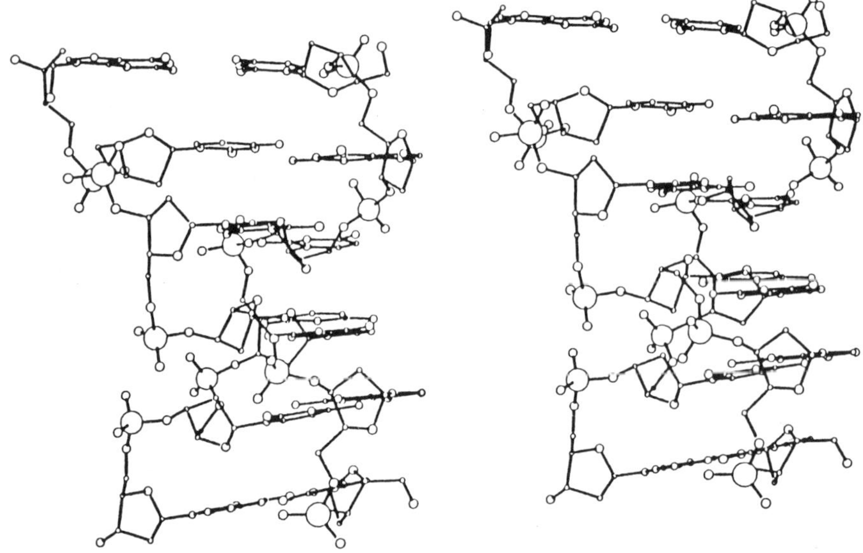

Fig. 10.3 The left-handed polymorph, Z-DNA shown by d(CGCGCG) (Wang *et al.* 1979). Note the zig-zag appearance of the phosphate-sugar backbone.

Such interactions can be accommodated in unit cells with a variety of symmetry operators, including sixfold, fourfold and twofold screw axes (Table 10.1). The same packing motif is utilized by a hybrid DNA-RNA decamer which forms an orthorhombic crystal (Wang *et al.* 1982a). Left-handed, Z-DNA helices, when crystallized, pack essentially the same way with end-to-end contacts in an orthorhombic space group (hexamers), and two closely related space groups (tetramers) (Table 10.2). The same sequence crystallizing in two different space groups has been reported for two sequences (Shakked *et al.* 1989; Sundaralingham and Jain 1989).

Table 10.3 lists selected references to deoxyoligonucleotides containing mismatched (non-Watson-Crick) base pairs and extra bases.

10.2 Techniques of X-ray analysis

10.2.1 Synthesis, crystallization, and data collection

Techniques of synthesizing deoxyoligonucleotides in the quantities required for single-crystal X-ray analysis (5–10 mg) have advanced greatly since the original solution methods (Arentzen *et al.* 1979; Gait *et al.* 1982) which resulted in larger quantities but which were very labour-intensive. Chapter 1 provides a review of X-ray diffraction methods. Modern automatic solid-phase equipment based on the phosphoramidite technique (McBride and

Table 10.1 Right-handed double helices

Sequence	Global form	Reference
d(pATAT)		Viswamitra *et al.* (1978, 1982)
d(IoCCGG)	A-form	Conner *et al.* (1982)
d(CCGG)	A-form	Chattopadhyaya and Chakrabarti (1988)
d(GGTATACC)	A-form	Shakked *et al.* (1981, 1983)
d(GGBrUABrUACC)	A-form	Kennard *et al.* (1986)
d(GGCCGGCC)	A-form	Wang *et al.* (1982b)
d(GGGGCCCC)	A-form	McCall *et al.* (1985)
d(GGGATCCC)	A-form	Lauble *et al.* (1988)
d(CTCTAGAG)	A-form	Hunter *et al.* (1989)
d(GGGCGCCC)	A-form	Rabinovich *et al.* (1988); Shakked *et al.* (1989)
d(GCCCGGGC)	A-form	Heinemann *et al.* (1987)
d(CCCCGGGG)	A-form	Haran *et al.* (1987)
d(GGmCCGGCC)	A-form	Frederick *et al.* (1987)
d(GTGTACAC)	A-form	Jain *et al.* (1987, 1989)
d(GGATGGGAG)	A-form	McCall *et al.* (1986)
d(ACCGGCCGGT)	A-form	Frederick *et al.* (1989)
d(CpsCGTACGTACGG)	A-form	Bingman *et al.* (1989)
r(GCG)d(TATACGC)	A-form	Wang *et al* (1982a)
r(UUAUAUAUAUAUAA)	A-form	Dock-Bregeon *et al.* (1988)
d(GpsCGpsCGpsC)	B-form	Cruse *et al.* (1986)
d(CGCGAATTCGCG)	B-form	Wing *et al.* (1980); Drew *et al.* (1982)
d(CGCAAATTTGCG)	B-form	Coll *et al.* (1987)
d(CGCAAAAAAGCG)	B-form	Nelson *et al.* (1987)
d(CGCATATATGCG)	B-form	Yoon *et al.* (1988)
d(CCAGGCCTGG)	B-form	Heinemann *et al.* (1989)
d(CGCAAAAATGCG)	B-form	DiGabriele *et al.* (1989)
d(CGCGAm6ATTCGCG)	B-form	Frederick *et al.* (1988)
d(ACCGGCGCCACA)	B-form	Timsit *et al.* (1989)
d(CGATCGATCG)	B-form	Grzeskowiak *et al.* (1989)

IoC = 5-iodocytosine; BrU = 5-bromouridine; mC = 5-methylcytosine; ps = Rp-phosphoro-thioate; m6A = 6-methyladenine.

Caruthers 1983), allows for the rapid synthesis of deoxyoligonucleotides up to 20–24 base pairs in length and of reasonable homogeneity. Purification, involving both ion-exchange and reverse-phase high pressure liquid chromatography and, on occasion, gel-chromatography, is currently the rate-determining step in the preparation of the starting material. Experience has shown that stringent purification is an essential preliminary for successful crystallization, and it is not uncommon to spend 24 hours on synthesis and one or two weeks on purification.

The resulting compounds, which generally consists of self-complementary sequences, are usually crystallized by vapour-diffusion (McPherson 1982; Holbrook and Kim 1985). A precipitating agent is diffused into a solution containing the oligonucleotide with associated cations and, in many cases,

Table 10.2 Left-handed double helices, Z-form

Sequence	Comment	Reference
d(CGCG)	high-salt	Drew *et al.* (1978, 1980, 1981)
d(CGCG)	low-salt, disordered	Crawford *et al.* (1980)
d(CGCGCG)	spermine, Mg and Na	Wang *et al.* (1979)
d(CGCGCG)	Mg and Na	Gessner *et al.* (1989a)
d(CACGTG)	2 forms	Coll *et al.* (1988)
d(BrCGBrCGBrCG)		Chevrier *et al.* (1986)
d(BrCGATBrCG)		Wang *et al.* (1985)
d(mCGATmCG)		Wang *et al.* (1985)
d(BrCGTABrCG)		Wang *et al.* (1984)
d(mCGTAmCG)		Wang *et al.* (1984)
d(mCGmCGmCG)		Fujii *et al.* (1982)
d(CG)r(CG)d(CG)		Teng *et al.* (1989)
d(CG)(araC)d(GCG)		Teng *et al.* (1989)
d(CGCGCGCG)	disordered	Fujii *et al.* (1985)
d(CGCATGCG)	disordered	Fujii *et al.* (1985)
d(CGTACCTACG)	disordered	Westhof *et al.* (1985); Brennan *et al.* (1986)

BrC = 5-bromocytosine; mC = 5-methylcytosine; araC = arabinose cytosine.

Table 10.3 Double helices with non-Watson–Crick base pairs or extrahelical bases

Sequence	Comment	Form	Reference
d(TGCGCG)	G.T	Z-form	Kennard (1985)
d(BrUGCGCG)	BrU.G	Z-form	Brown *et al.* (1986b)
d(CGTDCG)	T.2-aminoadenine	Z-form	Coll *et al.* (1986)
d(CDCGTG)	T.2-aminoadenine	Z-form	Coll *et al.* (1986)
d(CGCGFUG)	G.FU	Z-form	Coll *et al.* (1989)
d(CGCGXG)	G.X	Z-form	van Meervelt *et al.* (1989)
d(CGCGTG)	G.T	Z-form	Ho *et al.* (1985)
d(GGGGTCCC)	G.T	A-form	Kneale *et al.* (1985)
d(GGGGCTCC)	G.T	A-form	Brown *et al.* (1985); Hunter *et al.* (1986c)
d(GGGTGCCC)	G.T	A-form	Rabinovich *et al.* (1988)
d(GGIGCTCC)	I.T	A-form	Cruse *et al.* (1989)
d(CCAAGATTGG)	G(*anti*).A(*anti*)	B-form	Prive *et al.* (1987, 1988)
d(CGCGAATTTGCG)	G.T	B-form	Hunter *et al.* (1987b)
d(CGCGAATTAGCG)	G(*anti*).A(*syn*)	B-form	Brown *et al.* (1986a); Hunter *et al.* (1986b)
d(CGCAAATTCGGG)	C.A	B-form	Hunter *et al.* (1986a, 1987a)
d(CGCIAATTAGCG)	I.A	B-form	Corfield *et al.* (1987)
d(CGCAAATTGGCG)	A(*anti*).G(*syn*)	B-form	Brown *et al.* (1989)
d(CGCAGAATTCGCG)	extra A	B-form	Joshua-Tor *et al.* (1988)
d(CGCGAAATTTACGCG)	extra A	B-form	Miller *et al.* (1988)
d(CGCGCGTTTTCGCGCG)	hairpin	Z-form	Chattopadhyaya *et al.* (1988)

FU = 5-fluorouridine; D = 2-aminoadenine; X = N4-methoxycytosine.

spermine. Alternatively, water is diffused out of the nucleic acid solution by a desiccant such as hexane-1,6-diol or 2-methyl-2,4-pentanediol.

Conditions for crystallization are very critical, and many parameters have to be varied and optimized. Oligonucleotide crystallization techniques have not changed over the decade and still rely more on experience and intuition than on rational science. The average success rate for a random self-complementary sequence is less than 10 per cent. There are additional problems with stoichiometric mixing of non-self-complementary sequences.

When crystals are formed they generally contain about 50 per cent nucleic acid and 50 per cent solvent of crystallization. Consequently the crystals are often disordered and either do not diffract or diffract to a resolution of 2–3 Å. Crystals of fragments containing only Watson-Crick base pairs are generally stable at ambient temperature, while those with mismatched base pairs tend to be less stable.

Diffraction measurements are frequently carried out at a reduced temperature, typically at 4°C, on samples sealed in glass capillary tubes to prevent loss of solvent. Recently several structures were determined at 115 K, without a capillary, using a fast-quenching technique (Eisenstein *et al.* 1988; Hoppe 1988).

The quality of the diffraction data and the resolution that can be attained are determined by the size and, most importantly, by the perfection of the crystals used for measurements. The Z-hexamers form the most perfect crystals, often diffracting to 1 Å or better. Shorter or longer Z-forming sequences tend to be disordered. In general, the right-handed deoxyoligo-nucleotide crystals diffract to a maximum resolution of 1.7–2.5 Å. The use of synchrotron radiation can lead to improvements in effective resolution (Kennard *et al.* 1986) and may be used more commonly in future to aid structure solution by anomalous dispersion methods (Hendrickson *et al.* 1985). A typical data collection on a diffractometer would result in some 2500 reflections to a resolution of about 2.0 Å which would be used in the structure analysis. Recent availability of area detectors designed for the simultaneous measurement of very large numbers of reflections is revolutionizing this aspect of the subject (Arndt 1985; Hamlin 1985).

10.2.2 Structure solution and refinement

A variety of experimental and computational techniques can be applied to solve oligonucleotide structures, such as molecular replacement, multiple isomorphous replacement and anomalous dispersion (Blundell and Johnson 1976). The object is to find the correct model for the fundamental structural motif, the asymmetric unit. In many cases the asymmetric unit consists of two independent oligonucleotide strands which coil about each other to form a duplex. Less frequently the motif is a single strand and the double helix is formed by a twofold symmetry operation. Occasionally, the asymmetric unit contains four or more strands. So far no solution of such structures has been reported.

Many of the successful structure determinations to date employed the molecular replacement method with a model based on fibre diffraction coordinates (Rabinovich and Shakked 1984; Chattopadhyaya and Chakrabarti 1988). A notable exception is the use of maximum entropy methods for a 15-mer (Miller *et al.* 1988).

Once the phase problem is solved, the refinement process begins. This stage of the analysis aims to reduce discrepancies between structure factors calculated from the model and the experimentally determined values. In cases where the crystals do not diffract to atomic resolution, constrained/restrained methods of refinement are employed, and certain variables, such as bond lengths and angles, are held constant in an attempt to make the best use of the experimental observations (Westhof *et al.* 1985). For details, consult the original references.

Solvent molecules are an essential part of oligonucleotide crystal structures. Therefore refinement will generally include the location of solvent molecules from various electron-density maps, often with the aid of computer graphics. Usually, around 3–8 solvent molecules are found for each nucleotide in the molecule. Counterions have only been identified positively in the Z-DNA hexamers and in the two B-DNA decamers. It is common practice to assign solvent positions as water oxygen atoms. The refinement process is terminated when there is optimal fit of the model structure with the electron-density maps, when no further solvent molecules can be located, and when there is no further improvement in the crystallographic residual R. This residual is defined as $R = \Sigma F_o - F_c / \Sigma F_o$ where F_o is the observed and F_c the calculated structure factor. Most oligonucleotide structures are refined to an R-factor of between 12 per cent for well-resolved structures and 25 per cent for poorly resolved structures.

Resolution limits the degree of structural detail that can be accurately determined. Coordinates for individual atoms can only be determined reliably for structures that diffract close to 1 Å. For most oligonucleotide structures, resolution limits the analysis to the accurate determination of structural features, such as sugar conformation, orientation of bases, and the principal torsion angles. Information from the position of solvent molecules, particularly in the first hydration shell, is sufficiently reliable to allow for some generalizations of the role of hydration in relation to global conformations. This is discussed below.

A variant of the above techniques was used to analyse structures with base-pair mismatches. In these structures the contributions of the atoms of the mismatched bases were initially omitted from the calculations. At an intermediate stage in the refinement the positions of the atoms in the base-pair mismatch were located from electron-density maps and their contributions were subsequently included in the calculations. In this way the analysis was not biased by a preconceived model of the base-pair mismatch.

10.3 Numeric data

Tables 10.4–10.11 give numeric information derived from X-ray diffraction analyses of oligonucleotide fragments, published up to mid-1989. Tables 10.4, 10.5, and 10.6 give selected data for some of the oligonucleotide fragments crystallizing as A, B, or Z helices. For each crystal structure the following information is quoted: temperature of X-ray measurement, space group, unit cell dimensions, the resolution of the analysis, the R-factor, the number of solvent molecules located, and the volume per base pair. In all these crystals two strands of the oligomer form the double helix and in many cases the double helix has no imposed crystallographic symmetry.

Table 10.7 lists similar data for DNA fragments containing T.G, I.A, I.T, BrU.G, FU.G, X.G, G.A, or C.A mispairs. Comparison of this table with Tables 10.4–10.6 shows clearly that within each conformation type (A, B, or Z) the structures containing mismatches crystallize isomorphously with the native structures containing Watson-Crick complementary base pairs only. Table 10.8 gives data for three structures with extra bases.

Table 10.9 contains the global helical parameters for A-and B-type helices observed in typical single-crystal studies and, for comparison, values derived from fibre diffraction. The table is restricted to helices with Watson-Crick base pairs only. The parameters include the helix twist, rise per base pair, base-pair tilt, propeller twist, groove widths, and the displacement of the base pairs from the helix axis. For definitions see Dickerson *et al.* (1989). Table 10.10 gives average values which allow for a comparison of A-, B- and Z-form DNA. Table 10.11 presents average phosphate-sugar torsion angles observed in a number of right-handed fragments and, for comparative purposes, the values from fibre diffraction measurements. Note that there are quite substantial variations for each of these parameters in the different DNA fragments analysed. These variations can be attributed to differences in base sequence, particularly when oligonucleotides crystallizing with the same structural motif are compared. The groove width is one of the parameters particularly affected. In A-type crystals the width of the minor groove ranges from 10.2–8.7 Å compared to the fibre value of 11.0 Å and the width of the major groove from 10.1–3.2 Å compared to the fibre value of 2.4 Å.

10.4 Sequence-specific model structures

10.4.1 Alternating B-DNA

The relationship between sequence and variations in local helical parameters has been analysed in detail by Calladine (1982), Calladine and Drew (1984,

Table 10.4 Crystallographic details of selected A-form structures

Sequence	Temperature °C	Space Group	Unit Cell Dimensions (Å)			Resolution (Å)	R-factor (%)	Number of solvents	Volume per base pair (Å^3)	Ref.
d(IoCCGG)	−2	P4$_3$2$_1$2	41.1	41.1	26.7	2.2	16	86	1409	1
d(GGTATACC)	RT	P6$_1$	45.0	45.0	41.5	2.2	20	66	1519	2
d(GGBrUABrUACC)[a]	RT	P6$_1$	45.1	45.1	41.7	1.8	14	84	1530	3
d(GGGGCCCC)	RT	P6$_1$	45.3	45.3	42.3	2.5	14	104	1566	4
d(GGGATCCC)	RT	P6$_1$	46.8	46.8	44.5	2.5	17	9	1615	5
d(GCCCGGGC)[b]	RT	P4$_3$2$_1$2	43.2	43.2	24.6	1.8	17	68	1435	6
d(GGGTACCC)[b]	RT	P4$_3$2$_1$2	43.3	43.3	24.7	1.7	16	88	1447	7
d(GTGTACAC)[b]	4	P4$_3$2$_1$2	42.4	42.4	24.8	2.0	12	86	1393	8
d(GGGCGCCC)	RT	P4$_3$2$_1$2	43.3	43.3	24.7	1.8	16	88	1444	9
d(GGGCGCCC)	−158	P4$_3$2$_1$2	42.7	42.7	24.6	1.7	21	62	1403	10
d(GGGCGCCC)	RT	P6$_1$	47.0	47.0	44.3	2.9	12	45	1763	10
d(GGGCGCCC)	−158	P6$_1$	43.6	43.6	41.5	1.9	16	100	1424	10
d(CTCTAGAG)[b]	RT	P4$_3$2$_1$2	42.5	42.5	24.3	2.2	15	70	1315	11
d(GGCCGGCC)[b]	RT	P4$_3$2$_1$2	42.0	42.0	25.1	2.2	18	80	1383	12
d(CCCCGGGG)[b]	RT	P4$_3$2$_1$2	43.4	43.4	24.8	2.2	15	82	1459	13
d(GGATGGGAG)	RT	P4$_3$	45.3	45.3	24.7	3.0	33	—	1410	14
d(ACCGGCCGGT)	10	P6$_1$22	39.2	39.2	78.0	2.0	18	36	1772	15
d(CpsCGTACGTACGG)	RT	P6$_1$22	46.2	46.2	71.5	2.5	14	71	1836	16
r(GCG)d(TATACGC)	0	P2$_1$2$_1$2$_1$	24.2	43.5	49.4	2.0	16	179	1300	17
r(UUAUAUAUAUAUAA)	RT	P2$_1$2$_1$2$_1$	34.1	44.6	49.1	2.2	13	86	1334	18

Io = 5-iodocytosine.

[a] Synchrotron radiation used in the analysis.

[b] The asymmetric unit consists of a single strand or four base pairs.

A twofold axis is coincident to the octamer duplex twofold axis of symmetry.

RT signifies that room temperature was specified in the original publication.

References: 1, Conner *et al.* (1984); 2, Shakked *et al.* (1983); 3, Kennard *et al.* (1986); 4, McCall *et al.* (1985); 5, Lauble *et al.* (1988); 6, Heinemann *et al.* (1987); 7, Rabinovich *et al.* (1988); 8, Jain *et al.* (1987, 1989); 9, Rabinovich *et al.* (1988); 10, Shakked *et al.* (1989); 11, Hunter *et al.* (1989); 12, Wang *et al.* (1982b); 13, Haran *et al.* (1987); 14, McCall *et al.* (1986); 15, Frederick *et al.* (1989); 16, Bingman *et al.* (1989); 17, Wang *et al.* (1982a); 18, Dock-Bregeon *et al.* (1988).

Table 10.5 Crystallographic details of selected B-form structures

Sequence	Temperature °C	Space Group	Unit Cell Dimensions (Å)			Resolution (Å)	R-factor (%)	Number of solvents	Volume per base pair (Å³)	Ref.
d(CGCGAATTCGCG)	RT	P2₁2₁2₁	24.9	40.4	66.2	1.9	18	80	1385	1
d(CGCGAATTCGCG)	−257	P2₁2₁2₁	23.4	39.3	65.3	2.7	15	83	1253	2
d(CGCGAATTBrCGCG)	7	P2₁2₁2₁	24.2	40.1	63.9	2.3	17	114	1293	3
d(CGCAAATTTGCG)	RT	P2₁2₁2₁	25.2	41.7	65.8	2.5	16	32	1440	4
d(CGCAAAAAGCG)	4	P2₁2₁2₁	25.4	40.7	65.8	2.5	20	27	1417	5
d(CGCATATATGCG)	6	P2₁2₁2₁	23.5	38.9	66.6	2.2	19	43	1268	6
d(CGCAAAAATGCG)	12	P2₁2₁2₁	24.5	40.3	65.9	2.6	20	23	1355	7
d(CGCGAm6ATTCGCG)	RT	P2₁2₁2₁	25.6	40.7	67.3	2.0	17	87	1461	8
d(GpsCGpsCGpsC)	RT	P2₁2₁2₁	34.9	39.2	20.6	2.2	14	72	1175	9
d(CCAGGCCTGG)	RT	C2 $\beta = 117°$	32.2	25.5	34.8	1.6	17	–	1275	10
d(ACCGGCGCCACA)	4	R3	65.9	65.9	47.1	2.8	15	20	1640	11
d(CGATCGATCG)	RT	P2₁2₁2₁	38.9	39.6	33.2	1.8	22	42	1278	12

BrC = 5-bromocytosine; m6A = 6-methyladenine; pS = Rp-phosphorothioate; RT signifies room temperature.

References: 1, Wing *et al.* (1980); Drew *et al.* (1982); 2, Drew *et al.* (1982); 3, Fratini *et al.* (1982); 4, Coll *et al.* (1987); 5, Nelson *et al.* (1987); 6, Yoon *et al.* (1988); 7, DiGabriele *et al.* (1989); 8, Frederick *et al.* (1988); 9, Cruse *et al.* (1986); 10, Heinemann *et al.* (1989); 11, Timsit *et al.* (1989); 12, Grzeskowiak *et al.* (1989).

Table 10.6 Crystallographic details of selected Z-form structures

Sequence	Cations present	Space Group	Unit Cell Dimensions (Å)			Resolution (Å)	R-factor (%)	Number of solvents	Volume per base pair (Å^3)	Ref.
d(CGCGCG)	(Na, Mg, spermine)	P2$_1$2$_1$2$_1$	17.9	31.5	44.6	0.9	14	68	1048	1
d(CGCGCG)	(Na, Mg)	P2$_1$2$_1$2$_1$	18.0	31.0	44.8	1.0	18	88	1043	2
d(mCGmCGmCG)	(Mg, spermine)	P2$_1$2$_1$2$_1$	17.8	30.6	45.4	1.3	16	98	1027	3
d(mCGTAmCG)	(Mg, Na)	P2$_1$2$_1$2$_1$	17.9	30.4	44.9	1.2	16	98	1021	4
d(CG)r(CG)d(CG)		P2$_1$2$_1$2$_1$	18.3	30.9	43.1	1.5	20	82	1015	5
d(CG)(araC)d(GCG)		P2$_1$2$_1$2$_1$	18.6	30.7	42.9	1.5	17	91	1020	5
d(CGCG)	(Mg and/or Na)	C222$_1$	19.5	31.3	64.7	1.5	21	84	1232	6
d(CGCGCGCG)	(Mg and/or Na)	P6$_5$	31.3	31.3	43.6	1.6	19	85	1025	7
d(CGCATGCG)	(Mg and/or Na)	P6$_5$	30.9	30.9	43.1	2.5	16	—	991	7

mC = 5-methylcytosine; araC = arabinose cytosine.

References: 1, Wang *et al.* (1979, 1981); 2, Gessner *et al.* (1989a); 3, Fujii *et al.* (1982); 4, Wang *et al.* (1984); 5, Teng *et al.* (1989); 6, Drew *et al.* (1980); 7, Fujii *et al.* (1985).

Table 10.7 Crystallographic details of some structures containing non-Watson–Crick base pairs

Sequence	Mispair	DNA-form	Space Group	Unit Cell Dimensions (Å)			Resolution (Å)	R-factor (%)	Number of solvents	Volume per base pair (Å^3)	Ref.
d(GGGGCTCC)	G.T	A	P6$_1$	45.2	45.2	42.9	2.2	14	52	1584	1
d(GGGGTCCC)	G.T	A	P6$_1$	44.7	44.7	42.4	2.1	14	104	1529	2
d(GGGTGCCC)	G.T	A	P6$_1$	45.6	45.6	41.0	2.5	15	63	1507	3
d(GGIGCTCC)	I.T	A	P6$_1$	45.1	45.1	45.5	1.7	14	81	1600	4
d(CCAAGATTGG)	G.A	B	C2, $\beta = 119°$	32.5	26.2	34.3	1.3	17	48	1278	5
d(CGCGAATTAGCG)	G.A	B	P2$_1$2$_1$2$_1$	25.7	41.9	65.2	2.5	17	83	1464	6
d(CGCAAATTGGCG)	G.A	B	P2$_1$2$_1$2$_1$	25.2	41.2	65.0	2.3	16	94	1406	7
d(CGCIAATTAGCG)	I.A	B	P2$_1$2$_1$2$_1$	25.8	41.9	65.1	2.5	19	64	1466	8
d(CGCAAATTCGCG)	C.A	B	P2$_1$2$_1$2$_1$	25.4	41.4	65.2	2.5	19	82	1428	9
d(CGCGAATTTGCG)	G.T	B	P2$_1$2$_1$2$_1$	25.5	41.2	65.6	2.5	18	71	1436	10
d(TGCGCG)	G.T	Z	P2$_1$2$_1$2$_1$	17.9	30.7	45.1	1.5	18	–	1033	11
d(BrUGCGCG)	BrU.G	Z	P2$_1$2$_1$2$_1$	17.9	30.9	49.9	2.2	16	64	1152	12
d(CGCGTG)	G.T	Z	P2$_1$2$_1$2$_1$	17.5	31.6	45.6	1.0	20	91	1048	13
d(CGCGFUG)	G.FU	Z	P2$_1$2$_1$2$_1$	17.8	31.3	45.4	1.5	20	64	1047	14
d(CGCGXG)	G.X	Z	P2$_1$2$_1$2$_1$	18.2	30.4	43.9	1.7	16	79	1012	15

BrU = 5-bromouridine; FU = 5-fluorouridine; X = N4-methoxycytosine.

References: 1, Brown *et al.* (1985); Hunter *et al.* (1986c); 2, Kneale *et al.* (1985); 3, Rabinovich *et al.* (1988); 4, Cruse *et al.* (1989); 5, Prive *et al.* (1987, 1988); 6, Brown *et al.* (1986a); Hunter *et al.* (1986b); 7, Brown *et al.* (1989); 8, Corfield *et al.* (1987); 9, Hunter *et al.* (1986a, 1987a); 10, Hunter *et al.* (1987b); 11, Kennard (1985); 12, Brown *et al.* (1986b); 13, Ho *et al.* (1985); 14, Coll *et al.* (1989); 15, van Meervelt *et al.* (1989).

Table 10.8 Crystallographic data on DNA structures containing extra bases

Sequence	Temperature °C	Space Group	Unit Cell Dimensions (Å)			Resolution (Å)	R-factor (%)	Ref.
d(CGCAGAATTCGCG)	−150	C2, $\beta=99°$	78.5	42.8	25.2	2.8	15	1
d(CGCGAAATTTACGCG)	4	I222	37.0	53.7	101.5	3.0	24	2
d(CGCGCGTTTTCGCGCG)	RT	C2, $\beta=95°$	57.2	21.6	36.4	2.1	23	3

References: 1, Joshua-Tor *et al.* (1988); 2, Miller *et al.* (1988); 3, Chattopadhyaya *et al.* (1988).

Table 10.9 Average helical parameters for selected right-handed structures

	Helix twist (°)	Rise per base pair (Å)	Base pair tilt (°)	Propeller twist (°)	Groove width (Å)		Displacement Da (Å)
					Minor	Major	
A-form							
d(GGTATACC)	32	2.9	13	10	10.2	6.3	4.0
d(GGGGCCCC)	32	2.9	13	10	10.1	6.8	4.1
d(GGGCGCCC)	32	3.3	7	12	9.5	10.1	3.7
d(GGCCGGCC)	33	3.1	12	13	9.8	7.5	3.5
d(IoCCGG)	34	2.8	14	16	9.5	4.8	3.5
d(CTCTAGAG)	32	3.1	10	11	8.7	8.0	3.6
r(GCG)d(TATACGC)	33	2.5	19	12	10.2	3.2	4.5
r(UUAUAUAUAUAUAA)	33	2.8	17	19	10.2	3.7	3.6
Fibre A-DNA[a]	33	2.6	22	6	11.0	2.4	4.4
B-form							
d(CGCGAATTCGCG)	36	3.3	2	13	5.3	11.7	−0.2
d(CGCGAATTBrCGCG)	36	3.4	−2	18	4.6	12.2	−0.2
d(GpSCGpSCGpSC)[b]	37/34	3.5/3.3	0	9	6.4	10.2	−0.6
Fibre B-DNA[a]	36	3.4	2	13	6.0	11.4	−0.6

IoC = 5-iodocytosine; BrC = 5-bromocytosine.

[a] Based on fibre diffraction data (Arnott and Hukins 1972).

[b] There are two conformational states in this structure.

Table 10.10 Comparison of A-, B-, and Z-DNA

	A-DNA[a]	B-DNA[a]	B'-DNA[b]	Z-DNA[c]
Helix sense	Right-handed	Right-handed	Right-handed	Left-handed
Base pairs per turn	11	10	10	12 (6 dimers)
Helix twist (°)	32.7	36.0	34.1, 36.8	-10, -50
Rise per base pair (Å)	2.9	3.4	3.5, 3.3	3.7
Helix pitch (Å)	32	34	34	45
Base pair tilt (°)	13	0	0	-7
P distance from helix axis (Å)	9.5	9.3	9.1	6.9, 8.0
Glycosidic orientation	*anti*	*anti*	*anti*	*anti, syn*
Sugar conformation	C3'-endo	wide range	C2'-endo	C2'-endo, C3'-endo[d]

[a] Numerical values for each form were obtained by averaging the global parameters of the corresponding double-helix fragments.
[b] B'-DNA values are for a double-helix backbone conformation alternating between conformational states, I and II (Cruse *et al.* 1986).
[c] The two values given correspond to CpG and GpC steps for the helix twist and P distance values, to cytidine and guanosine for the others.
[d] Two values correspond to the two conformational states.

Table 10.11 Average torsion angles (°) for selected right-handed structures

	α(P-O5$'$)	β(O5$'$-C5$'$)	γ(C5$'$-C4$'$)	δ(C4$'$-C3$'$)	ε(C3$'$-O3$'$)	ζ(O3$'$-P)	χ(glycosyl)
A-form							
d(GGTATACC)	−62	173	52	88	−152	−78	−160
d(GGGGCCCC)	−76	178	63	84	−156	−70	−158
d(IoCCGG)	−73	180	64	89	−157	−66	−157
d(GGCCGGCC)[a]	−67	170	58	80	−142	−76	−162
d(CCCCGGGG)[a]	−87	182	72	79	−153	−66	−162
d(CTCTAGAG)[a]	−76	170	52	79	−175	−54	−156
d(GCCCGGGC)[a]	−71	177	55	35	−165	−64	−159
d(GGGCGCCC)[a]	−71	178	62	80	−158	−70	−163
d(GGGATCCC)	−101	181	81	81	−164	−63	−164
r(GCG)d(TATACGC)	−69	175	58	82	−151	−75	−162
r(UUAUAUAUAUAUAA)	−61	169	53	80	−148	−79	−159
A-DNA[b]	−50	172	42	79	−146	−78	−159
A-RNA[b]	−68	178	54	82	−153	−71	−158
B-form							
d(CGCGAATTCGCG)	−63	171	54	122	−169	−108	−117
d(CGCGAATTBrCGCG)	−62	170	55	120	−169	−109	−120
d(GpsCGpsCGpsC)[c]	−53	154	46	138	−164/−114	−102/−192	−109
B-DNA[b]	−33	138	33	142	−141	−157	−139

[a] Excluding α and γ of the fifth nucleotide.
[b] Based on fibre diffraction data (Arnott and Hukins 1972).
[c] Two values correspond to conformational states, I and II.

1986), and by Shakked and Rabinovich (1986). The rules that are emerging are somewhat tentative, since they are based on relatively few examples, and may well be modified as more information accumulates. An interesting generalization is the existence of regular alternation in the local helical parameters of d(AT) tracts. Such alternations were first observed in the tetramer d(pATAT) (Viswamitra *et al.* 1978, 1982; Klug *et al.* 1979), which led to the proposal of an alternating B-DNA helix for poly d(AT). In such a model the helix twist angle is smaller at the A-T step than at the T-A step. Similar alternation was reported recently in the central AT region of d(CGCATATATGCG) (Yoon *et al.* 1988). Alternating backbone conformation was also observed in the phosphorothioate analogue of d(GCGCGC) (Cruse *et al.* 1986) and, when examined closely, in the CGCG segments of the dodecamer d(CGCGAATTCGCG). Nuclease footprinting experiments have indicated the important role of the sugar-phosphate backbone conformation in some protein-DNA recognition processes (Drew and Travers 1984). Theoretical studies with *lac* repressor also highlight this aspect of protein-DNA interaction (Klug *et al.* 1979).

10.4.2 *Homopolymer tracts d(A)-d(T)*

The special characteristics of d(A)-d(T) tracts were first demonstrated crystallographically by Nelson *et al.* (1987) who analysed the duplex formed by d(CGCAAAAAAGCG) and d(CGCTTTTTTGCG). Although the crystals are isomorphous with the Dickerson-Drew dodecamer, the double helix has some different and interesting structural features. The A.T base pairs of the central region have unusually large propeller twists (approx 25°) which permit the formation of a three-centred hydrogen-bonding network in the major groove between the N-6 atoms of adenines and the O-4 atoms of two thymines, one the Watson-Crick partner and the other the base in the 3' direction of the opposing strand. Similar three-centred bonds were subsequently found in d(CGCAAATTTGCG), both when crystallized by itself and when complexed with distamycin (Coll *et al.* 1987). The presence of such inter-strand hydrogen bonds, in addition to a marked purine–purine base overlap, confers a certain rigidity of the homopolymer tract and could well provide an explanation as to why long runs of adenines are not located in the more sharply curved tracks of chromosomes but are found at the end of the nucleosomal DNA with decreased supercoiling (Satchwell *et al.* 1986). The flexibility and mode of bending of DNA is a highly controversial subject, and a number of different theories have been proposed to account for the many experimental observations. Interested readers are directed to Calladine *et al.* (1988) for discussions of this aspect of DNA structure.

10.5 Ribo-nucleotide double helices

Structural studies on transfer RNA have been detailed elsewhere (Saenger 1984); our concern here is with duplex RNA. There are only two examples of well-characterized RNA containing double helices. They are the hybrid decamer r(GCG)d(TATACGC) studied by Wang *et al.* (1982b) and the 14-mer r(UUAUAUAUAUAUAA) (Dock-Bregeon *et al.* 1988).

The hybrid structure consists of two DNA-RNA segments on either side of duplex DNA. All three sections display a conformation similar to the fibre-based elevenfold RNA duplex. Water molecules bridge the ribose O-2′-hydroxyl group and the cytosine O-2 group on the minor groove side of the base pair. Such an interaction may contribute to the particular ribo-conformation.

The 14-mer RNA duplex allows for the most detailed analysis of an RNA helix. This structure can be treated as three segments separated by kinks in the sugar-phosphate backbone. These kinks influence the major groove dimensions and contribute to a structure quite distinct from the fibre models.

10.6 Base-pair mismatches

The structure of deoxyoligonucleotides containing non-Watson-Crick or mismatched base pairs was reviewed recently both in the solid state (Kennard 1985, 1987) and in solution (Patel *et al.* 1987). This section is a brief summary of the most important results published since then.

The formation of stable mismatched base pairs G.T, A.G, and A.C by hydrogen bonding between the functional groups of the bases, and without having to invoke the presence of minor tautomeric forms, is now well established. The mismatched base pairs can be readily accommodated in duplex DNA by small concerted changes in the torsion angles of the sugar-phosphate backbone. The overall characteristics of the double helix containing base-pair mismatches deviate but little from those of the corresponding helices with Watson-Crick base pairs only. Water molecules play an important role in stabilizing certain base-pair mismatches and are sufficiently ordered to have been observed in most of the mismatches analysed. Recently these studies have been extended to include mismatches with the base inosine. Despite the absence of the N-2 amino group, this base was found to behave structurally very like guanine and forms a wobble base pair with thymine (Cruse *et al.* 1989) and I(*anti*).A(*syn*) base pairs with adenine (Corfield *et al.* 1987).

G.A is structurally the most variable mispair so far studied and is also the mismatch most likely to escape detection by DNA polymerase III (Fersht *et al.* 1982). This mispair has now been found in three different geometrical arrangements:

20G(*anti*).A(*syn*) in d(CGCGAATTAGCG) (Brown *et al.* 1986a; Hunter *et al.* 1986b); G(*syn*).A(*anti*) in d(CGCAAATTGGCG) (Brown *et al.* 1989); G(*anti*).A(*anti*) in d(CCAAGATTGG) (Prive *et al.* 1987, 1988).

The structural formulae for these pairings are illustrated in Fig. 10.4. All three types of base pairs can be accommodated in the B-DNA double helix, with the G(*anti*).A(*anti*) pair causing the largest distortions of the backbone. The variation in the mutual orientation of the bases appears to be related to the sequence environment of the mispair and to pH. In solution the G(*anti*).A(*anti*) pair is observed at neutral pH (Kan *et al.* 1983) whereas the G(*syn*).A(*anti*) base pair with a protonated adenine is the dominant form at pH values of 4.0–5.5 (Gao and Patel 1988). The neighbour dependence can, to some extent, be rationalized by the hypothesis that this purine.purine mispair formation is strongly influenced by dipole-dipole interactions with the adjacent bases (Brown *et al.* 1989). The importance of sequence environment is clearest for the G(*anti*).A(*anti*) pair which is stabilized by an intra-base-pair hydrogen bond from the amino N-2 of the guanine to the O-2 of thymine on the opposing strand (Fig. 10.5). Without a thymine in this position the mismatch stability would be lowered and possibly another orientation would be adopted.

There is evidence (Fazakerley *et al.* 1986) of differential recognition of the G.A mispairs *in vivo* in different sequence environments. This may well be related to the very different functional groups that are free to interact with polymerase or a repair enzyme in the different geometrical arrangements of the G.A pair. Given this variability, it is perhaps not so surprising that G.A is the least well-corrected mismatch.

10.6.1 Extrahelical bases

In 1988 there were three reports of crystal structures of oligonucleotide fragments that contain extrahelical bases. Two of these were the sequences d(CGCGAAATTTACGCG) (Miller *et al.* 1988) and d(CGCAGAATTC-GCG) (Joshua-Tor *et al.* 1988). In both structures B-form double helices with 14 and 12 base pairs, respectively, were formed, with the adenines, at positions 11 and 4, looped out from the helices. Crystal-packing forces have been identified as playing a role in determining the positions of the extrahelical adenines. The sequence d(CGCGCGTTTTCGCGCG) was studied by Chattopadhyaya *et al.* (1988). It adopts a monomeric hairpin conformation, with a six base pair Z-DNA stem. The extra thymines form a loop which caps the left-handed helical section. This structure is presented in Figure 10.6.

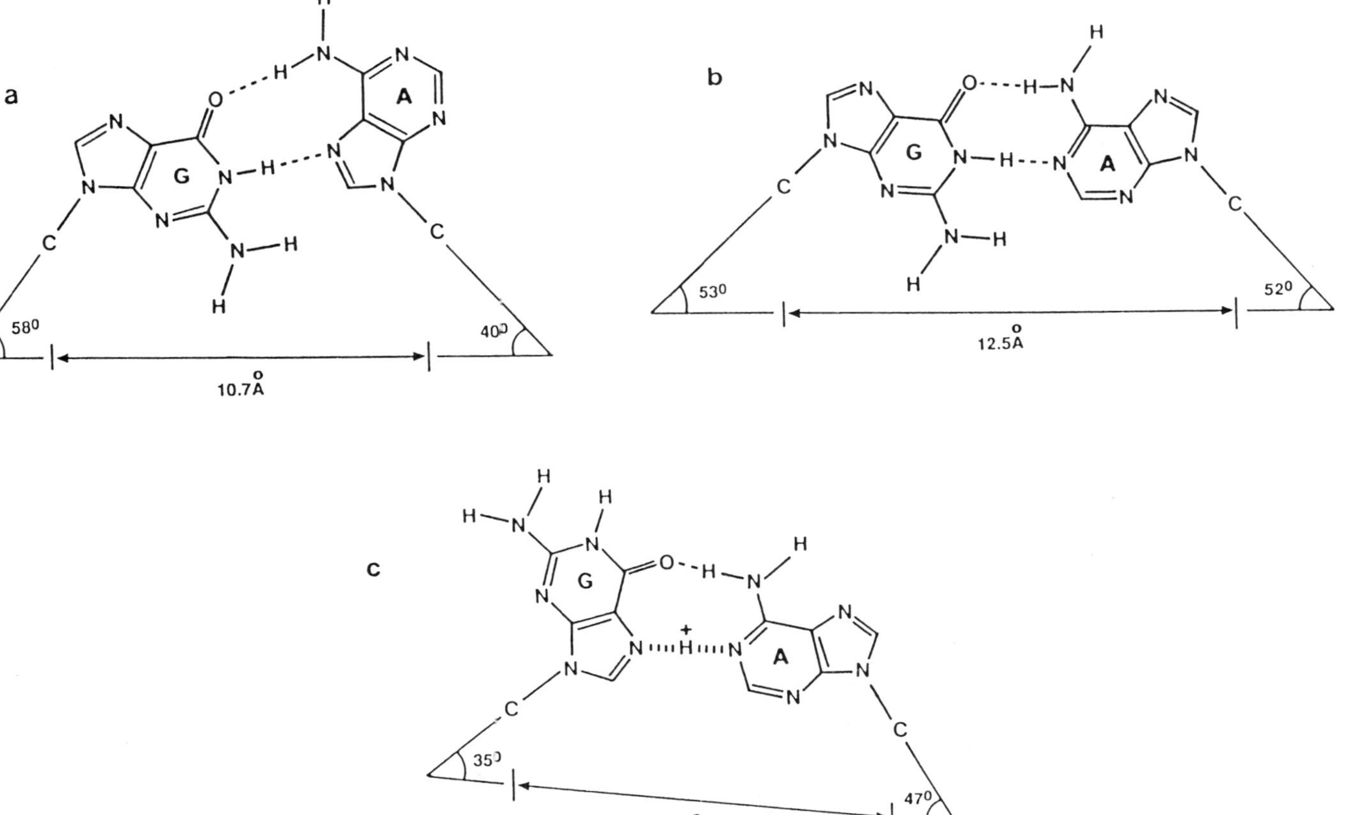

Fig. 10.4 Structural formulae and geometric details of three types of G.A pairing with the attendant C-1′-C-1′ distances and N(base)-

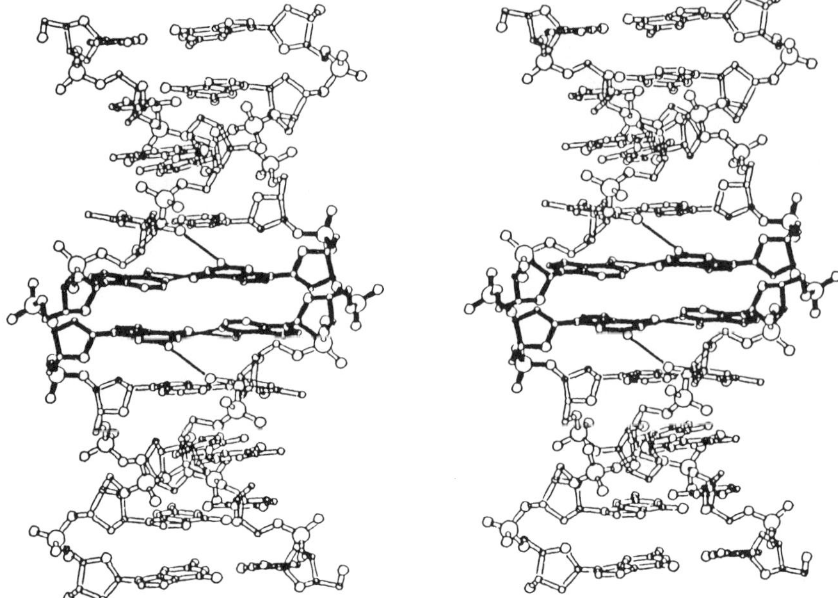

Fig. 10.5 The B-form decamer containing the G(*anti*).A(*anti*) base pairs (Prive *et al.* 1987, 1988). The mispairs are shown with dark bonds and the hydrogen bonds as thinner lines. Note the cross-strand hydrogen bonds adjacent to thymines.

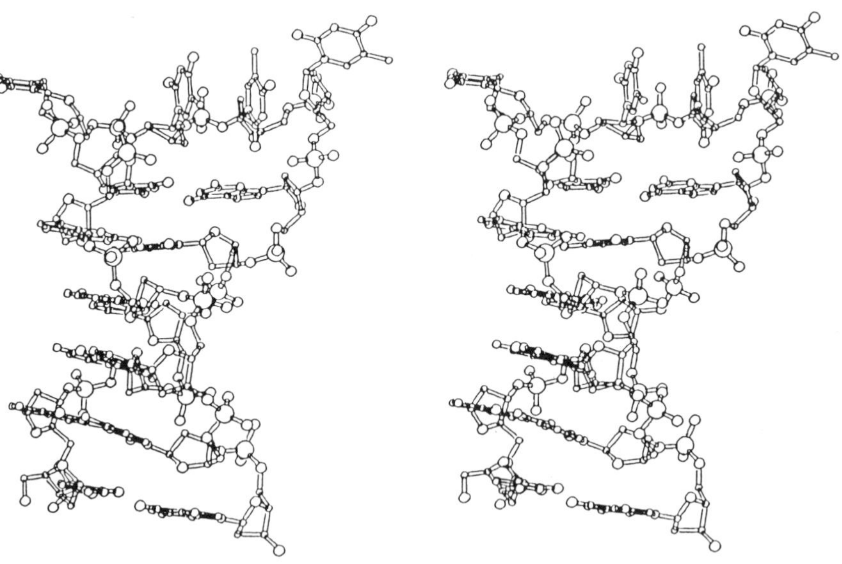

Fig. 10.6 Two stereoviews of the hairpin structure d(CGCGCGTTTTCGCGCG), analysed by Chattopadhyaya *et al.* (1988). The structure consists of a Z-DNA stem capped by four thymines.

10.7 Hydration

DNA crystals, as already mentioned, contain around 50 per cent solvent. While this extensive hydration contributes to the difficulties of obtaining well-ordered crystals and precise X-ray diffraction data, it does have the advantage of allowing crystallographers to study hydration experimentally and to extrapolate from the crystal structure results to *in vivo* structures, in much the same way as is common for protein structures.

In DNA, distinct hydration patterns are observed around the sugar-phosphate backbone and in the major and minor grooves. An examination by Saenger *et al.* (1986) of the water structure reported in numerous DNA fragments led to the proposal that in mixed sequence DNA the hydration of the backbone is related to the global conformation. The typical backbone hydration patterns of A-and B-DNA helices are illustrated in Figs 10.7 and 10.8. As can be seen, a continuous chain of water molecules can link the phosphate oxygens along the backbone in A-and Z-DNA. The water molecules are shared between adjacent phosphate groups, resulting in an effectively greater economy of hydration. In the B-form there are more water molecules around each phosphate group and virtually no water bridges between adjacent phosphate oxygens. These observations can be rationalized very simply by considering the O-1P/O-2P−O-1P/O-2P distances in the three global conformations. In A-and Z-DNA the distances are about 5.4 Å and 4.6 Å respectively, short enough to be bridged by a water molecule. The corresponding value in B-DNA is 6.5 Å or more, too great for a single water linkage (Fig. 10.9).

This pattern of backbone hydration can, however, be affected by local changes, which appear to be related to specific sequences. In two recently published structures, octamers with global A conformations, the backbone hydration pattern was disrupted (Heinemann *et al.* 1987; Hunter *et al.* 1989). The structures are observed to have sugar-phosphate backbone conformations that deviate markedly from standard 'textbook' A-DNA.

The hydration of the grooves is far more sequence dependent than the hydration of the backbone. Dickerson and colleagues were the first to observe a very regular pattern of hydration in the central AATT segment of the dodecamer d(CGCGAATTCGCG). A string of well-ordered water molecules is hydrogen bonded to nitrogen and oxygen atoms lining the minor groove, which is very narrow in this region (Drew and Dickerson 1981; Kopka *et al.* 1983). As the groove widens at the CGCG ends the regular arrangement of water molecules is disrupted (Fig. 10.10). This so-called 'spine of hydration' probably plays a significant role in stabilizing the B-DNA conformation, particularly of this specific sequence. In A-DNA, distinct hydration patterns have been observed in the major grooves. For d(IoCCGG) this consists of strings of water molecules across the groove linking the phosphates of each strand (Conner *et al.* 1984). The octamers

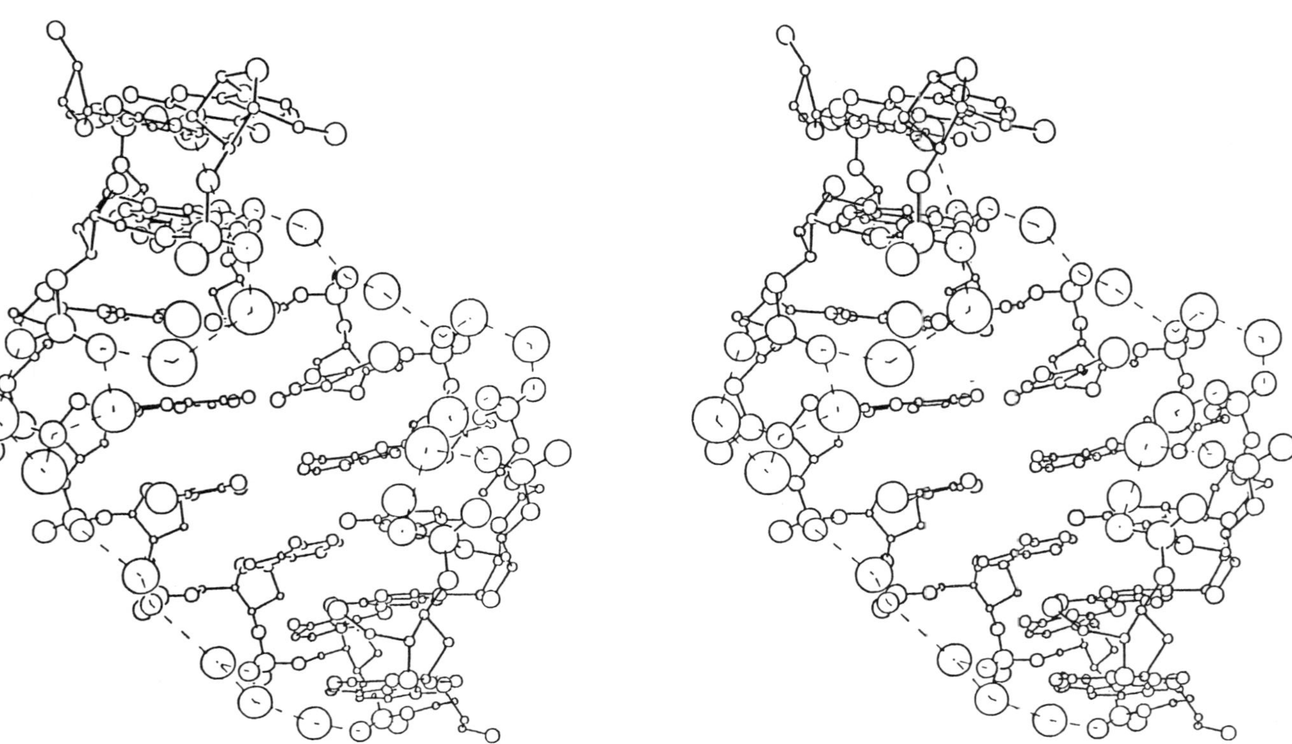

Fig. 10.7 The hydration of phosphate groups in the A-form fragment, d(GGBrUABrUACC). This stereo view shows the view into the major groove. Waters are represented as the largest spheres, with possible hydrogen bonds as dashed lines. Note how there is an almost continuous linking of phosphate groups by water molecules.

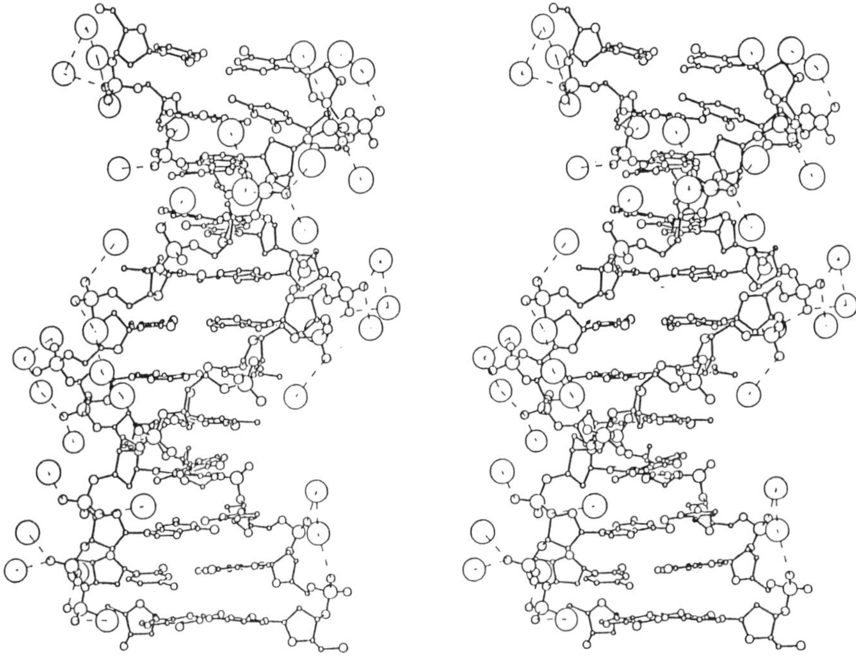

Fig. 10.8 The hydration of phosphate groups in the B-form dodecamer, d(CGCGAATTAGCG). As in Fig. 10.7, the waters are the largest spheres and attendant hydrogen bonds are shown by dashed lines. The phosphate groups tend to have clusters of water molecules around them rather than bridging waters.

d(GGBrUABrUACC) and d(GGTATACC) (Kennard *et al.* 1986) show a series of fused pentagons involving hydrogen bonds between water oxygen atoms and the functional groups of the bases (see Figs 10.8 and 10.11). This pattern is related to the base sequence and is disrupted in octamers of different sequences. More detailed accounts of hydration are given by Saenger (1984, 1987) and Westhof (1987, 1988).

10.8 Acknowledgements

We thank our colleagues, Peter Strazewski and Luc van Meervelt for discussions and comments on the manuscript. Figs 10.1, 10.2, 10.3, 10.5, 10.6, 10.9, and 10.10 used coordinates retrieved from the Brookhaven Protein Databank (Bernstein *et al.* 1977).

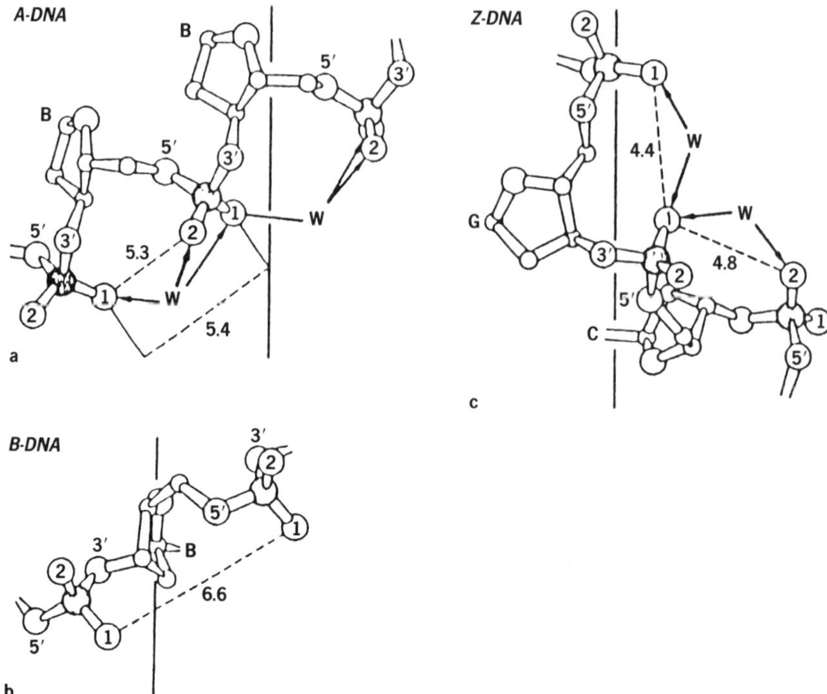

Fig. 10.9 A comparison of deoxyribose 3′,5′-diphosphate in (a) A-form, (b) B-form, and (c) Z-form DNA. Phosphate oxygen atoms are numbered. The vertical line is a representation of the helix axis calculated for each polymorph, the dashed lines and distances (Å) represent the shortest separations between free phosphate oxygen atoms. The base positions are given by B for the right-handed forms and by G,C for Z-DNA. W represents water molecles able to bridge free phosphate oxygens in A-and Z-form DNA.

Fig. 10.11 The major groove hydration of d(GGBrUABrUACC). Hydrogen bonds are shown as thin lines. Note the series of fused pentagons that fills the major groove of this DNA fragment.

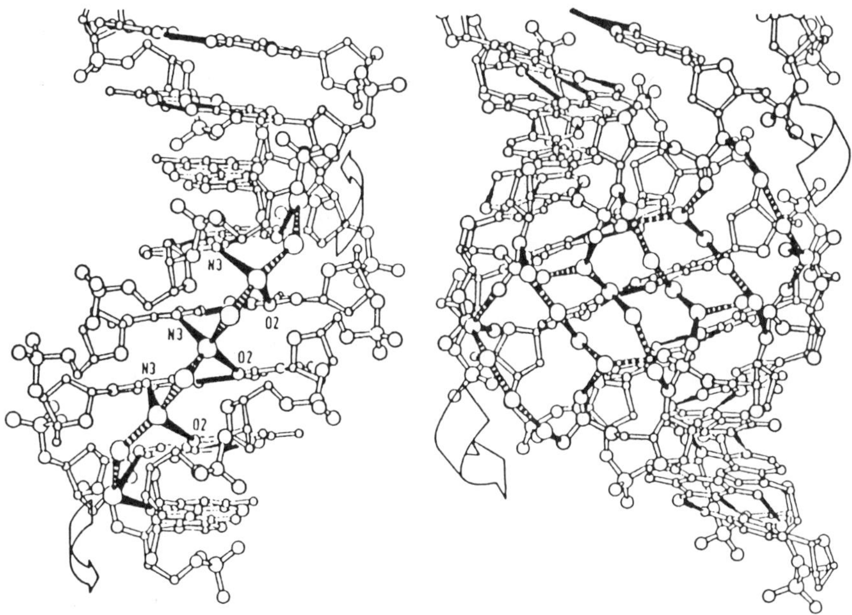

Fig. 10.10 Schematic illustration of the minor groove hydration, spine of hydration observed in the narrow AATT section of the B-form dodecamer (left), and major groove hydration observed in the A-form d(IoCCGG) (right). Broken lines show possible water–water hydrogen bonds, filled lines show hydrogen bonds formed between atoms on the DNA bases and water molecules. Note that in the B-form the purine N-3 and pyrimidine O-2 atoms form the hydrophillic attachments.

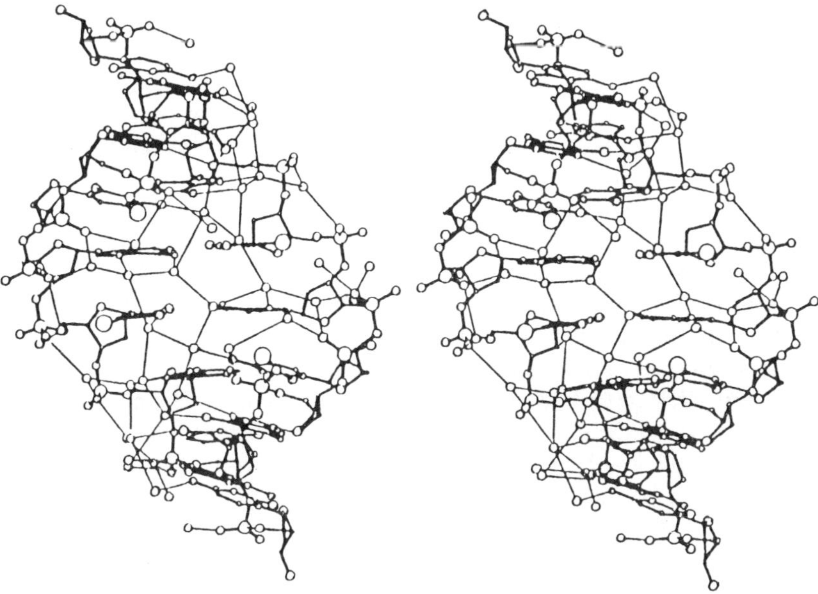

References

Arentzen, R., van Boecke, C.A.A, van der Marel, G., and van Boom, J.H. (1979). A convenient phosphorylating agent for the synthesis of DNA fragments by the phosphotriester approach. *Synthesis* 137–9.

Arndt, U.W. (1985). Television area detector diffractometers. In *Methods in Enzymology*, (ed. H. Wyckoff, C.H.W. Hirs, and S.N. Timasheff), Vol. 114, pp. 472–85. Academic Press, New York.

Arnott, S. and Hukins, D.W.J. (1972). Optimised parameters for A-DNA and B-DNA. *Biochemical Research Communications* **47**, 1504–9.

Bernstein, F.C., *et al.* (1977). The protein data bank: a computer based archival file for macromolecular structures. *European Journal of Biochemistry* **80**, 319–24.

Bingman, C.A., Jain, S., Jebaratnam, D., and Sundaralingam, M. (1989). Crystal and molecular structure of an A-DNA dodecamer. *Sixth Conversation in Biomolecular Stereodynamics, State University of New York at Albany, 6–10 June, 1989*, p. 28.

Blundell, T. and Johnson, L.N. (1976). *Protein crystallography* Academic Press, New York.

Brennan, R.G., Westhof, E., and Sundaralingam, M. (1986). The structure of a Z-DNA with 2 different backbone conformations. Stabilisation of the decadeoxyoligonucleotide d(CGTACGTACG) by $[Co(NH_3)_6]^{3+}$ binding to the guanine. *Journal of Biomolecular Structure and Dynamics* **3**, 649–65.

Brown, T., Kennard, O., Kneale, G., and Rabinovich, D. (1985). High resolution structure of a DNA helix containing mismatched base pairs. *Nature* **315**, 604–6.

Brown, T., Hunter, W.N., Kneale, G., and Kennard, O. (1986a). Molecular structure of the G.A base pair and its implications for the mechanism of transversion mutations. *Proceedings of the National Academy of Sciences USA* **83**, 2402–6.

Brown, T., Kneale, G., Hunter, W.N., and Kennard, O. (1986b). Structural characterisation of the bromouracil.guanine base pair mismatch in a Z-DNA fragment. *Nucleic Acids Research* **14**, 1801–9.

Brown, T., Leonard, G.A., Booth, E.D., and Chambers, J. (1989). Crystal structure and stability of a DNA duplex containing A(*anti*).G(*syn*) base pairs. *Journal of Molecular Biology* **207**, 445–50.

Calladine, C.R. (1982). Mechanics of sequence-dependant stacking of bases in B-DNA. *Journal of Molecular Biology* **161**, 343–52.

Calladine C.R. and Drew H.R. (1984). A base centred explanation of the B-to-A transition in DNA. *Journal of Molecular Biology* **178**, 773–82.

Calladine, C.R. and Drew, H.R. (1986). Principles of sequence-dependant flexure of DNA. *Journal of Molecular Biology* **192**, 907–18.

Calladine, C.R., Drew, H.R., and McCall, M.J. (1988). The intrinsic curvature of DNA in solution. *Journal of Molecular Biology* **201**, 127–37.

Chattopadhyaya, R. and Chakrabarti, P. (1988). Solving structures by MERLOT. *Acta. Crystallogr.* **B44**, 651–7.

Chattopadhyaya, R., Ikuta, S., Grzeskowiak, K., and Dickerson, R.E. (1988). X-ray structure of a DNA hairpin molecule. *Nature* **334**, 175–9.

Chevrier, B., *et al.* (1986). Solvation of the left-handed hexamer d(5BrCG5BrCG5BrCG) in crystals grown at two temperatures. *Journal of Molecular Biology* **188**, 707–19.

Coll, M., Wang, A.H.-J., van der Marel, G.A., van Boom, J.H., and Rich, A. (1986). Crystal structure of a Z-DNA fragment containing thymine/2-aminoadenine base pairs. *Journal of Biomolecular Structure and Dynamics* **4**, 157–72.

Coll, M., Frederick, C.A., Wang, A.H.-J., and Rich, A. (1987). A bifurcated hydrogen bonded conformation in the d(A.T) base pairs of the DNA dodecamer d(CGCAAATTTGCG). *Proceedings of the National Academy of Sciences USA* **84**, 8385–9

Coll, M., *et al.* (1988). Structure of d(CACGTG), a Z-DNA hexamer containing AT base pairs. *Nucleic Acids Research* **16**, 8695–705.

Coll, M., Saal, D., Frederick, C.A., Aymami, J., Rich, A., and Wang, A.H.-J. (1989). Effects of 5-fluorouracil/guanine wobble base pairs in Z-DNA. Molecular and crystal structure of d(CGCGFG). *Nucleic Acids Research* **17**, 911–23.

Conner, B.N., Takano, T., Tanaka, S., Itakura, K., and Dickerson, R.E. (1982). The molecular structure of d(IoCCGG) a fragment of right-handed double helical A-DNA. *Nature* **295**, 294–9.

Conner, B.N., Yoon, C., Dickerson, J.L., and Dickerson, R.E. (1984). Helix geometry and hydration in an A-DNA tetramer: IoCCGG. *Journal of Molecular Biology* **174**, 663–95.

Corfield, P.W.R., Hunter, W.N., Brown, T., Robinson, P., and Kennard, O. (1987). Inosine.adenine base-pairs in a B-DNA duplex. *Nucleic Acids Research* **15**, 7935–49.

Crawford, J.L., *et al.* (1980). The tetramer d(CGCG) crystallises as a left-handed double helix. *Proceedings of the National Academy of Sciences* **77**, 4016–20.

Cruse, W.B.T., Salisbury, S.A., Brown, T., Eckstein, F., Cosstick, R., and Kennard, O. (1986). Chiral phosphorothioate analogues of B-DNA: The crystal structure of Rp-d(GpsCGpsCGpsC). *Journal of Molecular Biology* **192**, 891–905.

Cruse, W.B.T., Aymami, J., Kennard, O., Brown, T., Jack, A.G. C., and Leonard, G.A. (1989). Refined crystal structure of an octanucleotide duplex with I.T mismatched base pairs. *Nucleic Acids Research* **17**, 55–72.

Dickerson, R.E., *et al.* (1989). Definitions and nomenclature of nucleic acid structure parameters. *EMBO Journal* **8**, 1–4.

DiGabriele, A., Sanderson, M., and Steitz, T.A. (1989). Crystal lattice packing is important in determining the bend of a DNA dodecamer containing an adenine tract. *Proceedings of the National Academy of Sciences* **86**, 1816–20.

Dock-Bregeon, A.C., *et al.* (1988). High resolution structure of the RNA duplex [U(U-A)6A]2. *Nature* **335**, 375–8.

Drew, H.R. and Dickerson, R.E. (1981). Structure of a B-DNA dodecamer: geometry of hydration. *Journal of Molecular Biology* **151**, 535–56.

Drew, H.R. and Travers, A.A. (1984). DNA structural variations in the tyrT promoter. *Cell* **37**, 491–502.

Drew, H.R., Dickerson, R.E., and Itakura, K. (1978). A salt-induced conformational change in crystals of the synthetic DNA tetramer d(CGCG). *Journal of Molecular Biology* **125**, 535–43.

Drew, H.R., Takano, T., Tanaka, S., Itakura, K., and Dickerson, R.E. (1980). High salt d(CGCG): A left-handed Z-DNA double helix. *Nature* **286**, 567–73.

Drew, H.R., *et al.* (1981). Structure of a B-DNA dodecamer; conformation and dynamics. *Proceedings of the National Academy of Sciences USA* **78**, 2179–83.

Drew, H.R., Samson, S., and Dickerson, R.E. (1982). Structure of a B-DNA dodecamer at 16K. *Proceedings of the National Academy of Sciences USA* **79**, 4040–4.

Drew, H.R., McCall, M.J., and Calladine, C.R. (1988). Recent studies of DNA in the crystal. *Annual Review of Cell Biology* **4**, 1–20.

Eisenstein, M., Hope, H., Haran, T.E., Frolow, P., Shakked, Z., and Rabinovich, D. (1988). Low-temperature study of the A-DNA fragment d(GGGCGCCC). *Acta. Crystallogr.* **B44**, 625–8.

Fazakerley, G.V., *et al.* (1986). Structures of mismatched base pairs in DNA and their recognition by the *E. coli.* mismatch repair system. *EMBO Journal* **5**, 3697–703.

Fersht, A.R., Knill-Jones, J.W., and Tsui, W.C. (1982). Kinetic basis of spontaneous frequencies, proofreading specificities and cost of proofreading by DNA polymerases of *E. coli. Journal of Molecular Biology* **156**, 37–51.

Fratini, A.V., Kopka, M.L., Drew, H.R., and Dickerson, R.E. (1982). Reversible bending in a B-DNA dodecamer: CGCGAATTBrCGCG. *Journal of Biological Chemistry* **257**, 14686–707.

Frederick, C.A., Saal, D., van der Marel, G.A., van Boom, J.H., Wang, A.H.-J., and Rich, A. (1987). The crystal structure of d(GGmCCGGCC): the effect of methylation on A-DNA structure and stability. *Biopolymers* **26**, 145–60.

Frederick, C.A., Quigley, G.J., van der Marel, G.A., van Boom, J.H., Wang, A.H.-J., and Rich, A. (1988). Methylation of the *Eco* RI recognition site does not alter DNA conformation: the crystal structure of d(CGCGAm6ATTCGCG) at 2.O Å resolution. *Journal of Biological Chemistry* **263**, 17872-9.

Frederick, C.A., *et al.* (1989). Molecular structure of an A-DNA decamer d(ACCGGCCGGT). *European Journal of Biochemistry* **181**, 295-307.

Fujii, S., Wang, A.H.-J., van der Marel, G., van Boom, J.H., and Rich, A. (1982). Molecular structure of (m5dC-dG)3: the role of the methyl group on 5-methyl cytosine in stabilising Z-DNA. *Nucleic Acids Research* **10**, 7879-92.

Fujii, S., Wang, A.H.-J., van der Marel, G., van Boom, J.H., and Rich, A. (1985). The octamers d(CGCGCGCG) and d(CGCATGCG) both crystallise as Z-DNA in the same hexagonal lattice. *Biopolymers* **24**, 243-50.

Gait, M.J., Matthes, H.W.D., Singh, M., Sproat, B.S., and Titmus, R.C. (1982). Rapid synthesis of oligodeoxyribonucleotides VII. Solid phase synthesis by a continuous flow phosphotriester method on a kieselguhr-polyamide support. *Nucleic Acids Research* **10**, 6243-8.

Gao, X. and Patel, D. (1988). G(*syn*).A(*anti*) mismatch formation in DNA dodecamers at acidic pH: pH dependent conformational transition of G.A mispairs detected by proton NMR. *Journal of the American Chemical Society* **110**, 5178-82.

Gessner, R.V., Frederick, C.A., Quigley, G.J., Rich, A., and Wang, A.H.-J. (1989a). The molecular structure of the left handed Z-DNA double helix at 1.0 Å atomic resolution. *Journal of Biological Chemistry* **264**, (14), 7921-35.

Gessner, R.V., Dean Williams, L., Rich, A., and Frederick, C.A. (1989b). The interaction of spermine with DNA: high resolution single crystal structure analysis of DNA-spermine complexes. *Sixth Conversation in Biomolecular Stereodynamics, State University of New York at Albany, 6-10 June, 1989*, p. 104.

Grzeskowiak, K., *et al.* (1989). *Sixth Conversation in Biomolecular Stereodynamics, State University of New York at Albany, 6-10 June, 1989*, p. 120.

Hamlin, R. (1985). Multiwire area X-ray diffractometers. In *Methods in enzymology* (ed. H. Wyckoff, C.H.W. Hirs, and S.N. Timasheff), Vol. 114, pp. 416-52. Academic Press, New York.

Haran, T.E., Shakked, Z., Wang, A.H.-J., and Rich, A. (1987). The crystal structure of d(CCCCGGGG): a new A-form variant with an extended backbone conformation. *Journal of Biomolecular Structure and Dynamics* **5**, 199-217.

Heinemann, U., Lauble, H., Frank, R., and Blocker, H. (1987). Crystal structure analysis of an A-DNA fragment at 1.8 Å resolution: d(GCCCGGGC). *Nucleic Acids Research* **15**, 9531–50.

Heinemann, U., Alings, C., and Lauble, H. (1989). Structural features of a G/C-rich DNA going A or B. *Sixth Conversation in Biomolecular Stereodynamics, State University of New York at Albany, 6–10 June, 1989*, p. 138.

Hendrickson, W., Smith, J.L., and Sheriff, S. (1985). Direct phase determination based on anomalous scattering. In *Methods in enzymology* (ed. H. Wyckoff, C.H.W. Hirs, and S.N. Timasheff), Vol. 115, pp. 41–55. Academic Press, New York.

Ho, P.S., *et al.* (1985). G.T wobble pairing in Z-DNA at 1.0 Å atomic resolution; the crystal structure of d(CGCGTG). *EMBO Journal* **4**, 3617–23.

Holbrook, S.R., and Kim, S.-H. (1985). Crystallisation and heavy atom derivatives of polynucleotides. In *Methods in enzymology* (ed. H. Wyckoff, C.H.W. Hirs, and S.N. Timasheff), Vol. 114, pp. 167–75. Academic Press, New York.

Hoppe, H. (1988). Cryocrystallography of biological macromolecules: a generally applicable method. *Acta. Crystallogr.* **B44**, 22–6.

Hunter, W.N., Brown, T., Anand, N.N., and Kennard, O. (1986a). Structure of an adenine-cytosine base pair in DNA and its implications for mismatch repair. *Nature* **320**, 552–5.

Hunter, W.N., Brown, T., and Kennard, O. (1986b). Structural features and hydration of d(CGCGAATTAGCG): a double helix containing two G.A mispairs. *Journal of Biomolecular Structure and Dynamics* **4**, 173–91.

Hunter, W.N., Kneale, G., Brown, T., Rabinovich, D., and Kennard, O. (1986c). Refined crystal structure of an octanucleotide duplex with G.T mismatched base pairs. *Journal of Molecular Biology* **190**, 605–18.

Hunter, W.N., Brown, T., and Kennard, O. (1987a). Structural features and hydration of d(CGCAAATTCGCG): a double helix containing two C.A mispairs. *Nucleic Acids Research* **15**, 6589–606.

Hunter, W.N., Brown, T., Kneale, G., Anand, N.N., Rabinovich, D., and Kennard, O. (1987b). The structure of guanine.thymine mismatches in B-DNA at 2.5 Å resolution. *Journal of Biological Chemistry* **262**, 9962–70.

Hunter, W.N., Langlois d'Estaintot, B., and Kennard, O. (1989). Structural variation in d(CTCTAGAG): implications for protein-nucleic acid interactions. *Biochemistry* **28**, 2444–51.

Jain, S., Zon, G., and Sundaralingam, M. (1987). The potentially Z-DNA forming sequence d(GTGTACAC) crystallises as A-DNA. *Journal of Molecular Biology* **197**, 141–5.

Jain, S., Zon, G., and Sundaralingam, M. (1989). Base only binding of spermine in the deep groove of the A-DNA octamer d(GTGTACAC). *Biochemistry* **28**, 2360–4.

Joshua-Tor, L., Rabinovich, D., Hoppe, H., Frolow, F., Appella, E., and Sussman, J. L. (1988). The three-dimensional structure of a DNA duplex containing looped-out bases. *Nature* **334**, 82–4.

Jovin, T.M., Soumpasis, D.M., and McIntosh, L.P. (1987). The transition between B-DNA and Z-DNA. *Annual Review of Physical Chemistry* **38**, 521–60.

Kan, L.-S., Chandrasegaran, S., Pulford, S.M., and Miller, P.S. (1983). Detection of a guanine adenine base pair in a decadeoxyribonucleotide by proton magnetic resonance spectroscopy. *Proceedings of the National Academy of Sciences* **80**, 4263–5.

Kennard, O. (1985). Structural studies of DNA fragments: The G.T wobble base pair in A, B and Z-DNA. The G.A base pair in B-DNA. *Journal of Biomolecular Structure and Dynamics* **3**, 205–26.

Kennard, O. (1987). The molecular structure of base-pair mismatches. In *Nucleic acids and molecular biology* (ed D. Lilley and E. Eckstein), Vol. 1, pp. 25–52. Springer-Verlag, Berlin.

Kennard, O. and Hunter, W.N. (1989a). Crystal structures of oligonucleotides. In *Landolt-Bornstein Tables, Group VII*, Vol. 1, pp. 255–360. Springer-Verlag, Berlin.

Kennard, O. and Hunter, W.N. (1989b). Oligonucleotide structure: a decade of results from single crystal X-ray diffraction studies. *Quarterly Reviews of Biophysics* **22**, 327–79.

Kennard, O., Cruse, W.B.T., Nachman, J., Prange, T., Shakked, Z., and Rabinovich, D. (1986). Ordered water structure in an A-DNA octamer at 1.7 Å resolution. *Journal of Biomolecular Structure and Dynamics* **3**, 623–47.

Klug, A., Jack, A., Viswamitra, M.A., Kennard, O., Shakked, Z., and Steitz, T.A. (1979). A hypothesis on a specific sequence dependent conformation of DNA and its relation to the binding of the lac-repressor protein. *Journal of Molecular Biology* **131**, 669–80.

Kneale, G., Brown, T., Kennard, O., and Rabinovich, D. (1985). G.T base pairs in a DNA helix: The crystal structure of d(GGGGTCCC). *Journal of Molecular Biology* **186**, 805–14.

Kopka, M.L., Fratini, A.V., Drew, H.R., and Dickerson, R.E. (1983). Ordered water structure around a B-DNA dodecamer. *Journal of Molecular Biology* **163**, 129–46.

Lauble, H., Frank, R., Blocker, H., and Heinemann, U. (1988). Three-dimensional structure of d(GGGATCCC) in the crystalline state. *Nucleic Acids Research* **16**, 7799–816

McBride, L.J. and Caruthers, M.H. (1983). An investigation of several deoxynucleoside phosphoramidites useful for synthesising deoxyoligonucleotides. *Tetrahedron Letters* **24**, 245–8.

McCall, M., Brown, T., and Kennard, O. (1985). The crystal structure of d(GGGGCCCC) – A model for poly(dG).poly(dC). *Journal of Molecular Biology* **183**, 385–96.

McCall, M., Brown, T., Hunter, W.N., and Kennard, O. (1986). The crystal structure of d(GGATGGGAG): an essential part of the binding site for transcription factor IIIA. *Nature* **322**, 661–4.

McPherson, A. (1982). *Preparation and analysis of protein crystals*. Wiley, San Francisco.

Miller, M., Harrison, R.W., Wlodawer, A., Appella, E., and Sussman, J.L. (1988). Crystal structure of 15-mer DNA duplex containing unpaired bases. *Nature* **334**, 85–6.

Nelson, H.C.M., Finch, J.T., Luisi, B.E., and Klug, A. (1987). The structure of an oligo(dA).oligo(dT) tract and its biological implications. *Nature* **330**, 221–6.

Patel, D.J., Shapiro, L., and Hare, D. (1987). Conformation of DNA base pair mismatches in solution. In *Nucleic acids and molecular biology* (ed F. Eckstein, and D.M.J. Lilley), pp. 70–84. Springer-Verlag, Berlin.

Prive, G.G., Heinemann, U., Chandrasegaran, S., Kan, L.S., Kopka, M.L., and Dickerson, R.E. (1987). Helix geometry, hydration and G.A mismatch in a B-DNA decamer. *Science* **238**, 498–504.

Prive, G.G., Heinemann, U., Chandrasegaran, S., Kan, L.S., Kopka, M.L., and Dickerson, R.E. (1988). A mismatch decamer as a model for general-sequence B-DNA. In *Structure and expression*, (ed. R.H. Sarma and M.H. Sarma), Vol. 2, pp. 27–47. Adenine Press, Schenectady.

Rabinovich, D. and Shakked, Z. (1984). A new approach to structure determination of large molecules by multi-dimensional search methods. *Acta. Crystallogr.* **A40**, 195–200.

Rabinovich, D., Haran, T.E., Eisenstein, M., and Shakked, Z. (1988). Structures of the mismatched duplex d(GGGTGCCC) and one of its Watson-Crick analogues d(GGGCGCCC) *Journal of Molecular Biology* **200**, 151–61.

Rich, A., Nordheim, A., and Wang, A.H.-J. (1984). The Chemistry and biology of left-handed Z-DNA. *Annual Review of Biochemistry* **53**, 791–846.

Saenger, W. (1984). *Principles of nucleic acid structure*. Springer-Verlag, New York.

Saenger, W. (1987). Structure and dynamics of water surrounding biomolecules. *Annual Review Biophysics and Biophysical Chemistry* **16**, 93–114.

Saenger, W., Hunter, W.N., and Kennard, O. (1986). DNA conformation is determined by economics in the hydration of phosphate groups. *Nature* **324**, 385–8.

Satchwell, S.C., Drew, H.R., and Travers, A.A. (1986). Sequence periodicities in chicken nucleosome core DNA. *Journal of Molecular Biology* **191**, 659–75.

Shakked, Z., and Kennard, O. (1984). The A form of DNA. In *Biological macromolecules and assemblies*. Vol. 2, *Nucleic acids and interactive proteins* (ed. A. McPherson and F. Jurnak), pp. 2–36. Wiley, New York.

Shakked, Z. and Rabinovich, D. (1986). The effect of the base sequence on the fine structure of the DNA double helix. *Progress in Biophysics and Molecular Biology* **47**, 159–95.

Shakked, Z. *et al.* (1981). Crystalline A-DNA: the X-ray analysis of the fragment d(GGTATACC) *Proceedings of the Royal Society of London* **B 213**, 479–87.

Shakked, Z., Rabinovich, D., Kennard, O., Cruse, W.B. ., Salisbury, S.A., and Viswamitra, M.A. (1983). Sequence dependent conformation of an A-DNA double helix: the crystal structure of the octamer d(GGTATACC). *Journal of Molecular Biology* **166**, 183–201.

Shakked, Z., Guzikevich, G., Frolow, F., Eisenstein, M., and Rabinovich, D. (1989). *Sixth Conversation in Biomolecular Stereodynamics, State University of New York at Albany, 6–10 June, 1989*, p. 304.

Sundaralingam, M. and Jain, S. (1989). *Sixth Conversation in Biomolecular Stereodynamics, State University of New York at Albany, 6–10 June, 1989*, p. 341.

Teng, M.-K., Liaw, Y.-C., van de Marcel, G.A., van Boom, J.H., and Wang, A.H.-J. (1989). Effect of the O2′-hydroxyl group on Z-DNA conformation: Structure of Z-RNA and (araC)-[Z-DNA]. *Biochemistry* **28**, (12), 4923–8.

Timsit, Y., Westhoff, E., Fuchs, R., and Moras, D. (1989). Unusual DNA: Structure of a frameshift mutation hot spot. *Sixth Conversation in Biomolecular Stereodynamics, State University of New York at Albany, 6–10 June, 1989*, poster presentation.

van Meervelt, L., Moore, M. H., Kong Thoo Lin, P., Brown, D.M., and Kennard, O., (1989). Unpublished results.

Viswamitra, M.A., *et al.* (1978). Structure of the deoxytetranucleotide d(pATAT) and a sequence dependant model for poly(dA-dT). *Nature* **273**, 687–8.

Viswamitra, M.A., Shakked, Z., Jones, P.G., Sheldrick, G.M., Salisbury, S.A., and Kennard, O. (1982). Structure of the deoxytetranucleotide d-pATAT and a sequence-dependant model for poly(dA-dT). *Biopolymers* **21**, 513–33.

Wang, A.H.-J., *et al.* (1979). Molecular structure of a left-handed double helical DNA fragment at atomic resolution. *Nature* **282**, 680–6.

Wang, A.H.-J., *et al.* (1981). Left-handed double helical DNA: Variations in the backbone conformation. *Science* **211**, 171–6.

Wang, A.H.-J., Fujii, S., van Boom, J.H. van der Marel, G.A., van Boeckel, S.A.A., and Rich, A. (1982a). Molecular structure of r(GCG) d(TATACGC): A DNA-RNA hybrid helix joined to double helical DNA. *Nature* **299**, 601–4.

Wang, A.H.-J., Fujii, S. van Boom, J.H., and Rich, A. (1982b). Molecular structure of the octamer d(GGCCGGCC): Modified A-DNA. *Proceedings of the National Academy of Sciences* **79**, 3968-72.

Wang, A.H.-J., Hakoshima, T., van der Marel, G.A., van Boom, J.H., and Rich, A. (1984). A.T base pairs are less stable than G.C base pairs in Z-DNA: the crystal structure of d(m5CGTAm5CG). *Cell* **37**, 321-31.

Wang, A.H.-J., Gessner, R.V., van der Marel, G.A. van Boom, J.H., and Rich, A. (1985). Crystal structure of a Z-DNA without an alternating purine-pyrimidine sequence. *Proceedings of the National Academy of Sciences* **82**, 3611-5.

Westhof, E. (1987). Hydration of oligonucleotides in crystals. *International Journal of Biological Macromolecules* **9**, 185-92.

Westhof, E. (1988). Water: an integral part of nuclcic acid structure. *Annual Review Biophysics and Biophysical Chemistry* **17**, 125-44.

Westhof, E., Dumas, P., and Moras, D. (1985). Crystallographic refinement of yeast aspartic transfer RNA. *Journal of Molecular Biology* **184**, 119-45.

Wing, R., *et al.* (1980). Crystal structure analysis of a complete turn of B-DNA. *Nature* **287**, 755-8.

Yoon, C., Prive, G.G., Goodsell, D.S., and Dickerson, R.E. (1988). Structure of an alternating B-DNA helix and its relationship to A-tract DNA. *Proceedings of the National Academy of Sciences USA* **85**, 6332-6.

Glossary

Amino acid abbreviations
(one-letter and three-letter): A and Ala, alanine; R and Arg, arginine; N and Asn, asparagine; D and Asp, aspartate; C and Cys, cysteine; G and Gly, glycine; E and Glu, glutamate; Q and Gln, glutamine; H and His, histidine; I and Ile, isoleucine; L and Leu, leucine; K and Lys, lysine; M and Met, methionine; F and Phe, phenylalanine; P and Pro, proline; S and Ser, serine; T and Thr, threonine; W and Trp, tryptophan; Y and Tyr, tyrosine; V and Val, valine.

Apoenzyme
Enzyme molecule without co-enzyme or prosthetic group.

CAMELSPIN
Cross-relaxation appropriate for minimolecules emulated by locked spins.

Chemical shift
The shift of an NMR resonance frequency due to its chemical environment (the electron density surrounding the nucleus). Usually measured relative to a reference compound and reported in p.p.m.

Coherence
The time and phase relationship between spins via through-bond interactions.

COSY
Correlated spectroscopy. A 2D NMR experiment in which cross-peaks are observed between coupled spins.

$d_{\alpha N}$
The distance between the αH of one residue and the amide of the next residue in a peptide. This term is used when describing both sequential NMR assignments and secondary-structure determination. Short $d_{\alpha N}$ is indicative of extended conformation, such as that found in b-sheets. Sometimes it can refer to a longer-range distance when written as $d_{\alpha N}(i,j)$, where i and j represent non-sequential protons.

$d_{\beta N}$
The distance between the bH of one residue and the amide of the next residue in a peptide. This term is used when describing both sequential NMR assign-

ments and secondary-structure determination. Short $d_{\beta N}$ can be found in any secondary structure, but is found regularly in helices.

d_{NN}

The distance between the amide of one residue and the amide of the next residue in a peptide. This term is used when describing both sequential NMR assignments and secondary-structure determination. Short d_{NN} is indicative of helical conformation.

2D NMR

Two-dimensional nuclear magnetic resonance.

3D NMR

Three-dimensional nuclear magnetic resonance.

DQF

Double quantum filter.

FT

Fourier transform. A mathematical procedure that converts time-domain data into frequency domain or vice versa.

Geminal coupling

The scalar (J) coupling between atoms two bonds apart.

HETCOR

Heteronuclear correlated spectroscopy.

Heterotropic effect

Interactions between different ligands, i.e. effector ligands influencing active site ligand binding and catalytic rate.

HMQC

Heteronuclear correlated spectroscopy via multiple quantum coherence.

HOHAHA

Homonuclear Hartmann Hahn spectroscopy.

Holoenzyme

Complete functional enzyme with co-enzyme or prosthetic group.

Homotropic effect

Interactions between identical ligands, i.e. active site ligands influencing other active site ligand binding and catalytic rate.

Isotropic mixing
A process where the effect of chemical shifts is eliminated. This puts the spins in the strong coupling limit spreading coherences throughout the entire spin system. It is used in both the HOHAHA and TOCSY experiments.

J ij
The scalar coupling constant. The three-bond value can be used to obtain structural information since it depends on the dihedral angle.

Knowledge-based modelling
A term popularly used to describe the construction of model structures using information from the Protein Data Bank.

MCD
Main-chain-directed protein ^{1}H NMR assignment method.

MQ
Multiple quantum NMR.

NMR
Nuclear magnetic resonance.

NOE
Nuclear Overhauser effect. The changes in intensity of spins due to cross-relaxation (transfer of magnetization) between them.

NOESY
Nuclear Overhauser effect spectroscopy.

p.p.m.
Parts per million. Units used to describe the chemical shift of NMR resonances.

Patterson function

$$P(uvw) = \frac{2}{V} \sum_{h=-\infty}^{\infty} \sum_{k=-\infty}^{\infty} \sum_{l=0}^{\infty} |F_{hkl}|^2 \cos 2\pi(hu + kv + lw)$$

Because the Patterson function is based on the structure factor amplitudes rather than the structure factors themselves, it can be calculated directly from set of recorded diffraction intensities. It has peaks or maxima corresponding to vectors between atom pairs. The complexity of a protein with many hundreds of atoms makes the Patterson uninterpretable. The heavy atom Patterson, however, contains many fewer vectors and is usually interpretable, giving atomic positions for the heavy atoms.

PreTOCSY
A process that prevents saturation of α-protons under the water resonance.

Primary, secondary, tertiary, and quaternary structure.
There is a hierarchy of structure in proteins in which the sequence is referred
to as the primary structure. The arrangement of that sequence into helices
and sheets is referred to as secondary structure. The arrangement of secondary structure into stable domains is referred to as tertiary structure. The
aggregation of tertiary structures into assemblies is referred to as quaternary
structure.

R factor

$$R = \frac{\Sigma_{hkl} \left\| F_o \right| - \left| F_c \right\|}{\Sigma_{hkl} \left| F_o \right|}$$

where F_o = the observed structure factor;
F_c = the calculated structure factor.
The summations are over all reflections hkl.

Ramachandran plot
This is a plot of ϕ, ψ angle pairs for each residue in a protein. Such plots were
originally derived by Ramachandran *et al.* (1963). The allowed regions of the
plot have been determined from a consideration of steric and energy effects
in small peptides. α-Helices and β-sheets have distinct allowed regions on
a Ramachandran plot. Any pair of angles which falls outside the allowed
regions should be examined with care. Glycine and proline residues are
special cases which must be considered separately.

Synchrotron sources
Synchrotron radiation is the electromagnetic emission from relativistic
charged particles submitted to an acceleration. Synchrotron sources used for
macromolecular crystallography are high energy electron or positron storage
rings.

Thermal parameters
The thermal parameters modify the atomic scattering factors in the structure
factor expression

$$F_{hkl} = \sum_{j=1}^{N} f_j \exp - B_j \left(\sin^2 \theta / \lambda^2 . \exp 2\pi i \left(h x_j + k y_j + l z_j \right) \right)$$

where B_j, the atomic temperature factor, has units $\mathring{A}^2$ and is related to the
mean square displacement of the atom in question.

RELAY
Relayed coherence transfer spectroscopy.

ROESY
Rotating Overhauser effect spectroscopy.

S/N
Signal to noise ratio.

SCUBA
Stimulated cross-peaks under bleached alphas. A process that prevents saturation of α-protons under the water resonance.

Site-specific mutagenesis
Replacement of a single amino acid protein residue with a different amino acid.

Spin diffusion
A process where magnetization is transferred several steps along a chain of spins. This is a secondary NOE effect.

Spin system
A group of coupled spins.

T_1
The time it takes a sample to reach equilibrium in a magnetic field.

T_2
The time it takes for transverse magnetization to reach its equilibrium value of zero.

τ_m
The mixing time in an NMR experiment.

TOCSY or **HOHAHA**
Total correlation spectroscopy.

Vicinal coupling
The scalar (J) coupling between spins three bonds apart. Its value is related to the dihedral angle between them.

Index